Withdrawn from
Davidson College Library

Library of
Davidson College

The Architecture of War

THE ARCHITECTURE OF WAR

KEITH MALLORY AND ARVID OTTAR

PANTHEON BOOKS

A Division of Random House, New York

First American Edition

Copyright © 1973 by Keith Mallory and Arvid Ottar

All rights reserved under International and Pan-American Copyright Conventions. Published in the United States by Pantheon Books, a division of Random House, Inc., and simultaneously in Canada by Random House of Canada Limited, Toronto. Originally published in Great Britain as *Architecture of Aggression* by The Architectural Press Limited, London.

Library of Congress Cataloging in Publication Data
Mallory, Keith.   The Architecture of War.
First published 1973 under title: *Architecture of Aggression,* by the Architectural Press, London.
Includes bibliographical references.
1. Military architecture—History. 2. Fortification—Europe—History.   I.   Ottar, Arvid, joint author.   II.   Title.
UG460.M38 1973   623'.3   73-7012
ISBN 0-394-48825-3
ISBN 0-394-70997-7 (pbk.)

Manufactured in the United States of America

CONTENTS

Acknowledgements 7
Preface 9

PART ONE: 1900–18

1. 1914 Fortress Zone: French, German and Belgian fortresses on the Western Front: their precedent and effects 12

2. Defence in Depth: Development of the deadlock on the Western Front, the development of the German system of defence in depth and concrete fortifications 34

3. Wider Scope of Defence Construction: Allied defences built on the Western Front and coastal fortifications including the first use of submarine pens 54

4. Support Structures: Nissen huts, mobile Zeppelin sheds, Richthofen's circus 72

PART TWO: 1918–40

5. Maginot Line: Design and construction of French defence 90

6. The Autobahn and West Wall: The birth of the Todt Organization, the highway system in Germany and the military implications of the West Wall 108

PART THREE: 1940–45

7. Fortress Britain: Organization and construction of English defences 126

8. Atlantic Wall: Up to D-Day 150

9. The Armed Camp: Build up of invasion forces in England and the resultant vast hutting programme; aspects of prefabrication 180

10. Birth of Mulberry Harbour and Fall of Atlantic Wall: Design and military implications of the Mulberry Harbour Project 200

11. The Shelter Controversy: British air raid shelter provision 214

12. German Bomb-Proof Mania: Protection of civilian morale: the public shelters and flak towers. Protection of military projects: the underground and bomb-proof factories, submarine pens, the V sites and the 'National Redoubt' 238

PART FOUR: CONTEXT

13. Context: Trends in military architecture since 1945; the relationship between military and civil architecture 268

Reference notes 288
Credits 303
Index 305

ACKNOWLEDGEMENTS

The initial study which led to this book was started in January 1970 by a research group at the University of Bath. The vast amount of photographic material gathered during this pilot study formed the basis of our later work and we are indebted to our co-researchers in this original group: Roger Cowles, Terence Fowler, Robert Pickering and Peter Reeve. Thanks are also due to the people who have enabled us to develop this initial study into its present form:

the University of Bath and in particular Professor K. H. Panter of the School of Architecture and Building Technology, for the facilities which were made available to us.

Mr. J. N. Padbury of the University of Bath, the Architectural Press London, and Godfrey Golzen of the Architectural Press for their help and encouragement.

We are indebted to the following who have generously and patiently helped us throughout our research:

the Corps Library of the Institution of Royal Engineers at Chatham and its staff, in particular Lt. Col. J. E. South and the late Lt. Col. F. T. Stear the Ministry of Defence Library Whitehall and Lt. Col. W. B. R. Neave-Hill of the Historical Section

the Imperial War Museum London, in particular the staff of the Photographic Library and Printed Books Section, and specifically Mr. M. J. Willis for his valuable criticism and advice

Martin Pawley who provided us with valuable criticism and access to his own research.

In addition we are extremely grateful to the many others who have at various times assisted us:

U.S. Department of Defense Washington, Bundesarchiv (Militärarchiv) Freiburg, The Royal Artillery Association London, The Institution of Civil Engineers London, The Institution of Structural Engineers London, The Royal Institute of British Architects, Purnell and Sons Ltd, Her Majesty's Stationery Office London, Diana Stoddart, Douglas H. Robinson, Dr. A. A. Caenepeel, Paul Virilio, Claude Parent, Dr. R. Buckminster Fuller, Christopher Duffy, Cecil Newman, Siegwert Geiger, Gen. W. Warlimont, Capt. A. de Lumsdaine, Stewart McKenzie, Peter Clare, Colin Partridge, Alvin Boyarsky, Paul Oliver, Tore Brantenberg, and Paal Didrik Holm.

Our debts to the various sources on which we have drawn in preparing the text, and the many publishers and photographers who have provided us with illustrations are detailed more fully in the reference notes and illustration credits.

Last but not least we should like to thank Robert Hall of the Architectural Press, who has not only designed the book but also given considerable help in obtaining illustrations, and Siri, Espen, Cheryl, Su and Heli who have made the book practically possible.

PREFACE

During the first half of this century almost astronomical sums were spent on war or the preparation for war. A large proportion of this money was expended on military construction—on programmes which covered every conceivable aspect of building. Simultaneously, technical advances were made in the more general context of weaponry and military research, such as new materials and techniques, which were to affect subsequent developments not only in military construction, but also in post-war civil architecture.

The huge energy potential which was directed into military construction, both directly and indirectly, has inevitably had an important technological impact on the development of civil architecture during the last seventy years. One would assume therefore that the activities and influence of the military sector of the construction industry would be closely examined, yet surprisingly little attempt has been made to do this. Instead the history of modern architecture has been primarily recorded by art historians as a series of art movements and personalities, an attitude which is typified by Pevsner's dictum that cathedrals are architecture—bicycle sheds are not. This approach was applied specifically to military construction by the Honorary Secretary of the Royal Institute of British Architects, the reactionary M. Waterhouse, who in 1943 stated firmly:

"It cannot be called Architecture, either as we knew it—or as we know it ought to be. I don't know what it really can be called. I am tempted to define it as a combination of Organisation and Improvisation."

Military construction has therefore been largely unconsidered in modern architectural history. It is not identified as 'architecture' even though, as the illustrations in this book suggest, it can easily be interpreted as such. The problem has perhaps been the perspective of time. Historians find little difficulty in assessing the broad basis of the architecture of 500, 1000, or 2000 years ago, when it can clearly be seen in the context of such factors as the social, ethnic, climatic, political and military environments in which it was created. Yet the history of our own civil architecture over recent years appears to be still too close to us to have produced a balanced objective assessment of the lines along which it is developing—or how they have come about. One book to do this was Banham's 'The Architecture of the Well-tempered Environment', in which an attempt was made to relate the development of modern architecture with the development of modern heating and electrical services. The original intention of our book was to identify a similar relationship between the development of modern architecture and military development, in terms of both military construction and weaponry 'spin-off'. However we met with a major problem, namely the lack of any significant secondary sources in which the history of contemporary military architecture had been brought together.

Thus the objectives of this book were, of necessity, amended. What was needed, before any attempt was made to examine the relationship between military and civil construction, was a book which simply described the development of military architecture in the twentieth century. This, in itself, was an ambitious project and our book can hardly claim to be the complete definitive history of the subject, nor was it intended to be. Instead our attention has been limited to specific topics within the geographic limitation of North West Europe and chronological period 1900–45; both the supposed time and place of the birth of the modern movement in civil architecture and perhaps significantly, the focus of the most extensive and costly military operations the world has ever seen.

There is also another parameter. Military architecture embraces a variety of modes and purposes; for instance, the London barracks of the 18th and 19th centuries, with their Doric columns, can hardly be termed any more military than Nash's London terraces of the same period. But the topics of military architecture described in this study all have the same common factor: all are very directly related to aggression, all are undeniably military in design and purpose, responding to developments in both weapons and military techniques and to the pressures of military and political events.

Within this common factor certain divisions can be made. Before the twentieth century, when soldiers attacked on foot or horseback, military construction was mainly described in three categories: field fortification, permanent fortification, and the structures built to counter them—

the scaling ladders and movable towers of the 'siege train'. Allowing for technological mutations these same three categories can be generally applied to the twentieth century. The field fortifications on the Western Front in 1914–18, the permanent fortifications of the 1914 fortress zone or even the Maginot Line of the 1930's, the 'siege train' of the artificial harbour project in 1944, continue a tradition that goes back to the beginnings of organised warfare. No matter how much military structures have changed to adapt to new weapons and technologies the principles behind them have remained the same. That convenient division is however a military rather than an architectural one. From an architectural point of view the interesting thing about military architecture is its responsiveness and adaptability to events—some would say more so than its civilian counterpart. For this reason we have described in some detail the military and political climate in which this construction occurred. Only in the context of events does the nature of military architecture as a continually adaptive process emerge.

We have not, of course, been able to cover everything. Many interesting areas are unfortunately neglected: airship hangars in Britain 1900–18, the large scale camouflage of critical structures such as power stations and granaries during 1939–45, military bridges, are just three examples. But we hope that what we have described and discussed is sufficiently representative to make the point that military architecture not only had a considerable impact on military (and indeed political) events, but also has an important and hitherto largely neglected place in the history of twentieth century architecture as such.

Keith Mallory/Arvid Ottar Bath 1.5.73

PART ONE 1900→18

1 1914 FORTRESS ZONE

French, German and Belgian fortresses on the Western Front: their precedent and effects.

Events

1870 French defeat in Franco–German War.

1874 General Séré de Rivières draws up plan for French fortress zone.

1888 July Belgians start work on their Meuse fortresses at Liège and Namur.

1900–14 Germans construct two fortified zones facing France: the 'Moselstellung' and Strasbourg.

1905 Schlieffen devises German invasion plan evading French fortress zone: sweeping down through Belgium.

1914 Aug. 1 German troops enter Luxembourg: seize railways needed for invasion of Belgium.
Aug. 4 War declared.
Aug. 16 Fortress at Liège falls.
Aug. 26 Fortress at Namur falls.
Sept. 5–9 German advance halted in the Battle of the Marne.
Sept. 24 First Battle of Ypres completes the deadlock.
Oct. 10 Fortress of Antwerp falls.

1915 Aug. 5 French orders reduce status of fortresses after rapid fall of Belgian fortresses.

1916 Feb.–June Battle for the fortified region of Verdun: French successfully defend.

1918 Nov. 11 Armistice.

1919 Preliminary studies of French Defence problem by General Staff.

1928 Feb. 17 First three sections of the Maginot Line go out to tender: design based on success at Verdun.

---

1. *Typical geometry of a Vauban fortress of the late seventeenth century.*

● The permanent fortress in 1914 was the continuation of a basic military concept which stretched back several thousand years, and which evolved through changes in artillery power, developments in military techniques, and advances in building methods and materials. In Ancient Greece and Rome the wall had been its prime expression—a principle which was also seen in the Wall of China. When, in the 12th century A.D. the Normans developed the feudal castle, walls became higher in order to resist scaling ladders. More complex crenellated designs were introduced and round corners became common in efforts to resist mining attack. Then came the introduction and development of gunpowder and cannon in the middle ages, which forced walls to become lower and thicker in protection against cannon balls. The eventual result was the bastioned fort, almost a walled ditch, which was developed into increasingly elaborate forms. By the end of the seventeenth century Louis XIV's engineer, Marshal Sébastian de Vauban, developed systematic methods of attacking these bastioned fortresses and, given time, the fall of a fortress was inevitable. Despite this, the eighteenth century saw relatively little change in fortress design.

During the nineteenth century design and layout changed rapidly. Three artillery developments brought this about. From 1792–1852 increased mortar power meant that shells falling vertically had to be taken into consideration and the guns in fortresses were given overhead protection with casemates. The next change was an even greater one. In the 1860s rifled artillery appeared with enormously increased firepower and accuracy—detached forts were the result. These allowed a girdle of forts to be added to existing fortresses thus keeping enemy artillery out of range of the cities they protected. Known as 'girdle' or 'ring' fortresses, the diameter of the ring formed by the detached forts soon increased in proportion to the steady increase in artillery power. The third major artillery development in the nineteenth century occurred in 1885–90: the French development of the explosive shell. This forced replacement of masonry with concrete and the use of metal cupolas to protect fortress artillery. By the end of the nineteenth century the major European fortresses were thus 'girdle' fortresses composed of a ring of detached forts which had concrete overhead protection and guns mounted in metal cupolas. The size and expense of these fortresses meant that they were confined to points of major strategic significance. To form more continuous lines of defence 'barrier' forts, small groups of

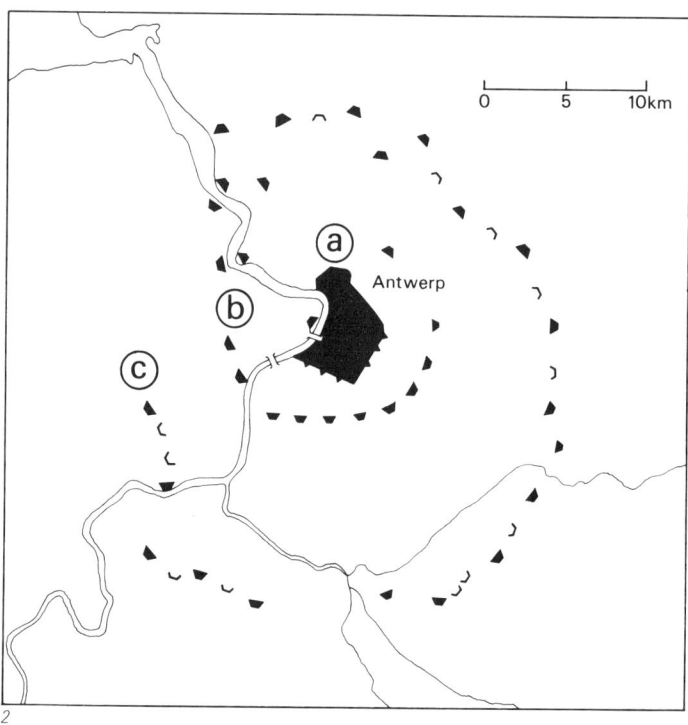

2. Diagram of the ring or 'girdle' fortress at Antwerp in 1914 showing three phases of development:
   a. the walled city
   b. the initial ring of detached forts built to meet growing artillery range.
   c. the outer ring of detached forts and intermediate ouvrages which was constructed when the outer ring became outclassed—it was scarcely completed in 1914.

3. The relationship between ballistics and detail design as illustrated in von Brunner's training manual of 1910.

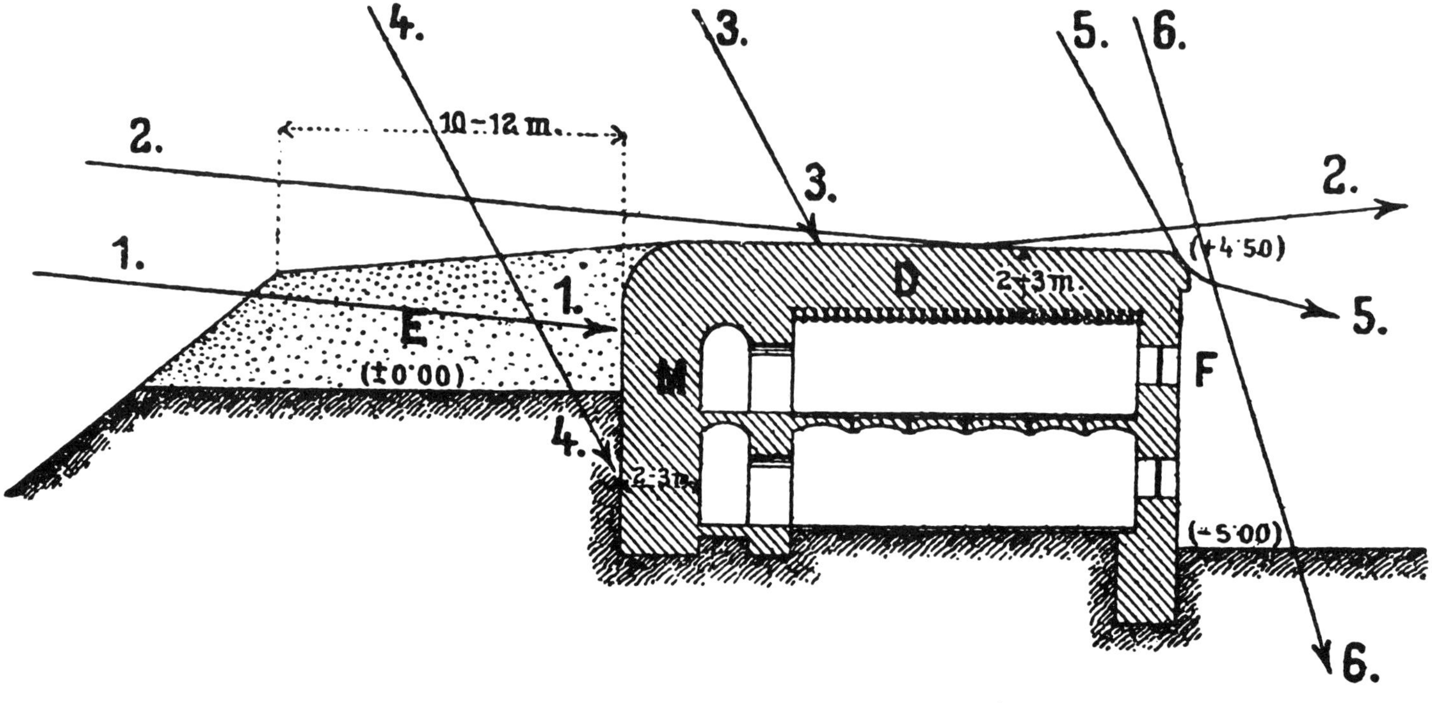

works or single forts, were built defending bridges, valleys, railway lines or other main lines of communication (1).

This very brief synopsis of fortification development indicates the striking part which considerations of artillery power had played in the evolution of the fortress during the period 1792–1914. While artillery thus dictated the basic form of fortress design at this time and the materials which were required, it also dictated their detail design. This fact was emphasised in a training manual by an engineer in the Austro-Hungarian army, Major Moritz Ritter von Brunner, which was translated and published by the British General Staff in 1910. It stated that the type and amount of hostile fire to be expected were the principal criteria in fortress design:

"The building materials, choice of armament and the mounting of guns are all dependent on the effect of this fire." (2).

Von Brunner went on to apply the theories of ballistics to the detail design of concrete fortifications:

"The majority of casemates are protected from gun fire in the following way (see illus. 3):
The front or escarp wall 'M' although several feet thick would soon be destroyed by continuous direct fire if left bare. It is therefore protected by a bank of earth 'E', which absorbs most of the striking energy of the shell (see Trajectory 1) . . . Shell with low trajectory which strike the crest of the earth parapet or the masonry roof of the casemate glance off and do no damage (see Trajectory 2) . . . A 'mortar' shell compared to one from a gun of the same calibre has a smaller striking energy but a far greater bursting charge . . . The damage caused by a shell hitting the roof 'D' of the casemate is due, firstly, to its striking velocity, and secondly, to the bursting of the charge immediately after the impact (see Trajectory 3); $6\frac{1}{2}$ to 10 feet of concrete is sufficient to withstand the shock of even two hits at the same spot. The roof is supported by arches or preferably by steel girders . . ." (3).

Although artillery power was undoubtedly as important as von Brunner pointed out and artillery developments forced changes in design it was new materials which made these changes possible. Brick and masonry were replaced first by concrete and later by reinforced concrete. After the cupola was introduced, offering superior protection to the casemate, the iron cupolas initially used were soon replaced by armour-plated steel. By 1900 all new French cupolas were made of this material. But the development of the cupola also suggested the extent to which materials and constructional developments, as quite separate factors from considerations of artillery power, themselves affected the development of fortress design. While the cupolas of 1880–90 were comparatively primitive, by 1910, when von Brunner's training manual was published in England, cupola design had become much more sophisticated. An indication of this sophistication was the hydro-pneumatic retracting carriage, a design which allowed a gun to emerge only when firing, giving much increased protection to both gun and crew. These developments were so fundamental that many of the advanced theories of fortress design used such cupolas as the basis of a new approach. This clearly indicated that such technical developments could affect fortification design just as much as the development of artillery power.

One very early theory based on the new cupolas was that of M. Mougin, a director of the St. Chamond firm, who in 1887 had proposed what was described by a contemporary English writer as a new system of fortification:

"His (Mougin's) fort is simple in conception, consisting of a huge block of concrete buried in the ground and carrying on the top three cupolas, each mounting two 15 cm guns. Four disappearing cupolas each containing two machine guns. complete the armament. Observation stations, electric lights, ventilating fans, and all the resources of science are included in the block of concrete, which it to be garrisoned by thirty or forty skilled mechanics." (4).

Mougin thus proposed a small concentrated fort but other authorities saw the cupola as the basis of a much more dispersed approach, eliminating the dangers caused by concentrating targets in forts. Von Brunner summarized two of the main theories which had become established by 1910:

"Several proposals to overcome this difficulty (of concentration) have been made. Some advocate the dispersion of all the component parts of the fort, others would replace the fort by several batteries of armour protected Q.F. (Quick Firing) and long-range guns. The first arrangement may be described as a system of 'Dispersed points d'appui' and the second as a system of 'armoured fronts'." (5).

The second of these systems was originally suggested by Lieutenant-Colonel Schumann, an engineer in the Prussian Army, and by Captain

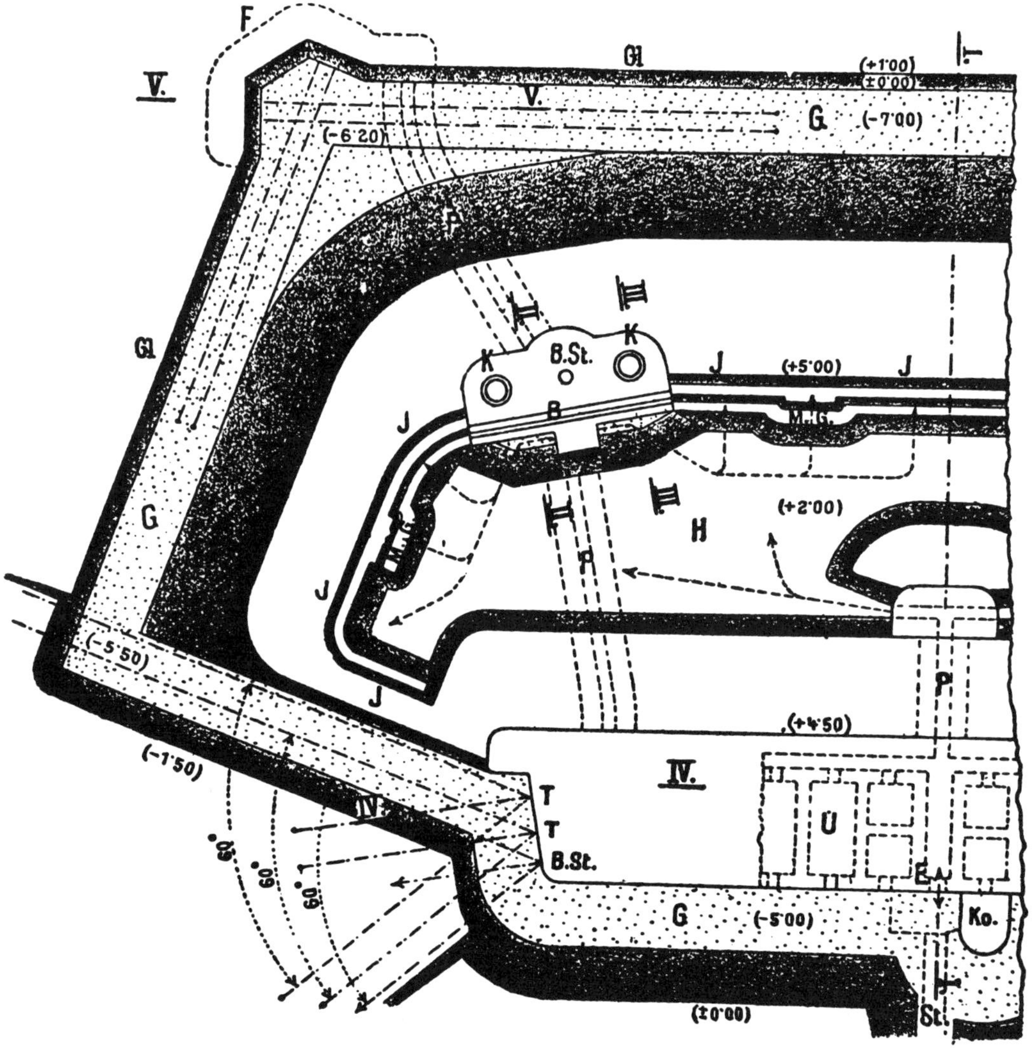

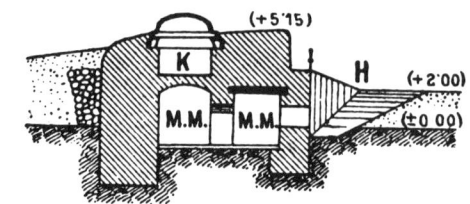

Section on III—III.

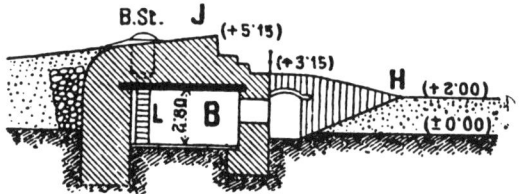

Section on II—II.

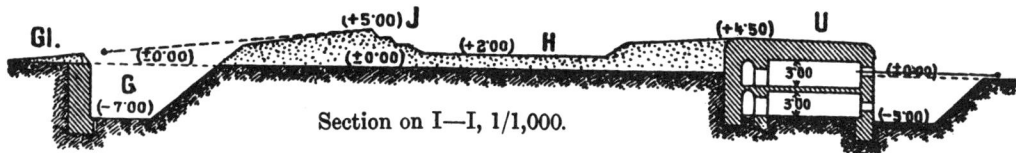

Section on I—I, 1/1,000.

4. Von Brunner's example of a detached works.

Meyer of the Swiss Army. It was based on the invention of a mobile armoured cupola. This 'cupola', an armoured quick firing gun on railway mountings, was also a part of Mougin's proposals and was in fact actually used at the turn of the century on the Sereth Line, a short fortified position stretching across the lowlands on the threshold of Moldavia.

Although cupola design became more sophisticated in this way and enabled new theories of fortification design to be developed, such theories were not generally applied in practice. While theories developed at the same accelerating rate at which artillery power increased the actual construction of fortresses found difficulty in keeping up. Construction was a slow and expensive affair and most of the European frontier fortresses had been started in the 1870's and 80's. Although the detached forts were being continually strengthened to meet new developments, a complete change-over to new ideas such as the 'armoured front' was not possible in practice. In addition there was the usual resistance to change. An English critic wrote in 1890 of General Henri Brialmont, the most influential of the 19th century European fortress engineers:

"All the authorities cited agree that changes are needed, and, with one exception, all advocate considerable innovations. General Brialmont, however, practically contents himself with sprinkling cupolas over his plans, and retains all the objectionable features of the stereotyped fort. In his view, the advances in the power of weapons merely call for increased protection. He accepts with effusion the designs of the ironmasters, and superimposes them upon all the tricks of drawing office fortification in its most aggravated form. The new ironmongery is in fact to be an addition to old methods, and cannot be trusted sufficiently to form the basis of a new system." (6).

The outbreak of war in 1914 thus found a situation in which the fortress zone consisted primarily of girdle fortresses and, although these did not reflect the latest theories, they were strengthened to varying standards with new materials and equipment such as the retractable cupolas. But some fortresses had not been brought up to date at all. They had become completely obsolete and officially declassified. The technical surprises of 1914, particularly the German 420 mm howitzer, were to make many more of the European fortresses obsolete—even those which had so far been thought impregnable; a fact which emphasised once more that fortress design had to be a continually evolving process.

The actual strategy behind the fortress zone of the French, Belgian and German frontiers varied from country to country and reflected their varied military attitudes. For France, her defeat in 1870 during the Franco–Prussian War and fears caused by subsequent German expansion had made the problem of defence assume considerable importance. The new frontier brought Germany within less than 320 km of Paris and, with Alsace-Lorraine now in German hands, there was no natural barrier such as the Rhine or the Vosges (7). The result was that General Séré de Rivières, a sapper, was given the task of constructing a large frontier defensive system. His plan, drawn up in 1874, called for three positions: firstly a frontier position; secondly an intermediate position with barrier forts and other small forts at Dijon, Langres, Besançon, Rheims, Laon and La Fère covering Paris at long range; thirdly a ring of detached fortifications around Paris (8). Along the Belgian frontier Séré de Rivières established barrier forts at Meziers, Maubeuge, Valenciennes and Lille to cover the principal lines of advance through Belgium. Along the German frontier large numbers of detached forts on the girdle principle were established at Belfort, Epinal, Toul and Verdun. A series of six barrier forts were constructed between Verdun and Belfort, and similar barrier forts were built between Epinal and Toul (9). In this way two definite fortress belts or defensive zones were constructed along the German border. Gaps were left between them at Stenay and Charmes, known as the Trouée de Dun and Trouée de Charmes, with the optimistic objective of channelling any German invasion into a position vulnerable to flank attack.

Séré de Rivières' plans were drastically cut. Improvements in artillery power and the use of concrete and armoured cupolas designed to counter them meant that reinforcement of these forts would be extremely expensive. Aided by a degree of post-war pacifism the forts of the intermediate position were not brought up to the new standards and many were eventually put out of service. Similarly, of all the forts along the Belgian border in the first position, only Maubeuge was to be brought up to standard with 'cupoles cuirassées' (10), armoured cupolas, but even this was incomplete when the war began. Only the defences actually along the border with Germany at Verdun, Toul, Epinal and Belfort were brought up to date and equipped with cupolas.

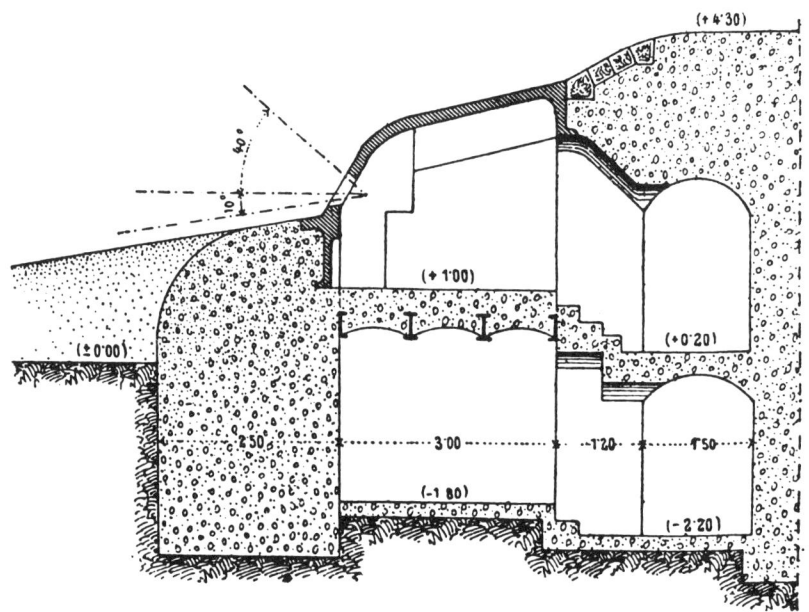

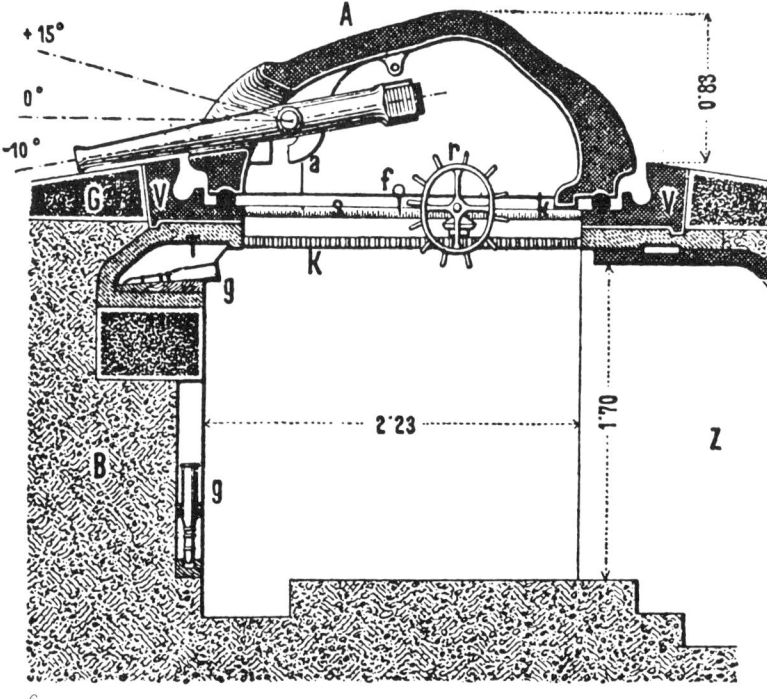

5. Armoured casemate for a howitzer.
6. Simple rotating cupola with centre-pivoted 3.5 inch quick-firing gun.
7. Rotating cupola with counter-balanced and muzzle-pivoted 12 cm gun.
8. Schumann's retracting cupola with 12 cm gun.
9. Mobile armoured cupola for a 53 mm gun as used on the Sereth Line.

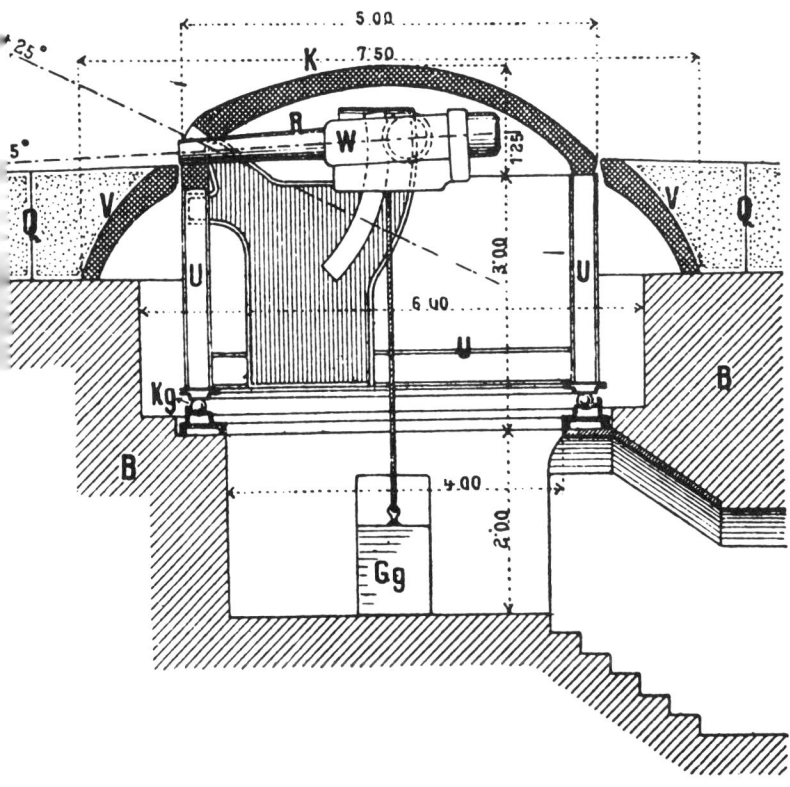

7

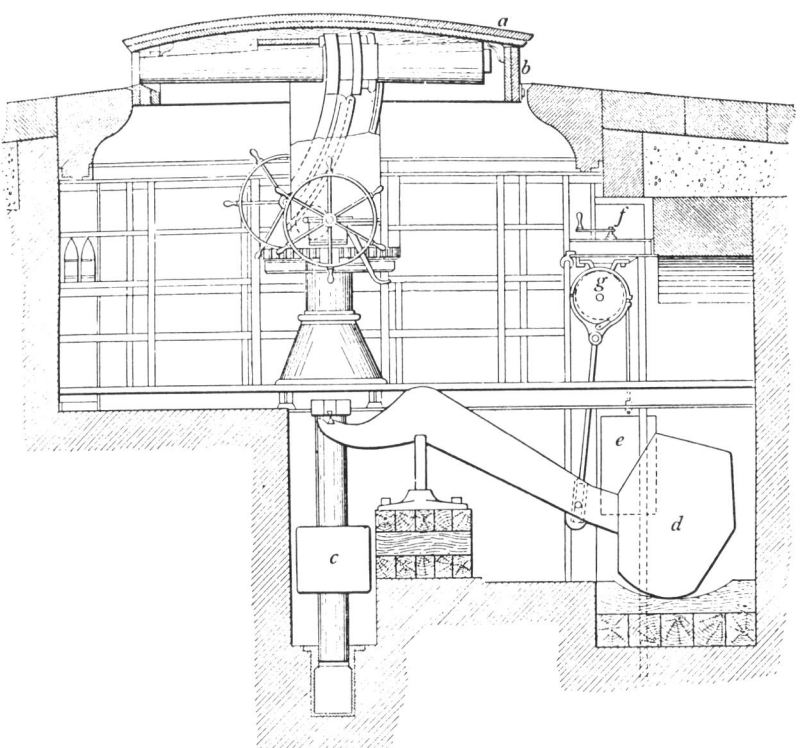

8

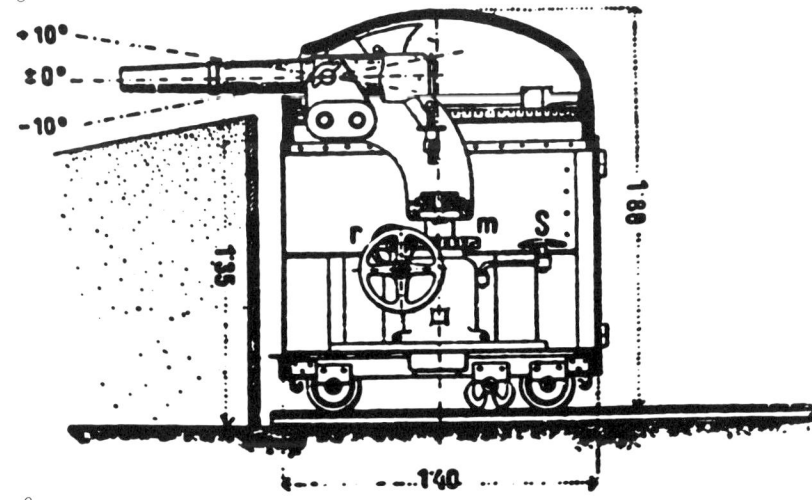

9

10, 11. Section and plan of the subterranean fort proposed by Mougin:
  a. artillery cupolas
  b. machine gun cupolas
  c. observation

At Toul the advanced works were strengthened by the construction of a defensive position known as the 'Couronné de Nancy' but this was incomplete in 1914. Thus, in August of that year, apart from the modernized defences of Paris and the incomplete Maubeuge, France's only modern fortresses of any value were on the German frontier itself—the two zones on the line Belfort–Verdun (11).

The modern German fortifications were similarly massed along the Franco-German frontier. Germany had traditionally guarded the Rhine but of the five fortresses: Wesel, Cologne, Koblenz, Mainz and Germersheim, only the last three had been given a ring of detached forts and only Cologne was up to date. But after the conquest of Alsace-Lorraine the old Rhine fortresses were in any case of less importance. The old French fortress at Strasbourg was turned into an enormous fortified zone barring 'La Vallée de la Bruche' by the construction of the modern 'Feste Mutzig', the Mutzig fortress, near Molsheim. North west, along the frontier, the fortress of Metz was also reinforced and enlarged with the modern 'Panzerfesten', armoured forts, and linked up with advanced forts at Thionville to form a second fortified zone (12) known as the 'Moselstellung' (13). Construction of these two main zones took place between 1900 and 1914 and they were consequently very modern in design—although German cupolas were generally inferior and weaker in construction, compared with both Belgian and French types. Their turrets were only of the simple turning type and, as in Belgium, only the small calibre guns were placed in the disappearing mounts. By comparison the French, who had changed their outdated cast iron cupolas for steel ones since 1900, had invested in hydraulic vanishing cupolas for even their large guns. Their armoured steel was twice the thickness of the German steel and the French concrete was, as the war soon proved, also just as superior to either Belgian or German equivalents (14).

Belgium, home of the military engineer Brialmont, had been faced with a different problem to either France or Germany in the design of her permanent fortifications. Her main problem was not the protection of Belgium itself but the possibility that, in the event of a Franco-German war, either France or Germany might choose an invasion route through Belgium (15). In the middle of the nineteenth century Belgium had thus abandoned a principle of national defence which relied on the protection of numerous small points—a principle which overstretched her Army. Instead the Belgians resigned themselves to barring the main invasion route through Belgium: the Meuse valley. Between 1880 and 1890 they constructed two enormous fortresses at Liège and Namur with this in mind, the forts at Liège being 14 km from the Dutch border, those at Namur 30 km from the French border. The individual forts were designed on either a triangular or pentagonal trace according to the terrain but all had the same characteristics: a low lying sunken central redoubt with thick concrete protection, a surrounding outer rampart composed of an earthen bank on one side and a thick concrete wall on the other with flanking galleries, with cupolas in the central mass of concrete housing 15 cm and 12 cm guns, 21 cm howitzers and 55 mm quick firing guns. The cupolas were of a design which had been determined by long experiments at Bucharest in the winter of 1885–86 (16). The two fortresses were 40 km apart but, with only the outdated defences at Huy in between them, they were the main Meuse defence. No barrier forts or field defences were constructed to link them up into a continuous defensive zone. As the cannon of these two fortresses had a range of only 8 km their strategy was subsequently open to criticism (17). In 1922 Général Benoit wrote that:

". . . it was impossible for them to efficiently guard the Meuse passage: they (the fortresses) could guard no more or less, in effect, than twenty permanent bridges; the nine other bridges, of which several were important, completely escaped them." (18).

Certainly in pre-war Belgium the Meuse fortifications and their designer, General Brialmont, were the subject of considerable criticism (19). The ex-War Minister, Chazal, pointed out that large garrisons would be essential. An increase in the standing army would be the result but the then War Minister, General Pontus, gave the assurance that:

"The proposed disposition does not call for any increase in our forces. The total strength of our army ought not to be increased because of its implementation, but some changes ought to be made in the composition and distribution of the artillery." (20).

As the Clerical Party strongly opposed any increase in the standing army this assurance by General Pontus probably enabled the scheme to get through the House of Representatives with a comfortable majority even though, as a British critic later pointed out, the assurance was 'radically unsound'. Despite this, work on the Meuse fortresses was

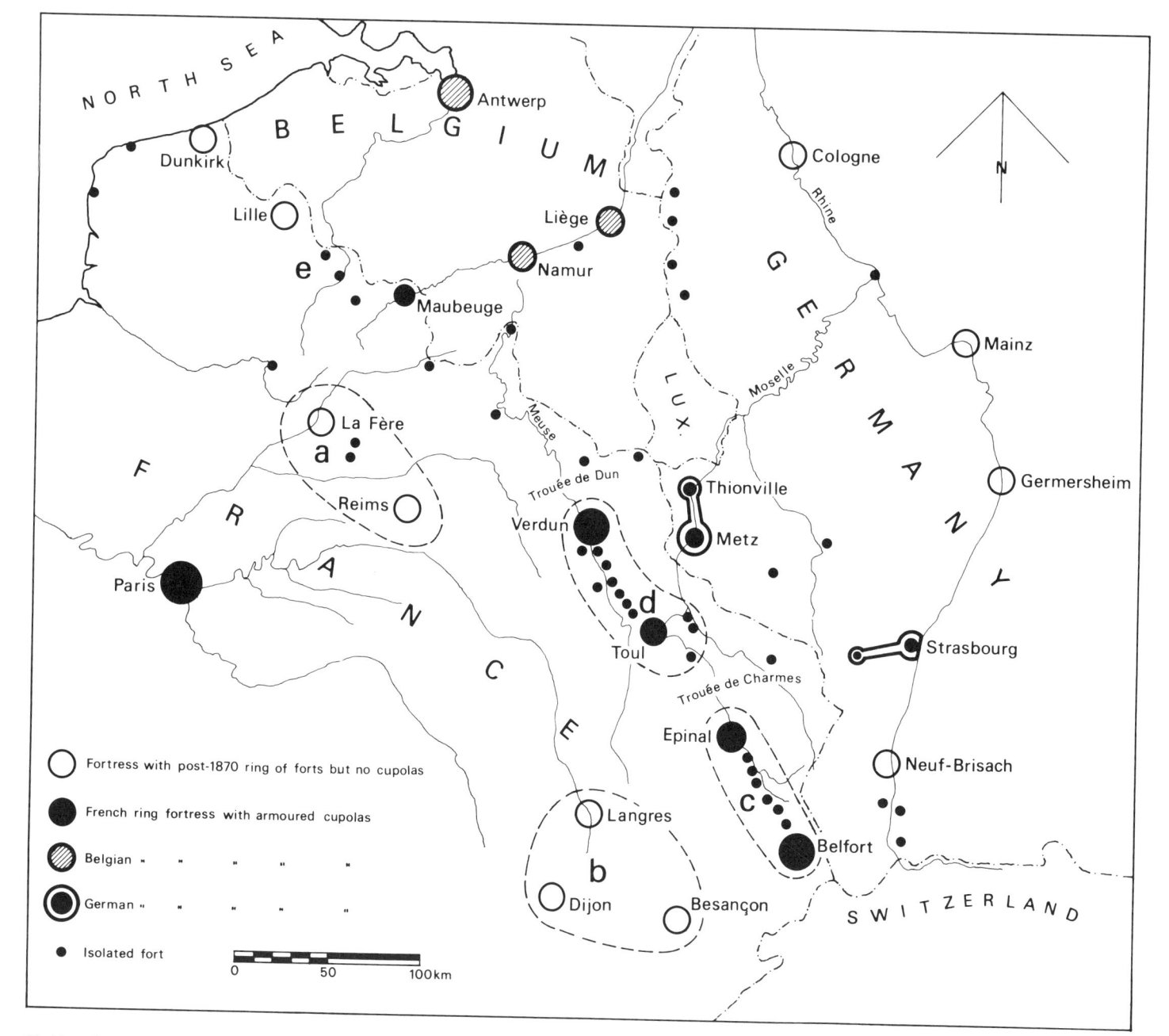

12. Map of the fortresses and fortified zones on the Western Front in 1914:
Zones a/b. were the remnants of de Rivières' plans
c/d. on the German frontier were France's strongest defences
e. on the Belgian frontier was the weakest zone and prompted Schlieffen's plan for an invasion route through Belgium.

started in July 1888. The same British critic, Major G. Sydenham Clarke, wrote two years later:

". . . that the conception of permanent defences of the Meuse was wrong. The crowding of artillery into small conspicuous areas was, I believed, a mistake. . . ." (21).

In addition to the two Meuse fortresses an even larger girdle fortress was constructed at Antwerp. This was also a controversial issue. Antwerp was planned as a pivot for the Belgian field army and a national redoubt in case of failure. A similar strategy had been devised twenty years earlier, in 1859, when Napoleon III had been the main threat, and Brialmont had fortified the city with a ring of thirteen forts. The circumference of this ring was 43 km but the city had grown and artillery range had increased. The new 1880's programme was for a new outer ring of thirty-one forts, 31–38 km in diameter and 108 km in circumference, forming a vast entrenched camp (22). Earthworks and brick casemates were to be replaced with concrete and steel. The fortress was to be much larger than either Liège, with its twelve forts, or Namur, with only nine (23) and even Brialmont opposed the plan on the grounds that the fortress would require not a garrison but an army to defend it. Despite Brialmont the outer ring was constructed (24). But although the three Belgian fortresses varied in size, all three were designed on similar principles: the ring of detached forts on a triangular or pentagonal trace, use of revolving cupolas, and use of concrete. Ultimately all three fortresses showed similar failings—in general summed up by Colonel Rébold, the French 'Chef du Bureau Fédéral de construction des ouvrages fortifiés' from 1906 until 1921:

"When war broke out, none of the three main Belgian defences fulfilled the requirements of a modern fortress. For a long time, the forts designed by Brialmont no longer complied with the new principles and techniques of construction. It was a failure which had ill-fated consequences." (25).

The varied strength and disposition of the French, German and Belgian fortresses were to have considerable effect on the military strategies of 1914, in particular the German invasion plan, and ultimately on the attitude to permanent fortifications during 1915. Even before the French cupolas were upgraded the German High Command was apprehensive of the strategic problems created by Séré de Rivières' fortified zones. In 1879 Kaiser William had written to Bismarck:

"The French frontier is almost hermetically sealed from Switzerland to Belgium. Even if we succeed in penetrating an unbroken line of fortresses, they would prevent any reinforcements being sent up, and exercise an enormous influence on the strategical advance of our forces. If we should be victorious, we should no longer be able to follow up the enemy as in 1870, and would have at once to lay siege to this girdle." (26).

It was not until 1905 that the solution to this problem was found. The plan, devised by the Chief of General Staff, General Graf von Schlieffen, was for a huge wheeling movement pivoted on the German 'Moselstellung', which would sweep round to the north of the French fortress zone. The main strength was to be in this right wing while the weak left was to lure the mass of the French army towards the Rhine, creating a 'swing-door' effect. The German army, having swept through Belgium would then be in a position to outflank the main French army. Ironically the French defensive plan at this time, the notorious 'Plan XVII', was for just the advance on the Rhine which von Schlieffen hoped to contrive (27).

If Dutch neutrality was to be respected, which avoided giving Britain the excuse to land troops in Holland, Liège was the key to the German advance. It lay on the line of the strong right wing of the Schlieffen Plan, blocking the gap between the 'Limburg appendix' of Holland and the Forest of the Ardennes (28). In 1911 Schlieffen's successor, Moltke, had prepared a memorandum which accurately foretold the events which were to come about three years later:

"Inconvenient though it is, the advance must proceed through Belgium without violating Dutch territory. A prerequisite is the possession of Liège. This fortress must therefore be taken at the outset. I believe that it is quite possible to reduce it by a 'coup de main'. The advanced forts are sited so badly that they cannot see or command the intervals. I have sounded all the routes of advance which lead between the works to the interior of the town. It is perfectly feasible to execute a push in several columns without being seen from the forts." (29).

Consequently, in August 1914, Liège was the first of the Belgian fortresses to feel the full force of modern warfare.

The twelve forts of the girdle fortress at Liège had been built in 1888–92. When Major-General Erich von Ludendorff, Chief of the Operational Section of the German General Staff from 1904–13, had

drawn up his plan of attack he had possessed exact details of their design (30). Brialmont had made a number of important errors: the most obvious being that the neat ring layout left a number of uncovered areas. But the main weakness in 1914 was the length of time which had elapsed since the forts had been designed. Over twenty five years nothing had been done to bring the forts up to modern standards (31). Designed for the relatively small armies and artillery power of the 1880's the forts could offer only limited resistance to the huge German army of 1914. Following the declaration of war on August 4th the town of Liège was captured after only three days. The forts fell, only eight days later on August 16th—a total resistance of eleven days. They did little to hold up the German invasion plan, which had allowed fourteen days for the capture of the fortresses.

The failure of Liège illustrated the fact that military developments had once more made certain standards of construction obsolete. But it also pointed out defects in design and in the use of the Liège fortress. In 1890 the British fortification expert, Sydenham Clarke, had reported that:

"The effective defence of the entrenched positions of Liège and Namur will depend mainly on a field force guarding the intervals." (32). But the Belgian field army was kept well back from Liège, behind the Get, and the fortress was left unsupported. King Albert of Belgium had put General Gérard Leman in command of the Liège fortress in January 1914. Though Leman realised the danger, he was forbidden by the General Staff to construct field works in the intervals for fear of offending Germany. Leman was only able to take last-minute measures. The infantry managed to dig only superficial interval defences, there was no time for minefields or extensive wire entanglements, and only the most vital fields of fire were cleared (33). Brialmont had himself been an opponent of prepared interval works (34). While the principle of the design of the Liège fortress was proved to be basically faulty in this respect other technical weaknesses were shown in its capture.

The Belgian fortresses were designed in the 1880's before any knowledge of the necessary concrete thicknesses was available. In their attempt to counteract growing firepower, the Belgians had not used the material effectively. It was not reinforced with metal as were the French forts, and the thickness of 1.5 m, laid on 40 cm of sand and 2 m of stone, was only half the general standard in 1914. With no appreciation of the effect of firepower on concrete the Belgians had made their situation worse by using only a weak type of concrete (35). In addition the concrete was laid in two layers, instead of being monolithic, causing lines of weakness (36), and being designed before the advent of the explosive shell this concrete could only withstand direct hits from 210 mm black powder shells (37). Fort Barchon, the first fort to fall on August 8th, was in fact only under 210 mm mortar fire for a short time (38). Though Krupps had secretly prepared 420 mm howitzers, popularly known as 'Big Berthas' after Frau Krupp von Bohlen, especially for the destruction of the Belgian fortress, they were not brought into action until August 12th—four days after the fall of Fort Barchon (39). The guns, weighing 75 tons, were slow to position, but even without this 'super' artillery the Liège forts had been proven unsuitable against modern artillery. Fort Pontisse, the first fort to come under 420 fire, surrendered after only 20 hours. The forts were completely outclassed, and during the next 48 hours Forts Embourg, Liers, Fléron, Boncelles and Lentin fell. By the 15th only three forts remained. An engineer at Fort Boncelles recorded the helplessness of the forts in his diary.

"Aug. 13. Arrival of German heavy artillery placed in such a way that we could not see them nor fire at them. Impossible therefore for us to render any help to Fort Embourg, which the enemy was now shelling....
Aug. 14. 6 p.m. Two shrapnels burst over our fort. The range probably being thus found, our fort was bombarded up to 8 p.m. Our telephone communications being smashed, all troops between the forts having been ordered to retreat on the 7th Aug., we were unable to defend ourselves. Orders were given to close the cupolas and await events....
Aug. 15. About 6 a.m. the concrete chambers where our cannons were began to fall in. Several of our cupolas turned no more." (40).
Fort Boncelles surrendered only 1½ hours later (41) and the remaining forts also by the following day.

While the forts mainly fell due to poor tactical siting, lack of support and obsolete protection, the conditions inside the forts increased their weakness. Ventilation was poor and air-tight doors were nonexistent (42): with explosive fumes filling the fort breathing was difficult and in some cases men were asphyxiated. Water cisterns cracked, and the observation posts were insufficient. Brialmont had taken little consideration of the conditions for the three to five hundred

13. One of the Belgian forts during construction.

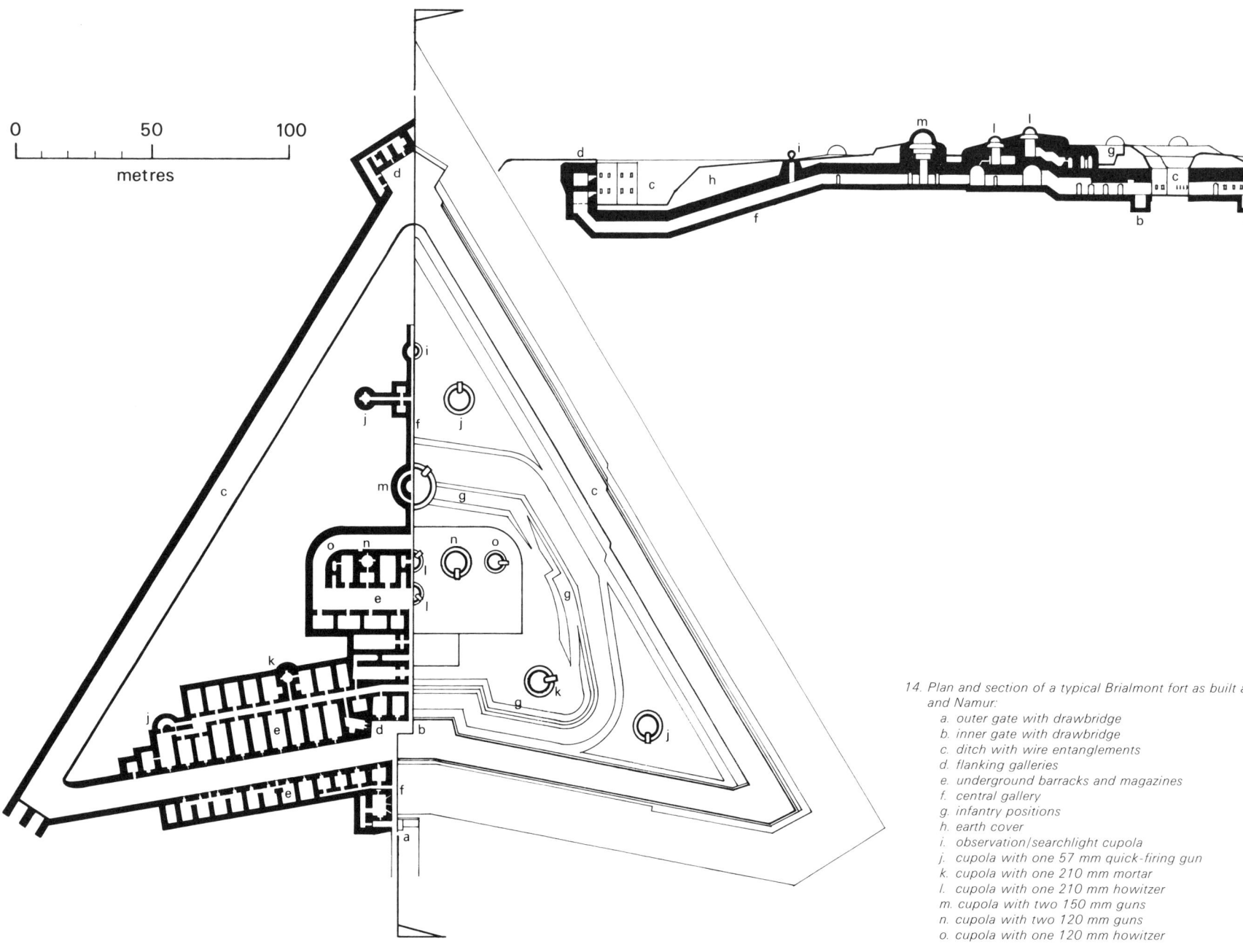

14. Plan and section of a typical Brialmont fort as built a and Namur:
   a. outer gate with drawbridge
   b. inner gate with drawbridge
   c. ditch with wire entanglements
   d. flanking galleries
   e. underground barracks and magazines
   f. central gallery
   g. infantry positions
   h. earth cover
   i. observation/searchlight cupola
   j. cupola with one 57 mm quick-firing gun
   k. cupola with one 210 mm mortar
   l. cupola with one 210 mm howitzer
   m. cupola with two 150 mm guns
   n. cupola with two 120 mm guns
   o. cupola with one 120 mm howitzer

men of the fort garrison. In 'La Rapport du Général Leman sur La Défense de Liège en Août 1914' Georges Hautecler states:

"... from the moment when enemy fire made it impossible to cross the ditched way, the men were separated from the stores and they were deprived of latrines (except at Loncin), in this way they were made to live without food in 'une atmosphère excrémentielle'." (43).

Général Leman is further quoted as saying:

"Brialmont's military genius had an academic bent, and he forgot that his works were made for human beings. He left out of account a natural function of mankind which does not cease during bombardment: quite the reverse." (44).

Having taken Liège the German route through Belgium was open. The fortress at Namur stood up to only four days bombardment and fell on August 26th. The Belgian army was besieged at Antwerp. An editorial in 'The Times' on October 1st, 1914 was optimistic:

"The Belgians are holding one of the strongest places in Europe, they must have at least 120 000 troops at their disposal, they possess an open seaport, they have only second-line troops aginst them, their outer defensive works are the newest and the most formidable of all.... We do not think that there is any need to worry about Antwerp." (45).

But, even though the outer ring of forts at Antwerp were the most modern in Belgium, hardly completed in 1914, they proved little better than the Liège forts. At Fort Wavre-Ste-Catherine a magazine exploded, 'the cupolas crumpled one by one' (46). On October 10th Antwerp fell.

In early September, therefore, the main German right wing was deep into Northern France despite the Belgian fortresses. The French minor forts and barrier forts in this sector, while oversized for blocking a route for a short time only, were not strong enough for an active defence, and caused no major obstacle to the advancing German army. It was the German Commander in Chief, Moltke, not the Allies, who was responsible for the halt of the German advance. By weakening the right wing and modifying Schlieffen's line of advance he exposed his exhausted army to flank attack from the Paris fortress: one of the few highlights for permanent fortifications in 1914. The Battle of the Marne followed. By the end of September a trench stalemate was established which was to last for four years.

When post-war French military writers were justifying the expenditure of large sums of money on a new generation of frontier fortifications, the success of the Belgian and French fortresses in 1914 were to be pointed out: they had at least prevented a total defeat by forcing the German army to adopt such an indirect and exhausting route. But during late 1914 and 1915 a general insecurity developed concerning the strength and value of permanent fortifications.

Verdun, the strongest part of the French fortifications, was still in French hands, forming the axis of a long salient where the French army had concentrated its main energy in Plan XVII. Now, however, Marshal Joffre persuaded the French Government to 'declass' Verdun as a fortress, using the failure of Liège and Namur as an excuse (47). Their resources and garrisons were to be put under him as Commander in Chief of the army. In a decree of August 5th 1915, the Government obliged. Four days later Joffre was able to issue the following instructions:

"Isolated forts liable to be invested have no longer a part to play.... Forts must on no account be defended for their own sake.... Fortress troops are field units exactly as are the other formations on the front...." (48).

While the report submitting the decree to the president for approval inferred that the defence of forts was of value, the Verdun forts were gradually disarmed. The engineers even received orders for the destruction of certain works. Ironically the Germans at this time had more faith in permanent fortifications: they were still completing the fortifications at Metz.

By the end of 1915 Verdun was drained of men and armaments. It was at this time that Falkenhayn, who had replaced Moltke a year earlier, decided to focus his attention on Verdun. Historians are still uncertain of his motives. His memorandum to the Kaiser in December 1915 put forward the intention of bleeding France to death:

"There remains only France.... If we succeed in opening the eyes of her people to the fact that in a military sense they have nothing more to hope for, that breaking point would be reached and England's best sword knocked out of her hand. To achieve that object the uncertain method of a mass break-through, in any case beyond our means, is unnecessary. We can probably do enough for our purposes with limited resources. Within our reach behind the French sector of the Western Front there are objectives for the retention of which the French General Staff would be compelled to throw in every man they have. If they do so

the forces of France will bleed to death—as there can be no question of a voluntary withdrawal—whether we reach our goal or not. If they do not do so, and we reach our objectives, the moral effect on France will be enormous. For an operation limited to a narrow front, Germany will not be compelled to spend herself so completely.... The objectives of which I am speaking now are Belfort and Verdun. The considerations urged above apply to both, yet the preference must be given to Verdun." (49). How Falkenhayn intended to prevent Germany 'bleeding to death' at the same time in his policy of attrition he did not say. Two months later in February 1916 the slaughter of Verdun began.

Verdun had been a fortified town for several centuries, sitting on the upper valley of the Meuse less than 65 km west of Metz. Following de Rivières' post 1870 plans it had received a ring of some 20 major forts and 40 intermediate 'ouvrages' constructed in the Brialmont system but as the forts were sited on the crest of hills they covered the valley and each other, an important feature lacking in the Belgian designs (50). The first belt of detached forts had been established between 1874 and 1880. Between 1880 and 1897 the external belt of forts, Douaumont, Vaux, Moulainville, and Bris-Bourns had been constructed. When the explosive shell was developed all the forts had been strengthened with reinforced concrete shelters, the French at this time having established a lead in reinforced concrete design. Between 1889 and 1914 the fortifications were being continually improved and added to (51).

The decision to 'declass' Verdun was therefore in many ways a surprising one. Although the Belgian fortresses had fallen with remarkable speed the French fortresses on the Franco-German border were much stronger in construction. In addition to being reinforced their concrete was both much thicker and of a better quality. Fort Douaumont, after having been reconstructed three times before 1914, had a thickness in the inner redoubt of 2.6 m—composed of a sandwich of concrete and shock absorbing sand. This was covered by 5.5 m of earth (52). Unlike the Belgians the French had carried out tests of the effect of firepower on concrete. In 1886 they carried out tests with 200 mm shells on their own Fort Malmaison which proved the weakness of construction at that time (53). The cupolas were also much stronger than those of the Belgians. While the 20 cm cast iron cupolas at Liège had been shattered and overturned, the French cupolas had been constructed from steel. Since 1900 cupolas had been constructed from laminated and toughened steel with a total thickness of over 30 cm (54), and a large number were retractable.

In early 1915 the Germans had in fact tested their 'Big Bertha' against the Verdun defences, in particular the renowned Douaumont. The fort successfully withstood direct hits from the 420s. The inhabitants, though mentally and physically shaken by the noise and impact of the shells were still ready to fight (55). Incredibly the garrison, who intially verged on mutiny, became acclimatized to the conditions. The structure of the fort was unharmed. In 1916, a year later, the onslaught on Douaumont was resumed following Falkenhayn's policy of attrition. Because of the hasty declassification of 1915, when the Germans arrived at Douaumont on February 25th they found only a small demolition party ready to blow up the fort, following Joffre's orders. It was taken without a shot being fired.

Unrest had already been growing in certain quarters about the state of the Verdun fortresses. On the day of Douaumont's capture General Pétain was given command and he immediately upgraded the permanent works. His order of March 1st stated:

"The experience of the recent fighting has proved the capacity for resistance of forts. These are naturally better organized than strong points constructed hastily during a battle ... and are no more shell traps than field works as they often cover an equally wide area. The forts, consequently can and should be used wherever they exist for the defence of sectors. The casemates will be re-armed, turrets repaired, charges placed for demolition will be removed and the personnel and material required will be indented for on armies forthwith." (56).

Fortunately most of the fixed armament in the forts, the retracting cupolas, had been too difficult to move during the drain on armament from the forts which took place in 1915. This did not help the already captured Douaumont but the fall of the fort marked the start of a long struggle for Verdun which lasted into the summer of 1916. During this period the combined German and French losses reached over 700 000 (57) and marked the birth of a legend in post-war France. For just as the defence of Verdun had become an important French political issue, as forecast by Falkenhayn in 1915, so its capture had become a German political issue of no less importance.

Following Pétain's reversal of policy in March 1916 the forts at Verdun—still in French hands—continued to withstand fire from the

◀ 15. One of the cupolas at Liège after it had been knocked out by German artillery.

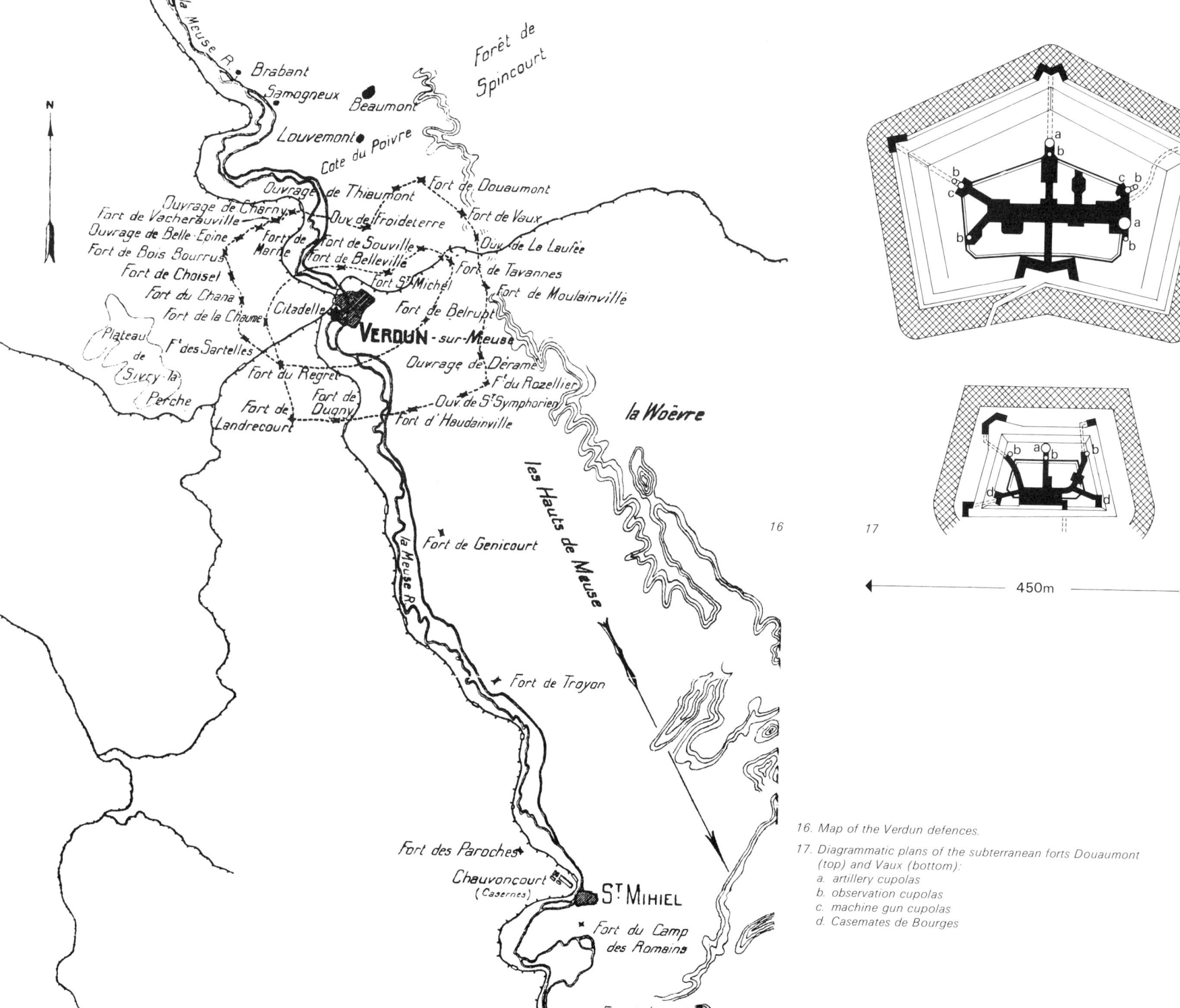

16. Map of the Verdun defences.
17. Diagrammatic plans of the subterranean forts Douaumont (top) and Vaux (bottom):
    a. artillery cupolas
    b. observation cupolas
    c. machine gun cupolas
    d. Casemates de Bourges

18 The outer ditch surrounding Fort Douaumont: in the distance one of the ditch flanking galleries.

19 External detail of a Verdun fort—the inscription reads: 'Better to be buried beneath the ruins of this fort than to surrender.'

420s. The armoured cupolas were almost all still in working order by 1918. The success of the Verdun forts under bombardment, and the 'hero' of Verdun—Pétain—were to become key influences in a post-war France once more attempting to decide on a strategy of frontier defence. In 1930 Pétain wrote his book 'Verdun' in which he quoted the conclusions drawn by General Descourtis, a commander of the engineers of the 11th Army:

"The war had demonstrated that the portions of our forts adapted to active combat, and the most important defensive element, defied the most powerful artillery. Except for the small works of Thiaumont, all the forts of Verdun are still, at this date, ready for action. Concrete, which may have proved unsatisfactory abroad, and which has been too hastily condemned as useless, has given good service in our case.... As to the armoured turrets the heaviest enemy projectile were unable to destroy any but a small number of machine gun turrets...." (58).

The implications were made obvious, and other post-war military writers drew similar conclusions. The result was that the 1930s once more saw France embarking on the construction of huge fixed frontier fortifications, known as the Maginot Line, which were to prove an expensive mistake in every possible sense.

While for post-war France the legend of Verdun was thus the most influential feature of the 1914–18 war, for Germany and Britain the lessons of the trench war were to be equally influential. The field fortifications which were developed during 1914–18 in fact proved stronger than the permanent fortresses. To a great extent this situation was caused by the defensive power of new weapons such as the magazine rifle and machine gun. A fortification engineer, Sydenham Clarke, had already anticipated this situation in 1890 when he explained the effect of the magazine rifle on fortress design:

"When the small arm was capable of being fired only about once a minute, this zone (of infantry attack) could be crossed with comparatively little difficulty.... Hence arose the vast ditches, the elaborate arrangements of flank defence the 'caponiers', countless galleries, etc., of the various systems of Fortification. The modern rifle has rendered all these expedients absolutely unnecessary in the future. The intensity of fire which a single line of men can now deliver upon a given area exceeds enormously the maximum formerly attainable by the combination of every conceivable system of cross-flanking.... There is no arm so potent in its influence on all questions of land defence, as the magazine rifle." (59).

During the four years of the stalemate new ideas of fortification and defence philosophy were thus rapidly developed, initially as field fortifications, which were so fundamental that they became integrated with the post-war development of permanent fortification. In addition, the increasing use made of the tank and the aeroplane in the siege situation on the Western Front was ultimately to cause more dramatic changes still in the design of permanent fortification.

20. Interior of Fort Douaumont.

2 DEFENCE IN DEPTH

Development of the deadlock on the Western Front, the development of the German system of defence in depth and concrete fortifications.

Events

1914 Aug. 4 Britain enters war.
Aug. 16 Liège falls: Germans sweep down towards Paris along line dictated by the Schlieffen plan.
Sept. 4 Galliéni's proposal for a flank attack from Paris sanctioned by Joffre.
Sept. 5–9 German advance halted in the Battle of the Marne.
Sept. 12 Germans retreat to R. Aisne and start to dig in.
Sept. 24 The First Battle of Ypres completes the deadlock.

1915 Jan.–Sept. Failure of Allied offensives to break through.
Sept. Battle of Champagne: first use of defence in depth by Germans.

1916 Feb.–June German offensive at Verdun.
July Battle of the Somme starts.
Aug. 29 Falkenhayn is replaced by Hindenberg with Ludendorff as First Quartermaster-General. Ludendorff, helped by Colonel Fritz von Lossberg sets about developing a flexible defence in depth.
Sept. 15 First premature use of British tanks.
Sept. 16 Hindenburg issues order for the building of the Siegfried Stellung.
Nov. Winter puts an end to the Battle of the Somme.

1917 March German withdrawal to Siegfried Stellung.
April 11 Lossberg asked to take over defence during British attack at Arras, where he develops principles of defence in depth further.
June 7 Third Battle of Ypres starts.
Dec. Third Ypres ends having proved the theory of defence in depth further.

1918 March German final offensive starts.
July Germans gain of defensive.
Nov. 4 Home front has been steadily deteriorating and now causes outbreak of revolution, starting at Kiel naval base.
Nov. 11 Armistice.

◀ *1. German troops constructing a large concrete dugout in 1916.*

● "At first, there will be increased slaughter on so terrible a scale as to render it impossible to get troops to push a battle to a decisive issue. They will try to, thinking they are fighting under the old conditions, and they will learn such a lesson that they will abandon the attempt. The war, instead of being a hand-to-hand contest in which the combatants measure their physical and moral superiority, will become a kind of stalemate, in which, neither army being able to get at the other, both armies will be maintained in opposition to each other, but never being able to deliver a final and decisive blow. Everybody will be entrenched in the next war; the spade will be as indispensable to the soldier as his rifle. . . ." (1).

This prophecy, written in 1897 by a Polish banker and economist named Bloch, was to prove remarkably accurate. In Bloch's view the reason for the slaughter and deadlock would be the magazine rifle. The American Civil War, the Franco-Prussian War, the South African War and the Russo-Japanese War had all proved the increasing power of defence which new weapons such as the machine gun were creating. The effect of the defensive power of new weapons on the type of wars which had been fought in the century before 1914 was to increase the value of protection, of field fortifications. In the war of 1854–5 in Russia, field fortifications had in fact proved themselves more effective than permanent ones against the new weapons. In the Russo-Turkish War of 1877 the Turks turned the Bulgarian village of Plevna into an earthwork fortress under the eyes of the Russians. The Russo-Japanese War of 1904–5 had similarly resulted in 'a digging match' along a hundred mile front. Yet despite all these precedents, the combatants in 1914 had not prepared for a trench war. The European General Staff refused to allow anything to affect their ideas of 'arme blanche' and mobility. As an indication of this, large numbers of cavalry were maintained in idleness throughout the war by both sides.

It was argued that these precedents for a trench war could all be ignored. Such stalemates could be avoided in the future by applying the 'offensive spirit'. Although field fortifications were still considered important and taught thoroughly, they were regarded as a 'local and temporary expedient' (2). The British 'Field Service Regulations' of 1914 summed up the general attitude of all the participants in the early days of war:

"Infantry in attack must not delay the advance or diminish the volume

2. German trench on the Western Front: the ideal situation.

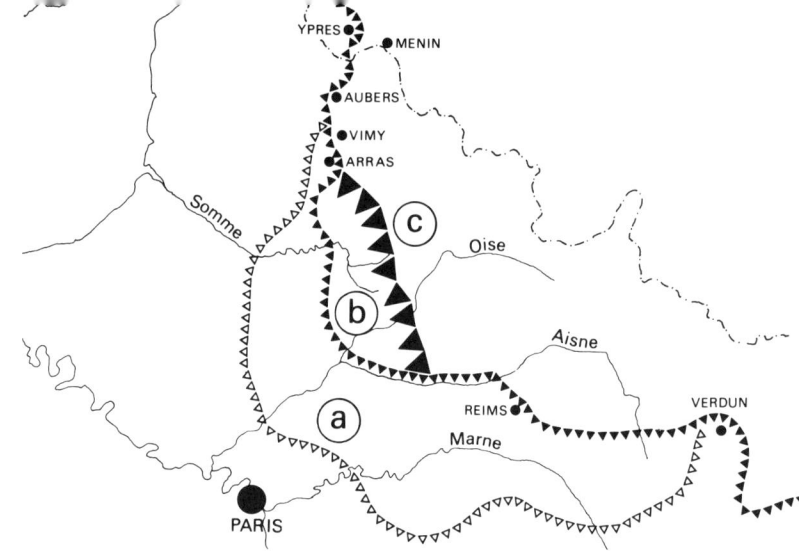

3. German positions during 1914–17:
   a. the maximum advance in 1914 before the Battle of the Marne
   b. the initial trench stalemate at the end of 1914
   c. the line at the end of the Hindenburg Retreat in Feb. 1917.

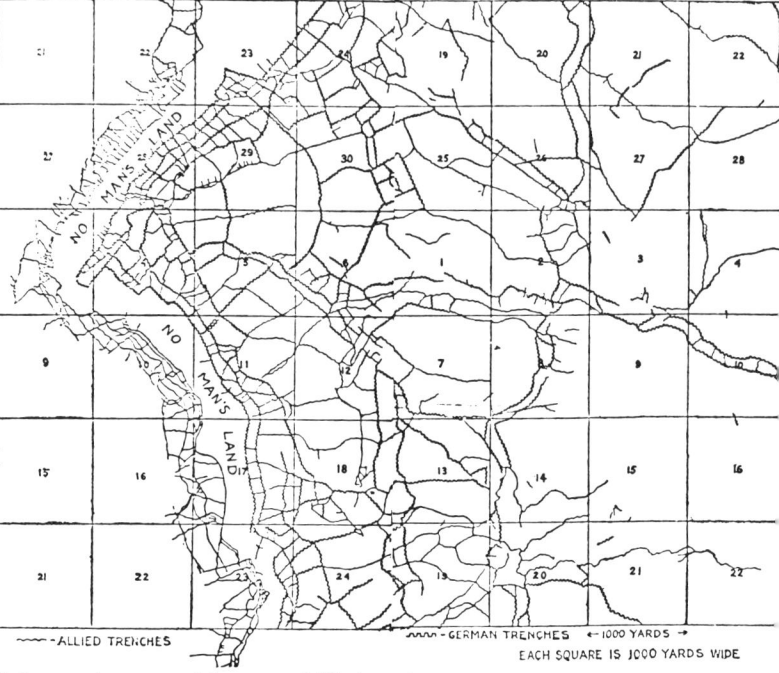

4. Comparative extent of German and Allied trench systems.

of fire by entrenching. Entrenchments in the attack are only used when, owing to further advance being impossible, the efforts of the attacking force must temporarily be limited to holding the ground already won. The advance must be resumed at the first possible moment." (3). Precedent, the influence of rapid fire and increased fire power were thus ignored.

In September 1914 Moltke's right wing, weakened by his amendments of Schlieffen's original plan, was brought to a halt on the R. Aisne and the first German defences, intended for a few days use only, subsequently developed into a vast entrenchment stretching 800 km from Switzerland to the Channel. Between 1914 and the German offensive in 1918 the front moved scarcely more than ten miles, except in one sector where the Germans made the strategic withdrawal to the Siegfried Line in 1917. Neither side had foreseen the stalemate or therefore found a way to avoid it. Instead more men, more equipment, more building effort were thrown into the Western Front, increasing the problem, but not solving it.

The trench stalemate on the Western Front became in effect a gigantic siege operation, although few of the Allied generals treated it as such. Instead, they favoured huge and costly frontal assaults on entrenched positions, preceded by long bombardments, interspersed by countless savage skirmishes involving infantry patrols. Under these conditions the Germans were to set a lead in defence construction, following the realisation by Falkenhayn, who replaced Moltke in September 1914, that a long war of attrition was inevitable. The German army was well equipped for such a war. Her siege equipment which had been developed for the seizure of the Belgian fortresses, such as hand grenades, mortars, signal light pistols, periscopes, searchlights and particularly the heavy howitzers, proved to be ideal for trench conditions. In addition her army was better equipped with the underrated machine guns. With her defensive initiative the German army thus developed the simple trenches of 1914 into deep systems of defence, at the same time developing in these systems concrete fortifications of a quite different kind to those at either Liège or Verdun.

In 1914, however, the general principles of trench theory and construction were basically the same in both the Allied and German armies. The German system in Belgium and France—prior to Ludendorff's take over on the Somme during 1916—was based on three successive trenches: 'Front' (Firing Line), 'Support' and 'Reserve'. The first of these trenches was to form the main line of defence. Details differed greatly and to some extent with national idiosyncracies, but the Allies often conceded that the German defences were superior to their own. 'The American Engineers in France' stated that:

"In systematic layout, in thoroughness of construction, and in elaborateness of design the German excelled all the others." (4).

The German Army had withdrawn in order to choose a better site for defence following the Battle of the Marne, pursued by the Allies. Generally this meant that the Germans found themselves defending highland and hills with the Allies below. Where the Allies did manage to capture a German position, they inevitably made the mistake of following them down into a new low ground position, recreating the same situation—one which was to govern the siting of the German trenches.

The Front trench was generally sited so as to be out of range of enemy mortars but not too far away to lose command of the 'No Man's Land' in front (5). It was as near as possible to the military crest (below the topographical crest) of the hill or rise from where it was able to command the entire slope. Eighty to two hundred metres behind this first trench the Support trench was sited, just below the topographical crest, in command of the Front trench and able to support it under fire. A good deal further behind this, about 200 to 500 metres, preferably placed on the other side of the topographical crest, came the Reserve trench, able to fire on an attacking column if it managed to break over the crest of the hill, but out of direct enemy fire and observation. All three of these trenches, incidentally, were traversed, bastioned or octagonal, zig-zag or wavy in plan, with the object of preventing enfilade fire and limiting the effect of shell explosions. Communication trenches named 'Wege', connected up to the front trenches at about 45 degrees—again with the possibility of being swept by enfilade fire in mind. They were also available as firing trenches and could be used as 'Switch' lines, forming a 'pocket' around an enemy break-in before the administering of a sharp and effective counter-attack. Well back from the front was the main artillery grouped in batteries of four guns, either placed out in the open without any attempt at protection, camouflaged, or in deep gun pits (6).

This three trench system was, except for topographical siting, similar to the system used by the Allies. A British Intelligence report dated 1916 stated:

5. Interior of German trench kitchen.

"They do not differ widely from our own trenches but they are slightly deeper, slightly narrower, and the dug-outs are almost universally well-constructed." (7).

By 1915, however, the German system of defence was already greater in depth than the Allied was ever to be. Whereas the Allies used the system of three trenches as their main line of defence, the Germans duplicated or even triplicated this system. Subsequently there might be two or three times as many lines of trenches, or 'Positions', in the overall German front.

The First Position, 'I Position', was continuous in 1915–16 and consisted of the usual Firing, Support and Reserve trenches. In this position were the bomb stores, pioneer advanced depots, company commander's dug-outs and advanced telephone stations. About 2000 metres behind this was the Second Position, 'II Position', which was also normally continuous and similar to the 'I Position', but not always (8). In extremely critical areas a 'III Position' was constructed and, although not always a continuous line, it was at least a number of reserve positions at intervals of two to four kilometres. As will be described later, however, the German system of defence was to evolve even further over the next two years.

Within these main principles governing the layout and construction of the German defensive system prior to 1916, there were many variations along the front. The trench front was certainly not a single ditch extending from the sea to the Swiss border. Changes in local conditions and topography called for different solutions. In places a system of trenches was impossible, such as in the area north of Verdun, and a series of strongpoints had to be built. Where the trench line reached the sea, in Flanders, with the water table less than a metre below ground level, a breastwork had to be built of sandbags, earth and brush hurdles.

In these different trench systems, varying not only according to types of ground, but also with the ideas of local commanders, a strong emphasis was soon put on the provision of underground facilities. Both sides used the traditional forms of mining attack which had been used in siege warfare for hundreds of years. Counter-mining, to find and destroy such enemy mines, was naturally used, and a war developed underground as well as above it. But if the British were the more successful in the 'underground offensive', the mined 'dug-out' and later the concrete one, for troop protection, were brought to a much higher level of sophistication by the Germans. At Ypres the Menin Road was under continual shelling day and night. It became so bad that the German engineers built a tunnel running under the road so that infantry and supplies could reach the front line under perfect cover (9). Their 'Keller' (basement) or 'Stollenkaserne' (tunnelled barracks) were numerous and well constructed, allowing German troops to withstand heavy bombardment with few casualties. At Aubers Ridge in May 1915, the German garrisons were able to emerge from their 'Wohngraben' (underground quarters) the moment the barrage lifted and take up their firing positions (10). The effectiveness of such underground facilities was increased by the lack of heavy artillery and howitzers in the Allied army. The French 75 mm and long-barrelled heavy guns had flat trajectories incapable of demolishing deep fortifications.

One initial reason, perhaps, for the wide application of dug-out techniques in the early months of war was that the Germans on high ground tended to dig down anyway, but the main reason was that troop protection formed an important part of German defensive theory. The German dug-outs and concern for their troops in this way paid good dividends both in keeping up morale (there was no mutiny like the one in the French Army) and creating the basis of a sound defence.

The construction and extent of dug-outs varied with the use they were put to. Rough shelters on the front might be simply dug into the banks of the trenches with enough concrete, timber, or earth cover to resist up to 77 mm shells. Very rough shelters, i.e. splinter-proof, became virtually obsolete in the German trench front by Nov. 1916 when the Prussian War Ministry stated that: 'Splinter-proof constructions have proved actually harmful. They not only fail to give protection, but block the trenches with their débris.' The use of stronger dug-outs was recommended in the same document, their strength depending on position:

"Every means must be used to provide shell-proof shelters, that is protection from continuous shelling by 6-inch guns; over and above this, every effort must be made, particularly in rearward lines, to secure bomb-proof shelters, that is, protection from continuous shelling by 8-inch calibres and heavy trench mortars, and single hits by heavier guns." (11).

In keeping with the policy, front line dug-outs were to be built for small garrisons, and larger dug-outs for rear positions, such as hospitals, headquarters, ammunition dumps. This sort of layout evolved only slowly, for even in 1917 many dug-outs were crowded into the front

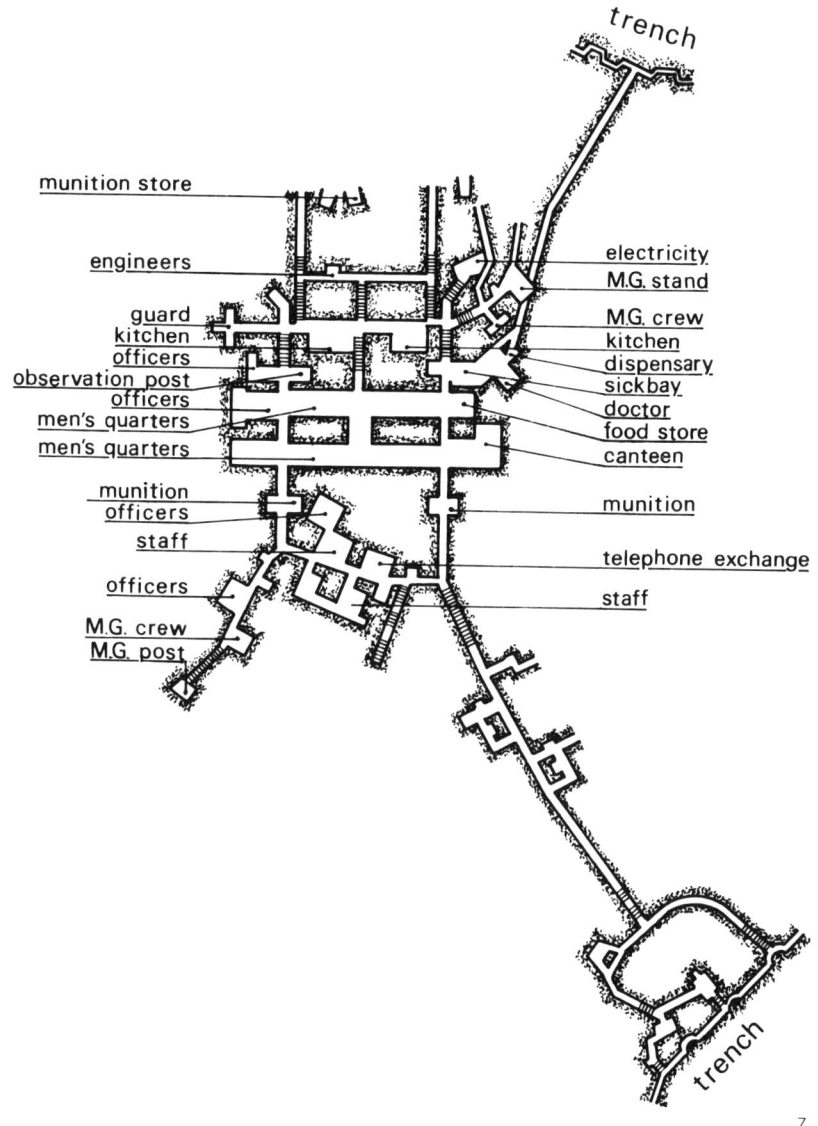

6. Entrance to captured German HQ dugout.
7. Plan of the dugout complex constructed by 126th Infantry Regiment (Württembergers).

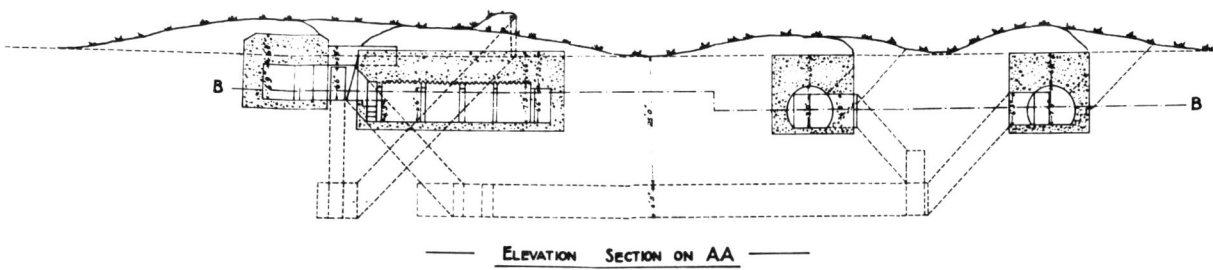
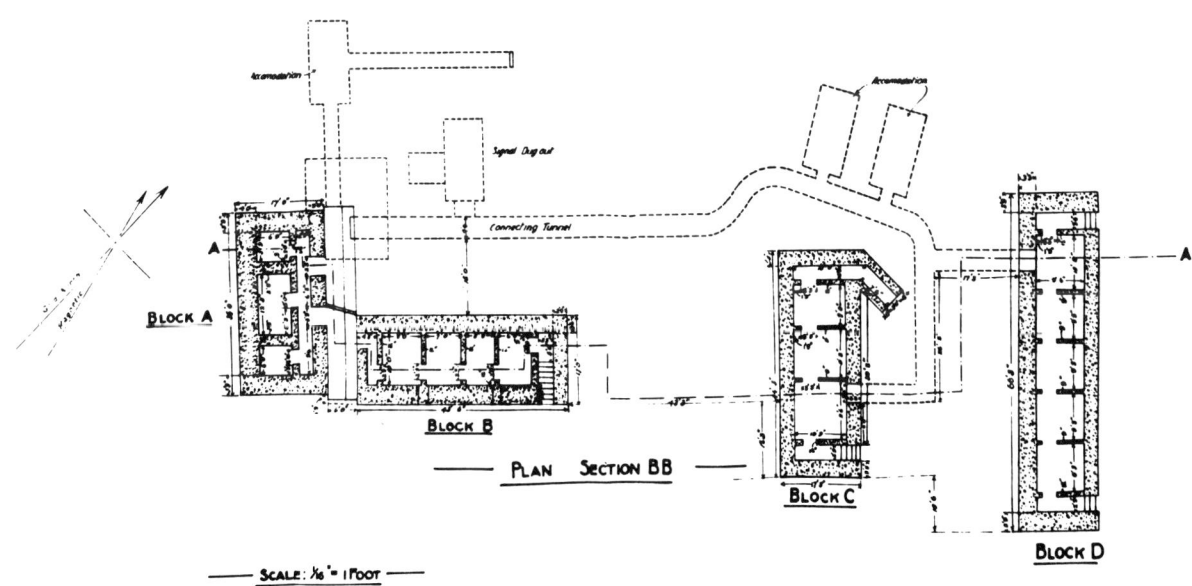

8. German Regimental HQ: concrete dugouts with mined communication tunnels.

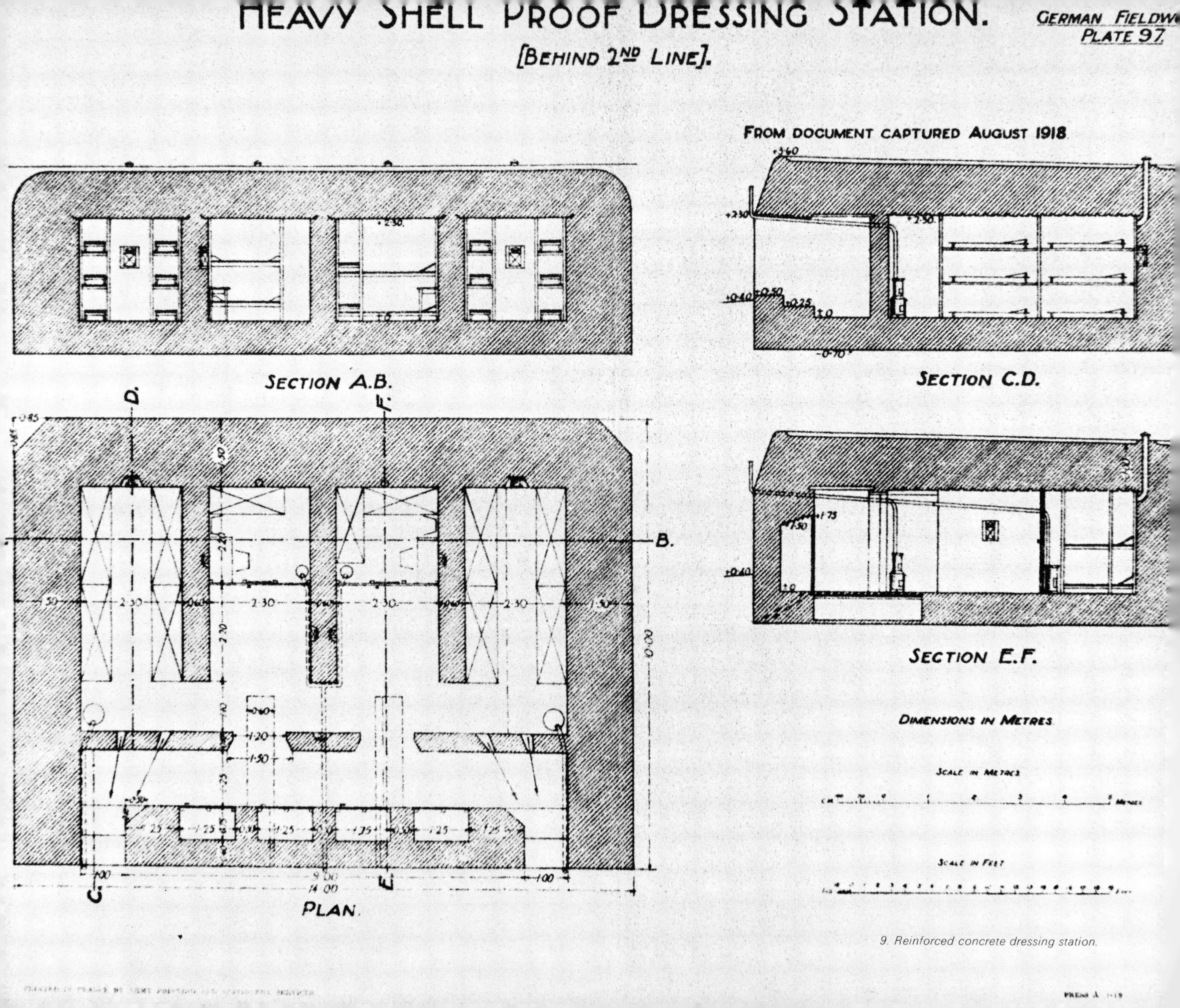

HEAVY SHELL PROOF DRESSING STATION.
[BEHIND 2ND LINE].
GERMAN FIELDW
PLATE 97.
FROM DOCUMENT CAPTURED AUGUST 1918.
SECTION A.B.
SECTION C.D.
SECTION E.F.
PLAN.
DIMENSIONS IN METRES.
SCALE IN METRES
SCALE IN FEET

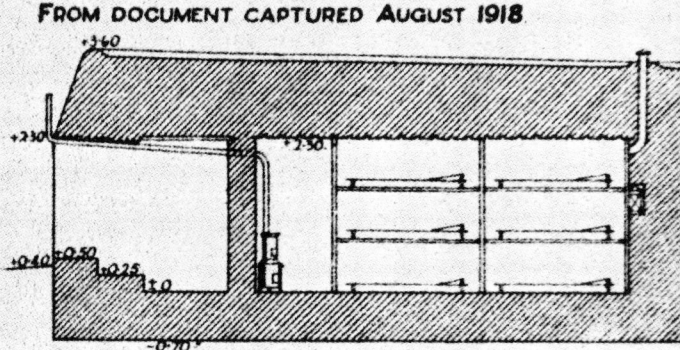

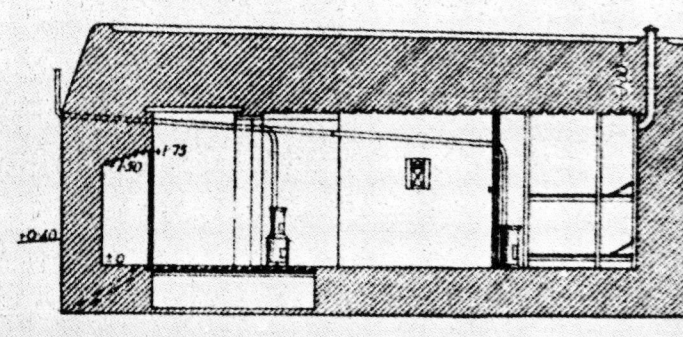

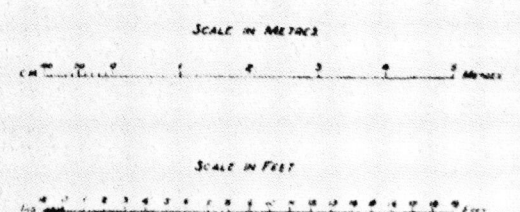

9. Reinforced concrete dressing station.

line at Flanders and other sectors which consisted of old positions. Design varied with the soil, topography, time available and the personal view of the engineer. Subsequently there were many different designs but because they were all underground they had certain similarities. One problem for example was access, or at least escape, and all dug-outs had at least two exits or more depending on size. Despite this sometimes all the entrances could be shelled in during a heavy barrage. 'The Construction of Field Positions' stated that:

"The larger the dug-outs, the more necessary it is for them to have convenient, concealed and well-protected exits. Underground communications between dug-outs and connections to the communication trenches are of advantage." (12).

Though some of the dug-outs were described as having electricity and comfortable furniture, including feather beds, the harsh reality of the Western Front was far from comfortable. Blankets soaked in chemicals hung over the entrances, providing gas protection. In the trenches duckboards were used as the trenches took over the functions of the original drainage system destroyed by artillery. In the deep dug-outs sumps were built and the water was pumped out manually. Though comfort was regarded as important, protection was the main objective.

The dug-outs did not always function successfully from the military point of view either. At Vimy Ridge in 1917 the short barrage and quick attack by the Canadians caught the inhabitants unaware. A captain in the 5th Imperial Division recalls:

"By ten a.m. the Canadians were in possession of the village of Thelus. Nearby we descended a deep shaft and to our surprise found that it led to a vast underground system of dug-outs and tunnels. So unexpected and rapid had been the advance that the inhabitants were blissfully cooking or awaiting their breakfast. They offered no resistance. Among them was an army colonel, who owned a nicely papered bedroom and a feathered bed with sheets." (13).

This type of failure led to changes in the design and use of German dug-outs. Most of the large dug-outs in the Fourth Army sector in Flanders, for example, had been ordered by Sixt von Armin to be barricaded off and only used for holding reserves for offensive action with the permission of responsible commanders (14). Concrete dug-outs became the rule as they allowed a shallower cover and garrisons could emerge more quickly. Dug-outs became smaller and more widely distributed and the eventual conclusion to this came in 1917–18, when the small dug-out and machine gun emplacement became a combined self-contained unit, as part of 'defence in depth' policy.

A brief description of the development of 'defence in depth' is necessary, as it was to form the basis of fortification design for the next thirty years. Although developed mainly after Ludendorff's takeover on the Somme in late 1916, that is in its correct meaning of 'elastic defence', the depth of the German defence zone had grown gradually over 1914–15—as has been mentioned earlier. From a single 'I Position' backed only by a single line of machine gun emplacements at Neuve Chapelle in March 1915, the defence had developed through the increasingly deeper defences of Aubers Ridge and Champagne. At the latter a first tentative approach to flexible defence had been made by withdrawal to the 'II Position' during the artillery attack. Still, however, much emphasis was put on the first trench of the 'I Position' by German commanders. General von Below of the First Army had ordered that 'any officer who gave up an inch of trench would be court-martialled, and that every yard of lost trench must be retaken by counter-attack' (15). When Ludendorff took over during the massive Allied offensive on the Somme in 1916, with the help of Colonel Fritz von Lossberg, First Army's new Chief of Staff, he reversed this method of resistance. Ludendorff no longer insisted on holding on to all ground, regardless of tactical importance, and stopped the continual small counter-attacks, hastily improvised and quickly defeated, which had been sapping the strength of his men. Instead the depth of defence was increased even further and by the end of September, although the Allies had gradually taken a strip some 480 km long and 11 km deep, the situation was saved.

The Somme was just as much a disaster for the Germans as for the Allies but out of the carnage they extracted their share of the lessons to be learnt. When winter put an end to the battle in November, the Prussian War Ministry was already suggesting that zones of defence should be generally increased (16). A series of manuals published in the winter of 1916–17 explained the principles and gave constructional details for 'defence in depth' (17). By mid-1917 much of the front had been changed over to the new method of defence. The lines no longer had the continuity of 1916. Behind a first continuous line a scattered grid of discontinuous trenches and machine gun positions were built

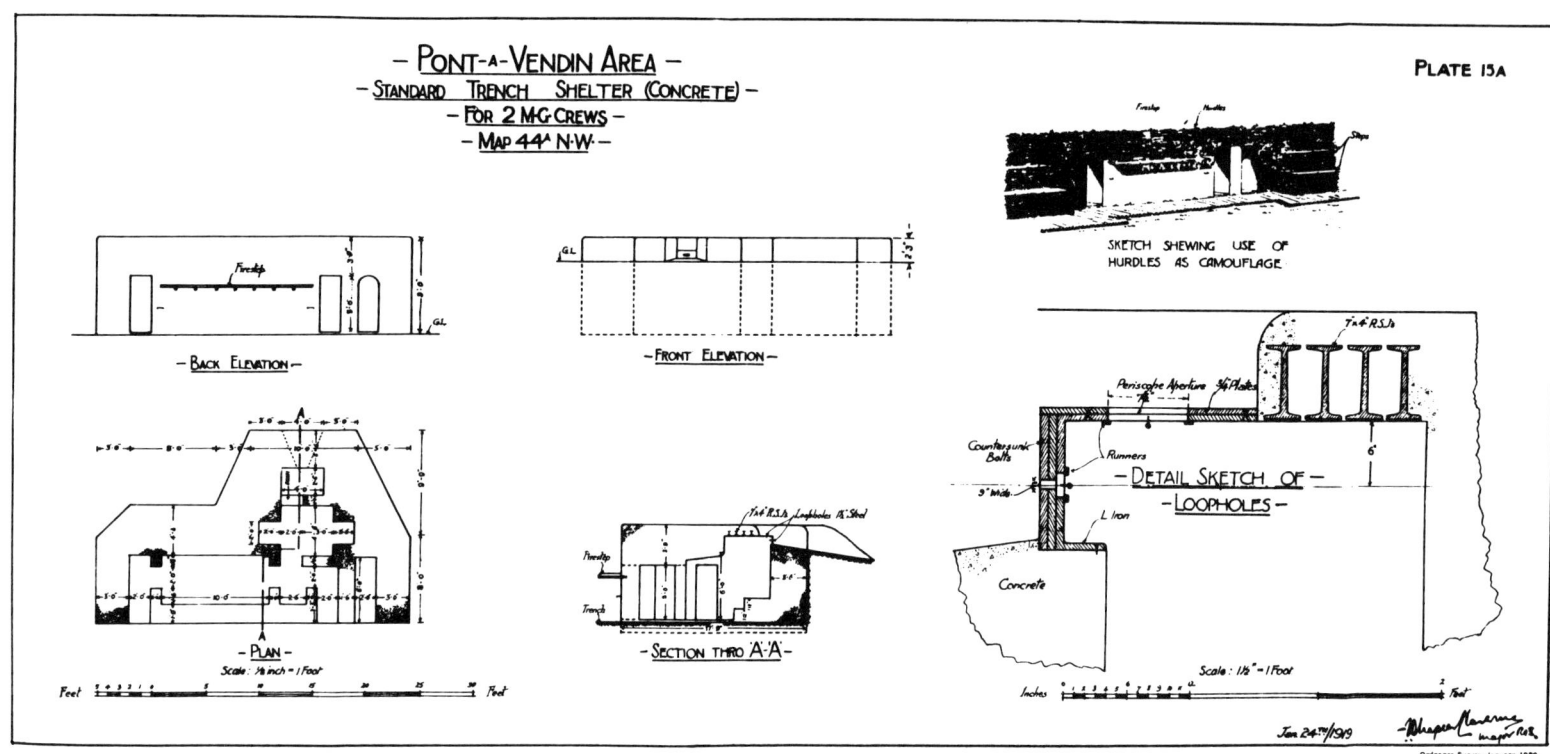

10. Standard German trench shelter.

to a depth of 2000 metres before a rear continuous line of last resistance (18).

The Battle of the Somme was also responsible for a further degree of German defensive initiative on the Western Front. When Hindenburg and Ludendorff reached Cambrai on 16th September, during the Somme crisis, an order was issued 'for the construction of a vast new line of defence which would greatly shorten the front and serve for a voluntary withdrawal in case of need. . . .' This was the Siegfried Stellung, later known to the Allies as the 'Hindenburg line' (19). Designed on 'in depth' principles the line shortened the front by 42 km stretching across the chord of the arc Lens-Noyons-Rheims. Ludendorff's withdrawal to it in early 1917 threw the Allied plans for attack on this front completely out of gear. The British were consequently forced to limit their offensive to the northern hinge of the line at Arras. Ironically it was this offensive which was to help develop 'in depth' theory even further.

On April 11th, two days after the attack started, Lossberg was appointed 6th Army's Chief of Staff in the defence of Arras. Vimy Ridge, north of Arras, was still organized on the old system of defence and Lossberg therefore executed a voluntary withdrawal from Vimy Ridge— forming a new front along the old 'III Position' on defence in depth principles. By using a weak front position, quickly improvised from shell-holes and dug-outs, with the bulk of the infantry well back out of reach of the enemy artillery (20), Lossberg hoped to draw the enemy infantry into a position where counter-attacks could be made unhindered. It meant in fact a complete change in defence for the Sixth Army. Only in September had they been ordered. 'The principal fighting line is the front line' (21), yet in a matter of days this situation had been reversed in the middle of the battle. Lossberg's success led to abandonment of the continuous front line and the establishment of the first principles of independent, yet mutually supporting positions, increasing in strength with depth. Two months later, in the Third Battle of Ypres, the use of such discontinuous positions was to be even further developed and, together with the extensive use of small concrete installations, the basis of future European fortification design was laid.

The Germans had been the first of the armies to use concrete in their field works, and reinforced concrete shelters and emplacements named 'Mebu' (22) were constructed from early 1915 onwards, reinforced concrete being preferred to mass concrete even then. The failure of the deep mined dug-out in 1916, when their garrisons had been unable to emerge sufficiently quickly, had led to the wide use of shallow concrete ones. A document issued by the German 6th Army on September 27th, 1916, entitled 'Supplementary Instructions as to the Construction of Defence' gives an impression of the way design had developed by that time:

"Concrete shelters are preferable to mined dug-outs in all circumstances. . . . From our experience on the Somme, mined dug-outs require from 23—26 feet overhead cover in hard chalk, and from 33— 36 feet in clay. Such dug-outs are impracticable in front line, as men cannot reach the fire-step in time from such depths. . . . Concrete shelters should be built in future of reinforced concrete, 5 feet thick (2 feet 8 inches hitherto). . . . In a gas cloud attack, braziers with low fires have kept the shelters completely free from gas. The shelters should be distributed over all three lines of trenches of a position, as required by the distribution of the formation in depth. . . . In general, too much cannot be done to provide secure shelters from shell fire on reverse slopes for the reserves. If there is not sufficient time to make complete dug-outs numerous recesses in the form of gallery entrances will answer the purpose." (23).

'The Construction of Field Positions (part 1)' published by the Prussian War Ministry a month later made the same basic point concerning the use of concrete for dug-out construction and from 1916 its use was preferred if not actually obligatory.

Concrete work was done by trained Fortress Engineers and Pioneers, and by Civil Engineers chosen for their peace-time experience in its use. Though the reinforced concrete in the first months of the war had been extremely crude, using scrap metal such as iron gates, rails, steel joists, and even screw pickets for reinforcement, this type of construction had soon proved unsatisfactory and been superseded by the use of standard steel rod reinforcement, well distributed and tied, or by correctly positioned and tied steel joists and rails. The sand and gravel used in the concrete was also locally obtained at first, but later Rhine gravel and sand were brought by barge and rail in order to improve the quality and strength of the concrete. Shuttering was mainly timber or corrugated iron (24), the wide use of the latter being due in part to its recommendation by the central organization, the War Ministry:

"Corrugated iron is specially suitable for the inner revetment of

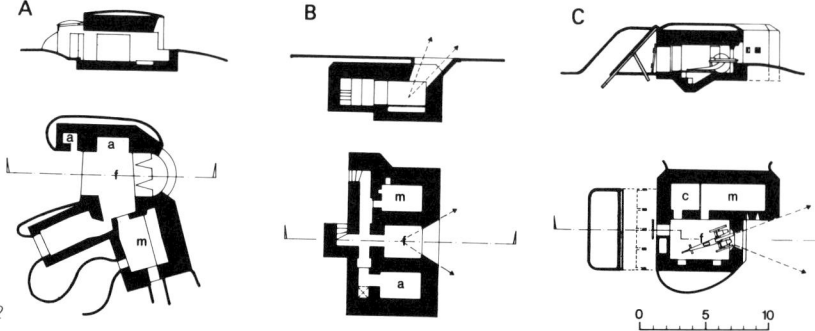

11. German reinforced concrete emplacement captured before completion: 1917.

12. Reinforced concrete gun emplacements:
    A. 77 mm battery emplacement at Aubers Ridge
    B. Mortar emplacement
    C. Emplacement for a flanking field gun.
    Key: f firing chamber, a ammunition, m men, c commander.

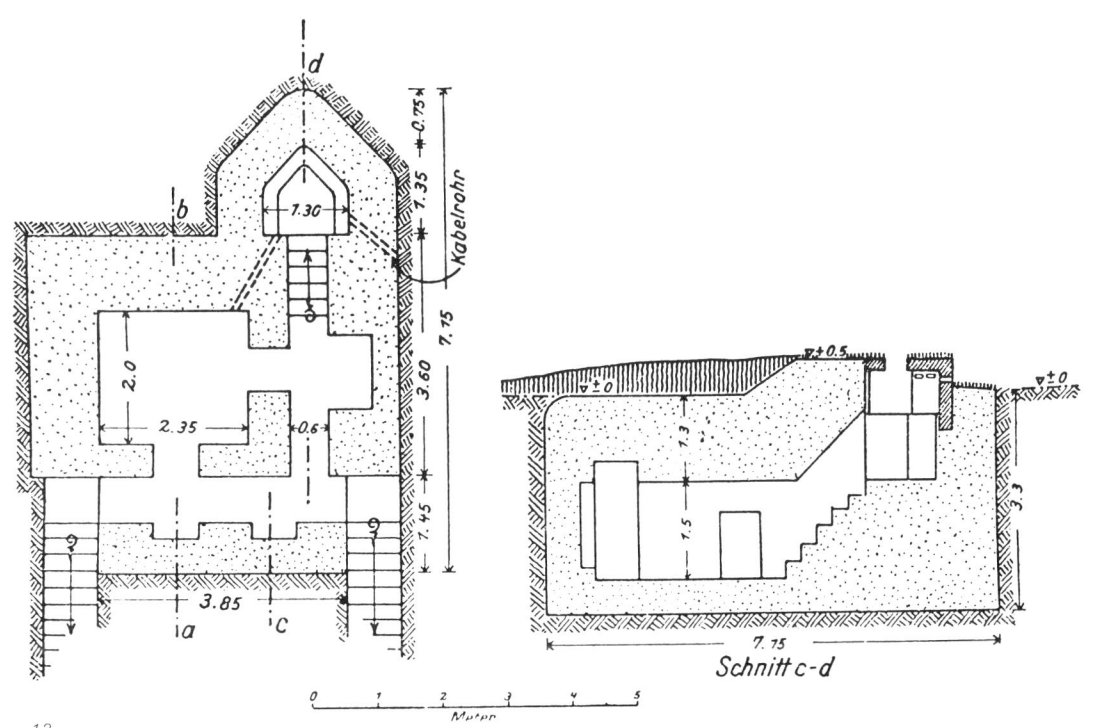
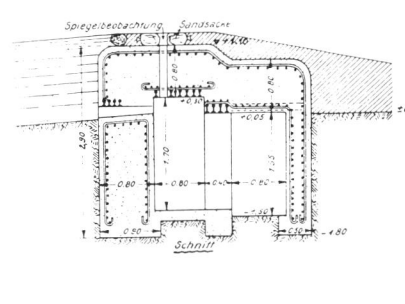
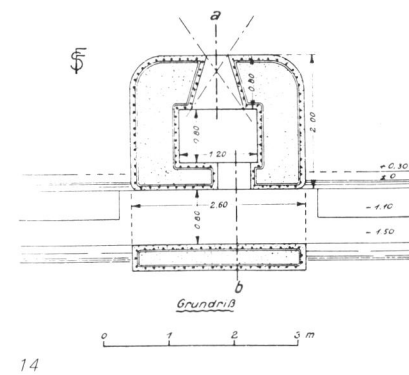
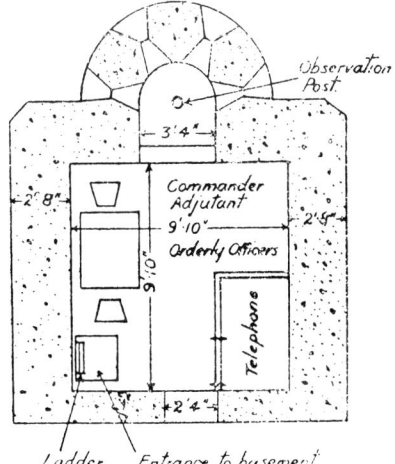

13. Type observation post 1916–17.

14. Front line observation post 1916.

15. Command post for a Battalion or Brigade Commander: from the British translation of the Prussian manual 1916.

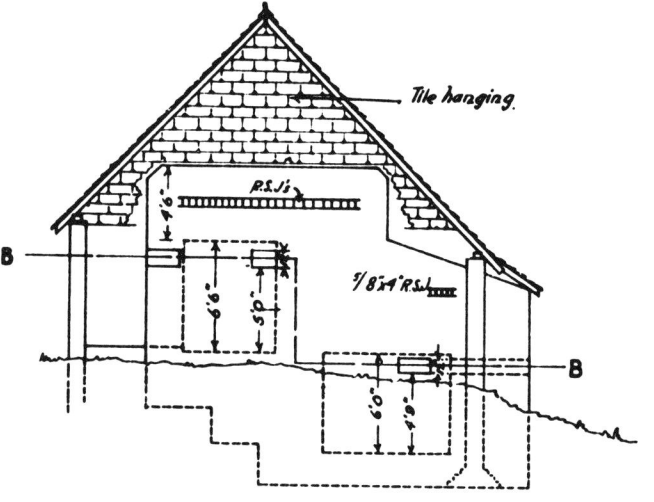
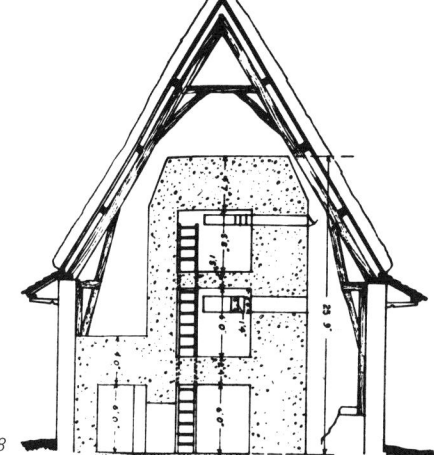

16. German concrete strongpoint built under existing farm buildings near Mouchy-au-Bois.

17. German light signalling station at Aubers camouflaged to match existing buildings.

18. German brigade headquarters constructed within existing farm buildings at Aubers Ridge.

concrete walls and structures and considerably increases their strength." (25).

Up to 1916 the recommended concrete cover for most shelters for 'shell-protection' was 800 mm and for 'splinter protection' 400 mm, but certain critical shelters such as Battle Headquarters, which were in the rear area susceptible to heavy shelling from the heaviest calibres, were graded 'bomb-proof' and were considerably thicker. To meet the increase of fire power in 1917 the thickness of reinforced concrete for dug-outs was increased, and 'shell-proof' by this time called for 900 to 1200 mm of concrete, machine gun posts and trench shelters being in this category, while 'bomb-proof' headquarters had a thickness nearer and sometimes exceeding two metres (26).

The use of concrete for the construction of field positions presented many problems which were not to be found in civil construction. Curved corners, however desirable by ballistic and text-book standards, were seldom possible under battle conditions. Silence had to be maintained and the approved method was to mix the concrete dry in the rear lines before fetching it up. As every material used in the front lines had to be brought up manually this in turn affected design:

"To reduce carriage of stores, the minimum sufficient thickness of roof and walls must be employed." (27).

The front line shelters, therefore, were seldom of 'bomb-proof' standard.

Another main consideration which affected design was that of concealment. Some structures such as trench shelters, concrete dug-outs and first-aid posts could be situated with their roof about 500 mm below ground level, and similarly trench mortar and searchlight emplacements could be built below ground level with an open roof aperture. Machine gun posts and front line observation posts had however to rise above ground level in order to function. Both farm buildings and villages, where in existence, were inevitably converted into strong points for this reason, either in first or second positions or as regimental or brigade headquarters. One type often situated in existing buildings was the observation post, which because of its function needed to rise well above the ground where possible. Consequently it was not unusual for church steeples and industrial chimneys, where available, to be heavily reinforced with concrete—literally filled—and used for this purpose (28). The Prussian Manual of Positional Warfare 1916 stated that:

"Elevated observatories in trees, towers, chimneys and such like are very often desirable as observation posts for the senior artillery commanders . . . buildings can be strengthened inside in an inconspicuous way by the use of concrete. . . ." (29).

The method of concealment inside buildings was of three main types: the use of existing buildings as an 'envelope' inside which a structure was designed to fit the space available, whilst at the same time remaining free of the walls of the existing building; the construction of emplacements or bunkers to a form which suited its purpose and the addition of a mock building façade of hung tiles etc. in a form which imitated neighbouring buildings; and, perhaps the most satisfactory method, the use of existing buildings as a kind of permanent shuttering. The latter was not only good camouflage during construction, but also left the concrete with rough brick impressions after the impact of several direct hits knocked the old brickwork off (30). The use of such camouflage, however, was not always possible and here the height above ground level was kept extremely low. The average height of machine gun posts and front line observation posts above ground level was about 900 mm. The rest of the structure was hidden below ground but, even though only a small amount of concrete was visible, it was camouflaged and this was in some cases critical as installations of this type were usually in vulnerable front line positions (31).

In September 1917 in the middle of the slaughter at the Battle of Ypres (Third Ypres), Fourth Army was ordered:

"Concrete structures, (dug-outs) which should at the same time, all be constructed for use as machine gun emplacements, must be of low profile, and the slopes of the earth covering them kept flat. . . ." (32).

Even with only a height of just over 900 mm there were times when this was too much. Reducing this height created a very difficult design problem. A standard machine gun post up to shell-proof standard was one metre above the ground and this was the minimum firing height without reducing the depth of protection. There were only two possible alternatives to this, the first being to reduce the thickness of the roof over the actual observation or firing position and strengthening it with heavy steel joists. This could reduce the height to 680 mm as in the standard trench shelter for two machine gun crews, which was used on Aubers Ridge. The second alternative, and one which could reduce the height even further, was the introduction of a special section over

19

20

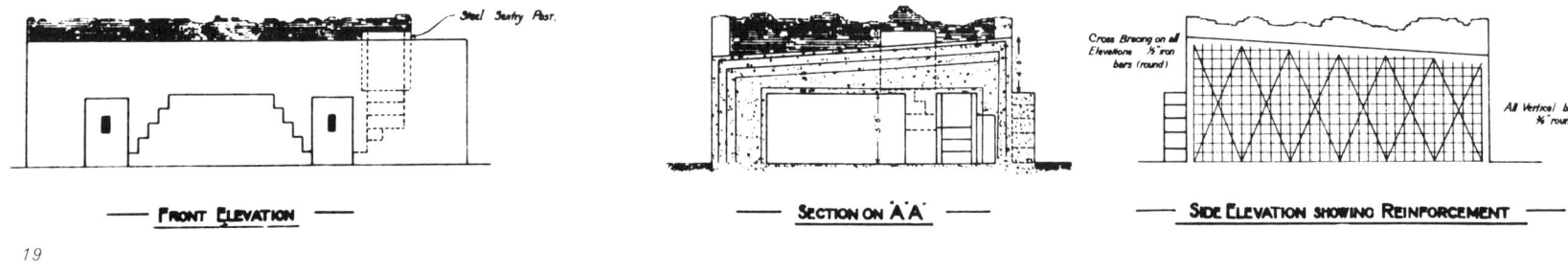

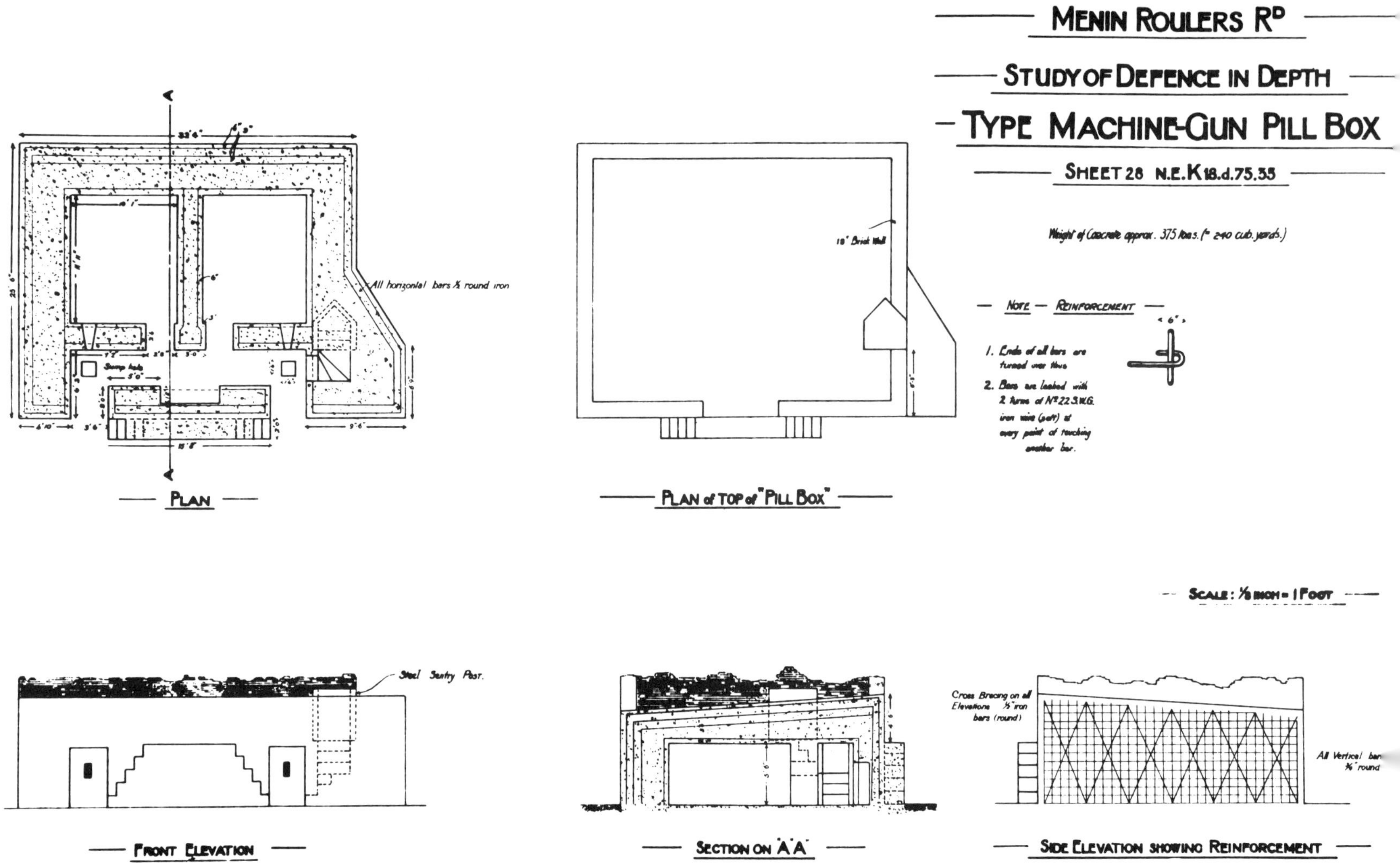

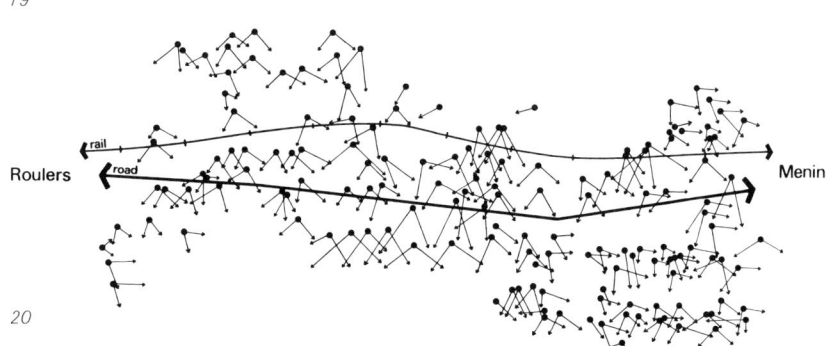

19, 20. Type machine gun pill box and diagram of distribution on a ten kilometre section of the Menin-Roulers Road 1917–18.

the observation or firing position, consisting of armoured steel allowing a height of 400 mm. Though the special armoured sections made this type of construction something of a luxury, the qualities of concealment were thought to make it worthwhile in certain cases. When available, cupola-type construction was therefore used in the same manner as it was to be used in the Atlantic Wall nearly thirty years later (33).

This variety of semi-underground construction was not always possible in places with a high water table. The type machine gun pill-box used on the Menin—Roulers Road for example, part of the defence in depth system in that sector, stood completely above ground and even with its minimal internal ceiling height of 1.7 metres (few concrete structures had internal heights of more than 1.98 metres) stood out strongly with only a token attempt at camouflage (34). The small concrete 'forts', which also played a part in the defences built in 1917, stood well above ground level with loop-holes and slits for machine guns. Heavily armed, if small, these forts depended, like the type machine gun pill-box, on a deeper concrete cover than normal. Ominously the tank forts built near Perenchies in 1917 with a cover of one metre of reinforced concrete were soon considered insufficiently strong for the job of tank warfare.

Two types of emplacement which were usually above ground level in order to function were the artillery battery and the battle headquarters, mostly situated well back from the front line. Any type of artillery emplacement was rare in the first year of the war but there was an increasing tendency to protect guns during 1916–17. Though concrete gun emplacements created severe problems for artillery, limiting the field of fire and making camouflage difficult, when within enemy artillery range it was essential. Paragraph 14 of the Manual of Positional Warfare 1916 stated that:

"Battery points must be irregularly situated with irregular intervals between the guns. The guns should be masked or be in concrete emplacements. Shell-proof and bomb-proof shelters for the detachments and ammunition are necessary." (35).

The Battle Headquarters, in the same way as the artillery, functioned better when placed above ground, allowing the direct observation of light signals or the front line itself. Like the artillery, they were in the area more susceptible to shelling from the heaviest calibres, and being the nerve centres of their areas of operation they had to be well protected. The Prussian War Ministry laid down the following requirements:

"The Battle Headquarters of sector and sub-sector commanders are to be located in shell-proof or, better bomb-proof buildings. They must be so situated that the commanders have an uninterrupted view forward, backward and to the flanks of the next command post, and over the country (para. 17). . . . Battle Headquarters, which must not be confused with staff 'living dug-outs', should be so small that they can readily be adapted to the ground, to provide shelter for subordinate personnel of a staff below the Battle Headquarters. If not, a shell-proof shelter of a simple kind should be provided near at hand. It also must be of minimum dimensions (para. 79). . . . The higher the formation to which the staff belongs the more accommodation it will want (para. 80)." (36). With a cover of 1.8 metres or more, and consisting of six or more component buildings, the Battle Headquarters were usually the most complex of the German concrete structures on the Western Front.

Concrete was undoubtedly the most important constructional feature of the German defences built on the Western Front, and the use of reinforced concrete in 1914–18 by the Germans was probably the first use of the material on such a vast scale. The obvious reason for its rapid growth in popularity was that it proved highly successful in resisting direct hits from 6-inch and even 8-inch howitzers when used in the correct depth. Thus photographs taken after heavy area bombardment show scenes of incredible desolation, often with water turning the surface into a morass, and with the straight edges of a pill-box the one thing still recognizable (37). Capturing such emplacements was an extremely difficult job and virtually the only method which showed any measure of success was the use of infantry who, crawling up to the emplacement, might manage to drop a bomb into the interior through a firing slit. A slightly safer method developed in later years was the use of a Mills bomb fired with a special adaptor from a .303 rifle (38) but although these methods were reasonably effective, concrete emplacements inevitably proved difficult and expensive obstacles to overcome.

From an architectural point of view it is tempting to examine such emplacements purely in terms of detail design but in order to understand the outside factors which influenced this they cannot be considered as isolated elements. As the use of concrete developed, so it was inte-

21. Pill box (ringed) following heavy bombardment at Trones Wood: September 1916.

grated into the German defence philosophy and the design of emplacements and defence philosophy interacted strongly, the one complementing the other. When defence in depth became an established principle in 1917 therefore, it called for large numbers of machine gun posts and small concrete dug-outs—a method of defence which saw its final major success of the war at the Third Battle of Ypres in late 1917. Not surprisingly Lossberg was again Chief of Staff, this time for Fourth Army under Sixt von Armin. By now the acknowledged expert on 'defence in depth', Lossberg's order of June 1917 on the construction of defences in Fourth Army sector was a direct development from his experience at Arras in April and May. It stated that:

". . . there must be a deliberate transition from the old pattern of positions, which is visible and will be shot to pieces by the enemy, to a zone of defence organized in depth. This must allow of offensive action by the defence from positions which are, as far as possible concealed, and are lightly held in front and more strongly held in rear." (39).

The following extract from the same order, refers even more obviously to the experience of Arras:

"During the battle, continuous trenches are no longer to be insisted on in the front line positions, but their place will be taken by shell hole nests, held by groups and single machine guns, distributed chequerwise. The shelter in the shell holes will be improved by the employment of mining frames or by joining of adjacent shell holes by tunnels. . . . Close behind the foremost shell hole line, strong points will be constructed for machine guns, assault troops and elements of the supports that have been brought forward. . . . Supports and reserves must work methodically at the construction of a continuous system consisting of several lines of trenches, which must be screened as much as possible from the enemy's observation (a reverse slope position). . . . This system will form a support for the defenders who are organized in depth in front of it. . . . This system will generally be the artillery protective line and will be about 1,650—2,200 yards from the foremost shell hole line." (40).

By the time of the Passchendaele offensive, the final failure of Third Ypres in October and November, the defence had a strong framework of machine gun posts in concrete emplacements around which the weak front defences could be based, drawing the enemy into the depth of the defence where counter-attacks were made. The concrete machine gun posts, many distributed above ground because of the waterlogged ground, proved ideal in these circumstances. The British method of attack, which made the waterlogged ground into an impassable quagmire, was just as unsuitable as the German defence proved suitable.

Third Ypres was the last German defensive battle before they switched from the defensive to the offensive in March 1918, and made the last desperate gamble for victory. But by July Ludendorff's armies were again on the defensive and began the gradual withdrawal leading to a final German defeat, already ensured by the blockade. Even in this final switch to the defensive Ludendorff put great hope in the German power of defence, although by now he must have been certain of defeat. In July he issued orders deepening the 'outpost zone' even further and calling for an even more elastic defence (41).

On July 22nd 1918, Ludendorff issued yet another order stressing the importance of fortifications in depth. This extract from it makes a suitable epitaph on German defensive history of 1914–18:

"In the defensive battle, distribution in depth and flanking fire are the points of most importance. . . . The construction of defences has been undervalued of recent times. In some quarters, it was even held that it was more difficult to attack an enemy who was not in a fortified position, than one fighting behind wire entanglements and trenches. In any case, the value of dug-outs was still acknowledged. . . . THIS IDEA MUST BE RIGOROUSLY DISCOURAGED. The construction of dug-outs is, undoubtedly of primary importance. At the same time, however, there CANNOT BE TOO MANY TRENCHES AND ENTANGLEMENTS FOR A DEFENSIVE BATTLE." (42).

By August Ludendorff was proposing that peace negotiations be opened before the situation deteriorated further. Yet the exhausted German forces still occupied more French soil than in 1917, with the Siegfried Line, the Hunding-Brunnhild and Hermann lines, and the Antwerp-Meuse line either ready or in preparation for them to withdraw to (43). Even when a peace settlement was reached at the end of 1918 they still occupied large areas of France and Belgium. Over four years the German army had developed ideas of defence and fortification which, far in advance of their contemporary Allied equivalent, were to be the basis of European defence construction for the next three decades.

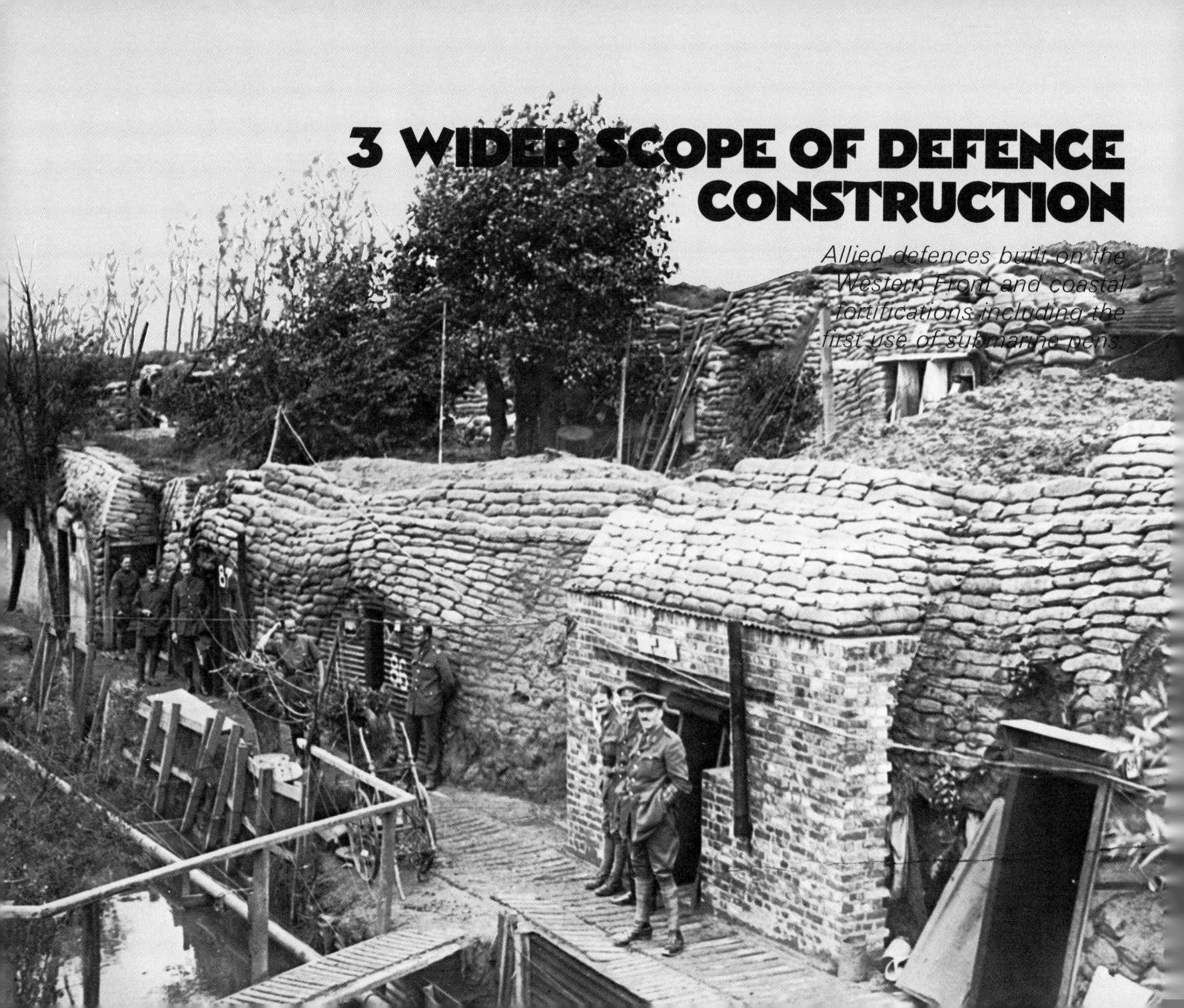

3 WIDER SCOPE OF DEFENCE CONSTRUCTION

Allied defences built on the Western Front and coastal fortifications including the first use of submarine pens

Events

1859 Royal Commission appointed to examine the question of defence.

1880's The power of the new rifled guns changes British coastal emplacement design.

1882 Decision that Navy should form first line of British defence.

1904 Committee of Imperial Defence established.

1908–9 Committee of Imperial Defence decides on policy of defence against German invasion.

1912 Admiralty urge for defences at Scapa Flow and Cromarty.

1914 Mobilized ships at Portland sail through Straits of Dover in secrecy following Austria's ultimatum to Serbia.
Aug. War.
Oct. Kitchener fears that deadlock will release German soldiers for invasion of Britain.
Oct. British naval aircraft raid Zeppelin sheds at Dusseldorf and Cologne.
Nov. Flanders Flotillas set up submarine pens at Bruges.
Dec. German Fleet raid West Hartlepool and Scarborough.

1915 Jan. Churchill renews promise that navy will prevent any more than 70 000 men reaching British coast.

1915–16 Zeppelins bomb England.

1916 German fear for invasion through Jutland.

1917 June 26 Fourteen twin engined Gotha aeroplanes bomb London causing 462 casualties (162 fatal).
June French Second Bureau start publishing reports on captured German documents dealing with defence in depth.
Dec. Both Haig and Pétain issue instructions on defence. Pétain lays the foundations for future French understanding of 'in depth' and anticipates German offensive in his Directive No. 4.

1918 March German offensive starts at St. Quentin. Germans penetrated 65 km after two weeks.
July Last German offensive near Rheims changes into Allied counter-offensive.
Nov. 11 Armistice: Along the British coast miles of trenches are filled in.

● The German theory of defence in depth and German concrete emplacement design on the Western Front were to be two of the most significant developments during the 1914–18 period. They had considerable influence on later fortification design (1). But other aspects of European defence were of importance and interest during this period. Three aspects can be defined: firstly on the Western Front the Allies followed the German lead in defensive theory but added their own characteristics—British prefabrication and the air-space theory; secondly both Germany and Britain feared invasion and put varying amounts of effort into coastal defence; finally the aeroplane forced the most important development of all and protection against air attack had become a vital consideration by 1918. For European civilians it was a taste of things to come.

On the Western Front the Germans, led by von Falkenhayn, had already prepared themselves for a war of attrition in 1914. They therefore naturally took the lead in this type of warfare but the Allied commanders in France stubbornly continued to believe in ultimate mobility. The war, they thought, would soon be over—an opinion which persisted despite their unsuccessful and costly attempts at a breakthrough in 1915 and 1916. Brigadier-General Charteris records that one of the few realistic prophets at GHQ in the first year of war was General Rice, the senior sapper:

". . . He predicts that neither we nor the Germans will be able to break through a strongly defended and entrenched line, and that gradually the line will extend from the sea to Switzerland, and the war end in stalemate. D.H. (Haig) will not hear of it. . . ." (2).

This failure to understand the new type of war, in addition to the need to hang on to every inch of French soil for reasons of prestige and

1 Dugouts on the banks of the Yser Canal, Ypres, 27th August 1917.

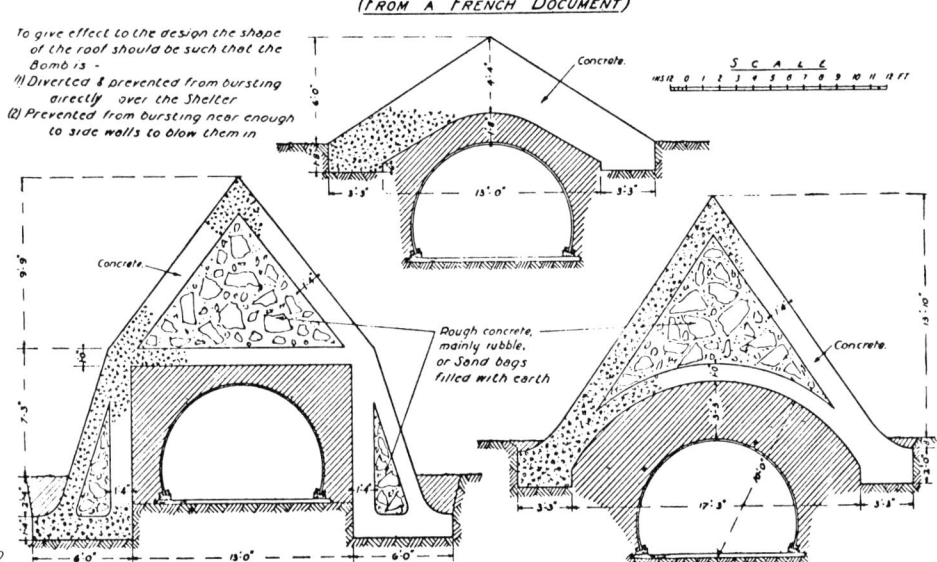
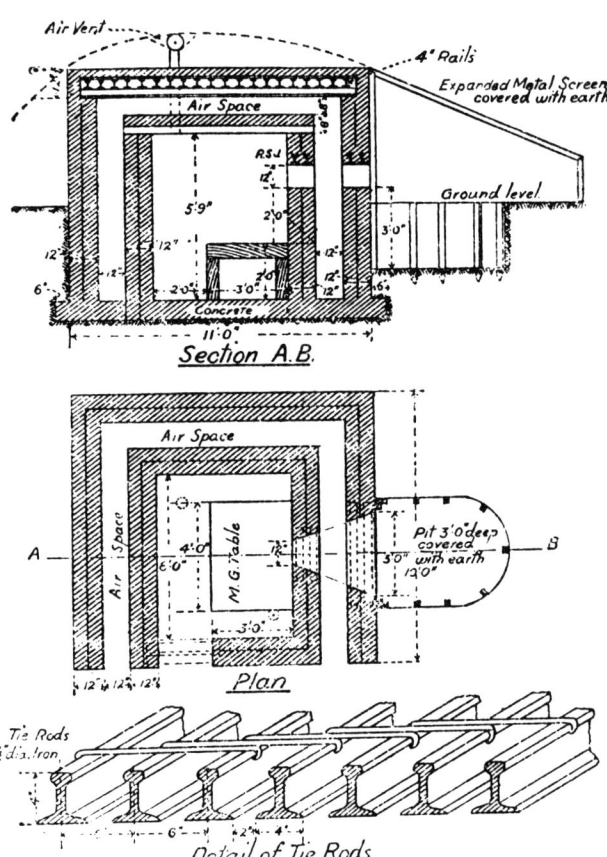

2, 3. Examples of Allied concrete design on the Western Front: the bomb deflecting shelter as proposed in the French Army (2), and the design for a machine gun emplacement based on the air space theory as published by the British General Staff in 'Notes on Trench Warfare for Infantry Officers', December 1916 (3).

morale, led the General Staff to adopt a defence which evolved from a belief in their own offensive power . . . from the pre-war doctrines of 'arme blanche' and mobility.

When the armies had stabilized themselves after the Battle of the Marne the first rapidly dug trenches, where the Allied assault groups had ended their advance, became the front trench. The front was therefore established along a line which was dictated by the strength of the enemy defences and one which inevitably left the Allied armies in the weaker position. Small areas of high ground or groups of buildings or other areas which offered defensive advantages projected out from the enemy line. Such positions made the British trenches vulnerable to enfilade fire. But the British commanders insisted that if a line was to be straightened it should be done by pinching out the strong enemy salients, not by withdrawal. In order to deny any territorial gain to the enemy the front trench, the firing trench, became the main defensive position. To make the situation worse this overcrowded front line, within reach of the German artillery, was not provided with strong dug-outs for the troops like its German counterpart.

Such provision was discouraged by the General Staff as a danger to the offensive spirit. In 1916 the 'Times History of the War' accurately reflected British military opinion on the Western Front where this subject was concerned:

"There is no doubt that in some cases this great security (of the German dug-out) was harmful to the defence as the defenders refused to come out to take an active part in the defence." (3).

Similarly, a British Intelligence report in 1916 on 'German Methods of Trench Warfare' stated that:

"While the Germans are experts in bombing and in machine gun tactics, they invariably show themselves inferior to any of the nations against whom they are fighting when it becomes a matter of 'cold steel.'" (4).

This belief in the power of the offensive persisted. Even when the German machine guns had proved their defensive strength in the first months of the war Haig insisted that: 'The machine gun was a much overrated weapon and two per battalion were more than sufficient' (5). By the end of the war forty per battalion was normal. Similarly, the tank, Churchill's 'trench-crossing armoured car', found its way on to the battlefield despite the General Staff. Lord Kitchener called it: 'A pretty mechanical toy. . . . The war would never be won by such machines' (6). Throughout the four years of war the Allied Staff continued to believe that a massive offensive, based on brute physical force, was the key to success.

The result of this attitude on the part of the British General Staff was that British defence provisions were minimal when compared with German measures. But, as Barrie Pitt writes: '. . . if the British trenches were in places shallow and unsafe, the French trenches were in places virtually non-existent' (7). Correlli Barnett puts the relatively poor condition of the French defences down to national characteristics:

"By contrast (to the Germans) the French soldier would not dig; would not take pains to make his defence strong, comfortable and clean; the French officer could not or would not master the intricate problem of mass organization posed by the new type of war." (8).

The result of French disregard for the safety of their troops, of too great a reliance on the headlong offensive, was the French mutiny of 1917. But although in the Allied armies defence measures were not given as much emphasis as in the German army, the task of designing and executing defence works was not ignored. Information in the form of a stream of pamphlets, which described how Allied and German companies had constructed their defences, were issued—although no central control by the General Staff was enforced until late 1917. The chaotic situation which resulted was further encouraged by the constant movements of the troops (9). Thus no overall conception of how to build, hold and defend a system of defences was developed.

In the area of defensive strategy the Allies were never to catch up with the German lead but in detailed construction they developed in a completely individual way. Colonel G. R. Pridham wrote in the Royal Engineers' Journal that:

"The details of work in ours (the British trenches) were probably better executed, and a lavish supply of materials made our trenches appear almost luxurious compared with those of the enemy, (though) . . . in layout and design they were probably inferior." (10).

This description was probably a little optimistic but it is quite clear from documents available that British developments in detail design showed considerable initiative.

In certain ways British detail design was similar to that of the Germans. On the Ypres Salient (11) both German and British engineers

built concrete observation posts inside the walls of existing farmhouses and buildings (12). There were minor differences: the Germans tended to use planks for formwork while the British engineers favoured corrugated iron. In the same area the British favoured the use of hollowed tree trunks lined with steel, or even 'steel trees' appropriately camouflaged, for observation posts (13). The difference between German and British construction was more basic than these idiosyncracies suggest however. While the Germans mainly used monolithic reinforced concrete construction for their installations, the Allies tended much more towards a form of composite concrete construction designed to resist against shell penetration. A sandwich construction had been used for the Verdun forts prior to 1914. Similar theories were developed further on the Western Front. The use of shell-bursting slabs and the air-space theory became characteristics of British construction. In essence the air-space theory was extremely simple: a bursting slab was to detonate the shell, behind this slab the air space or a soft absorptive material, such as earth or sand, cushioned the blast and generally reduced the effect of the explosion before the shock wave reached the inner and main protective shell. This theory was implemented throughout the British defences, both in timber and earth designs as well as in concrete. Designs for machine gun posts on these principles featured in 'Notes on Trench Warfare for Infantry Officers', issued by the General Staff in December 1916 (14). Surveys of actual constructions (15) confirm their general, if not universal, application in various forms. The burster principle, incidentally, was to be used both post-war, in the French Maginot Line, and during the 1939–45 war in German submarine pens and other bomb-proof structures.

If the burster principle was one feature of British defence construction then prefabrication was the other. The shortage of materials and labour on the continent led to the introduction of prefabrication in the British hutting programme. In the sphere of defence construction, however, the criteria of speed and silence were probably more important reasons for its use. Prefabrication never found general application in defence construction but by 1918 the tendency was firmly established.

The ultimate in this tendency towards prefabrication was probably the 'Moir' pill-box, which owed its origins to experience on the Western Front and featured in the 'Manual of Field Works' published by the War Office in 1921 (16). The 'Moir' was a circular machine gun post formed from interlocking concrete blocks. On top of these blocks was a cupola type mounting. This was a sophisticated structure and the type of prefabrication which found successful application on the Western Front was much cruder and more adaptable to improvisation. In this context the 'Ryes' O.P. Plates were the most outstanding example, offering considerable flexibility with only a single type of component (17). These plates were designed to meet the demands by the Royal Artillery for an Observation Post capable of being easily erected to a suitable height, giving protection to the observers, and resisting destruction by gun fire. The resulting conditions which the design satisfied were: simplicity in manufacture and erection, portability, strength combined with lightness, ability to resist destruction by shell fire, capability of being strengthened to an unlimited extent and of being adjusted to any height. But its flexibility was still wider. The plates, in 2740 and 1830 mm lengths, could be bolted together with one type of bolt to form a variety of building plans and sections. They were not only suitable for observation posts but also for pill-boxes, palisades and bomb shelters, used either on their own, being of structural shape, or reinforced with concrete, brick rubble, stone or earth. They could also form individual building components such as box girders or flange girders within larger structures. The usefulness of such a component in a war of constant improvisation can well be appreciated but to what extent the plates were used is difficult to assess. More successful and more readily available for improvisation were the sheets of 'elephant iron' or corrugated steel. These were used extensively by both sides as formwork and as reinforcement in dug-out work.

Though the Allies often excelled in detail design, 'no far reaching departure or invention can be claimed' by the Allies having 'important bearing on actual Defence' (18)—that is defensive strategy. When the first modifications of rigid defence started in 1914 and a small amount of flexibility was introduced, different modifications of the defence system occurred within each company border. Thus the 'strongpoint' or 'island' idea was in great favour until it was found that strongpoints, if overdone, were easily crushed by concentrated artillery fire and the whole defence collapsed. When the Allies tried to develop a more flexible and deeper defence in 1917, the result was only a poor copy of German strategy based largely on translations of captured German documents.

◀ *4. Men of the Border Regiment resting in a front line trench: Thiepval Wood August 1916.*

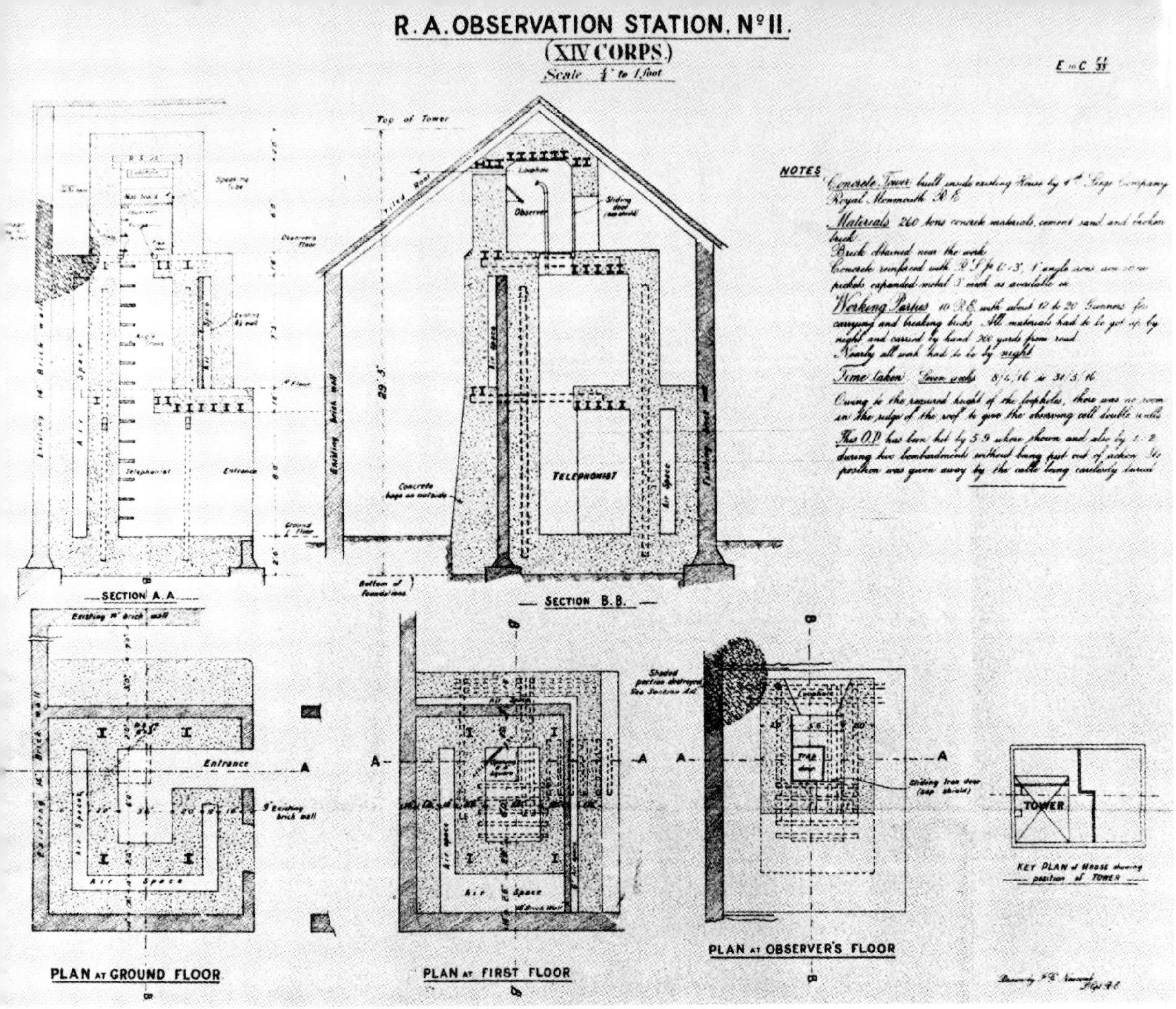

5, 6. British observation posts constructed within existing buildings in Flanders: drawing showing use of air space construction (5), and more recent photograph showing a post with external walling blasted off to reveal the use of corrugated steel shuttering (6).

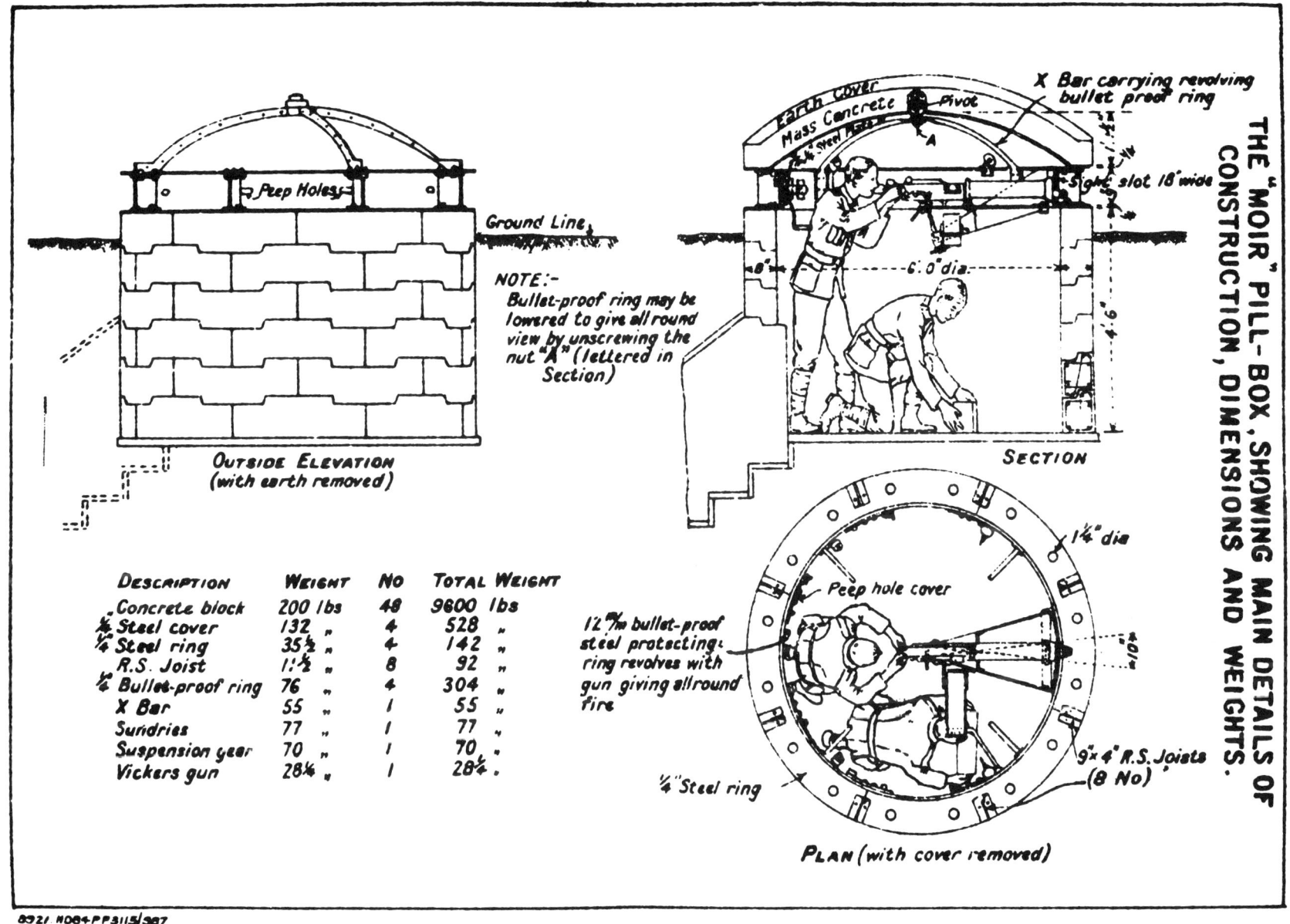

7. The British 'Moir' pill-box: a design based on the use of interlocking concrete blocks.

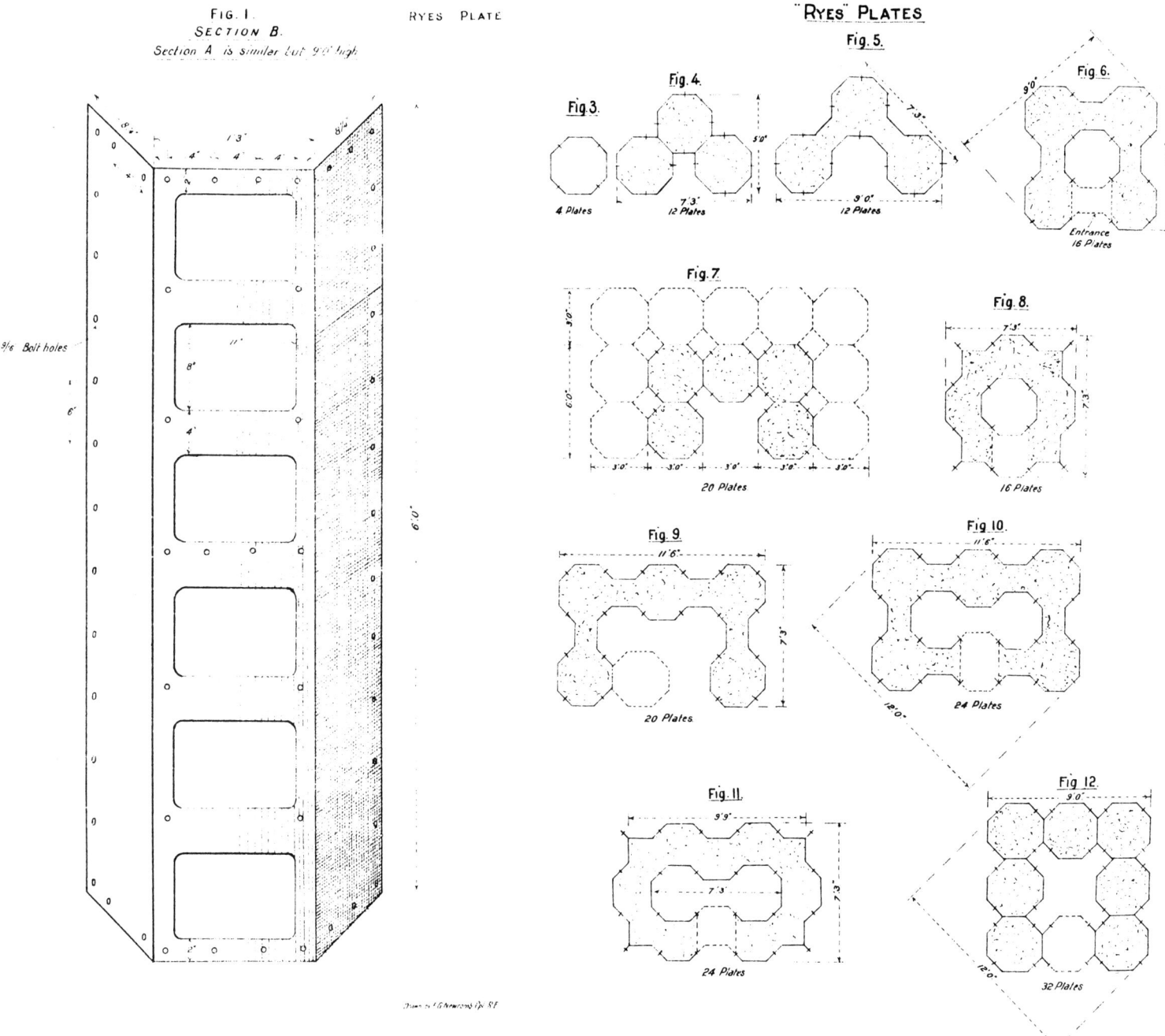

8. The 'Ryes' plate and plans showing its inherent flexibility.

From June 1917, the French Second Bureau started publishing several reports of such documents: 'German Infantry Bulletin', dated June 10th, contained an appendix about 'German ideas on positional warfare'; 'German Instructions on the Counter-attack in Depth', dated January 31st; finally a translation of the German instructions on construction for the defensive positions used at Third Ypres was published on September 15th (19). Except for an early instruction by Pétain that the second position was to be constructed at 'such a distance from the first that it cannot be subjected to preliminary bombardment at the same time as the first', prior to this no Allied instructions on defence had been issued on comparable principles to those developed by Ludendorff and Lossberg. In December 1917, however, a series of instructions was issued from Field Marshal Haig's headquarters in which, for the first time, the emphasis was on the defensive rather than the offensive. Similarly Pétain signed instructions embodying what was to become the French theory of defence in depth: 'Instructions on the Defensive Action of Large Formation in Battle.'

The armies were slow to fully understand and implement the principles involved. When Ludendorff launched his main offensive in March 1918, using Captain Geyer's theory of infiltration, the Germans broke through the British defences with comparative ease. The implementation of the new theories of defence in the British Army failed, having been based on dispersion rather than depth. After a front line, a battle zone of a depth of up to 400 metres had been built using discontinuous trenches and pill-boxes, and this had been followed by a third zone. But the main principle of German 'in depth', the use of weakly held front positions and positioning the main body of the infantry out of enemy artillery range, had been forgotten. Infantry were distributed evenly throughout and German artillery could cover the battle zone. Within the British 3rd Army sector, corps commanders had been tempted by the strength of the forward defensive positions into packing them tightly with troops. The results were disastrous. Later Haig introduced a certain amount of flexibility when he withdrew his front line on April 16th and 17th, thus taking the edge of Ludendorff's convergent blow either side of Ypres, April 18th. But the only Allied commander to fully understand at least the basic theory of 'in depth' was France's Pétain (20). Following the principles laid down by his 'Instructions on the Defensive Action of Large Formation in Battle', in his preparation for a German attack at Rheims, he was the only Allied commander who was able to oppose the German 1918 offensive plan with great success. The action proved Pétain's theories, and gave him a position which led to his strong influence on the construction of France's post-war defences. Gouraud received personal fame and Foch, who followed the break-up with his long planned tank counter-attack, was now in a position to turn the tide.

With the Germans once again in defence, the defensive theory of 'in depth' was not to be tried again by the Allies. A weakening German home-front, together with Ludendorff's failure to achieve a decisive victory, had achieved what the Allied offensive had attempted to do in 1915, 1916, and 1917. But although few of the army staff on either side realised it at the time, the decision had been forced at sea through the blockade—not in the trenches.

The war at sea leads on to the second aspect of this 'wider scope of defence construction', namely coastal defence. The preparation of coastal fortifications in Britain had always been regarded as being secondary to the development of naval power itself. The last coastal defence programme of any size had been undertaken in 1860, following the report of a Royal Commission directed against France as the potential enemy. Portsmouth, for example, acquired ring forts under this programme. Massively constructed sea forts were built in main estuary approaches. It was not until the end of the century that it was realised that the enemy might not be France, but Germany. By the time war broke out the disposition of defences was still being debated. On March 29th 1913, Churchill, the First Lord of the Admiralty, had summed up the situation as follows:

"With the exception of Chatham, no naval or military harbour exists (on the front against Germany). Chatham itself has no graving docks for the later Dreadnoughts, and the depth of the Medway imposes serious limitations of tides and seasons upon great vessels using the dockyard. Harwich affords anchorage only to torpedo-craft (and light cruisers), and is lightly defended. The Humber and the Tyne are unsuitable for large battle fleets, and are but lightly defended. Rosyth will not be ready even as a war repairing-base till 1916 at the earliest. Defences are being erected at Cromarty and a temporary floating base is in the process of creation at that point. Only improvised emergency arrangements are contemplated for Scapa Flow and the Shetlands are quite

9. Map of Anglo-German naval bases in 1914, including the sites for the later German submarine bases.

unprotected. The only war bases available for the fleet along the whole of this front are Rosyth, Cromarty, and Scapa—the more remote being preferred, although the least defended." (21).

In spite of the emphasis on ships rather than fortified bases it was, ironically enough, only because of a test mobilisation which had been previously arranged that the fleet happened to be at sea in the summer of 1914. Then came Austria's ultimatum to Serbia. Instead of dispersing the fleet Churchill sent it through the Straits of Dover to take up positions on the East Coast. As it happened, the open sea was safer than its bases; and it was because of this, rather than deployment on strategic tasks that the Grand Fleet continued to spend most of its time at sea until the second half of 1915.

Meanwhile a frantic programme to secure the safety of the bases was being carried out. A desperate appeal from Jellicoe, whom a submarine alarm had chased away from Scapa Flow for the second time in October 1914, produced this reply from the Admiralty:

"Every effort will be made to secure your rest and safety in Scapa and adjacent anchorages. Net defences (anti-submarine nets) hastened utmost, will be strengthened by successive lines earliest. . . ." (22).

By mid-1915 Scapa Flow had been equipped by 4-inch and 6-inch batteries, searchlights, submarine booms in the principal entrances and block-ships in some of the others; it could be regarded as a safely protected harbour for the Grand Fleet. The main worry of the navy was now no longer unprotected bases but the siting of these very bases.

The Channel and the western exit from the North Sea were covered by the Second Fleet, based at Portland and the Dover Patrol. The north was protected by the Grand Fleet at Scapa Flow and the battle cruisers at Rosyth. The East Coast, however, was vulnerable to hit-and-run raids as was shown by the German bombardment of Hartlepool and Scarborough in December 1914. This weakness arose because the strategy set out by the Committee of Imperial Defence in 1908–9 had been based on the assumption that if there were a foreign attack on British shores it would be a full scale invasion by not less than 70 000 men. Smaller raids were not properly taken into account and the only provision made for defence against them consisted of a variety of obsolete naval vessels, old submarines, about 300 naval guns from 4-inch to 9.2-inch calibre and a strategic mobile force of 120 000 men stationed behind the thin coastal crust (23).

When the deadlock was realized, however, the fear that Germany might attempt to strike a blow at England, while the armies in France were neutralized in the trenches, led to additional measures being taken. The shore defences were strengthened, the entrances to undefended ports were closed by block-ships and mines, and plans were made for entrenching the coast generally. This last measure was delayed, for reasons described by Lord Hankey, then Secretary of the Home Ports Defence Committee:

"The coast had not yet been entrenched. Plans had been prepared, but except at defended ports it had not been thought worthwhile to cause alarm by digging entrenchments during the summer season at the seaside. . . ." (24).

Thus London's first defences in the direction of the Essex estuaries, where no serious natural obstacles existed, were left in a rudimentary state until October 1914. There were eventually three trench systems: the outermost running north of Chelmsford by Maldon and Danbury Hill and the innermost by Ongar and Epping (25). Large parts of the rest of the East Coast were similarly entrenched and in November 300 000 half-trained troops were deployed along the East Coast. To these first field fortifications semi-permanent concrete pill-boxes were later added (26). By 1918 the coastal crust, held by an army varying between 300 000 and 500 000 men, was further strengthened by: sixteen squadrons of fighter aircraft, 480 anti-aircraft guns and 706 searchlights as well as numerous coastal observers, radio and telephone communication links.

The chaos surrounding the provision of British coastal defence was to be repeated little more than twenty years later. The standardized concrete pill-boxes built in the Thames and Medway area were not in fact unlike those built for a similar purpose in 1940. But in one aspect of detail design the coastal fortifications differed from future designs and those of the immediate past: this was the provision of overhead protection for artillery intended to engage naval craft. Small pill-boxes, designed to combat the embarking invader, were given overhead protection against small howitzers and trench mortars. But the coastal artillery had no such protection.

Due to the rapid development of mortars and howitzers during the wars of 1792–1815, enemy projectiles had been as likely to arrive vertically as they were horizontally. The artillery in permanent fortifica-

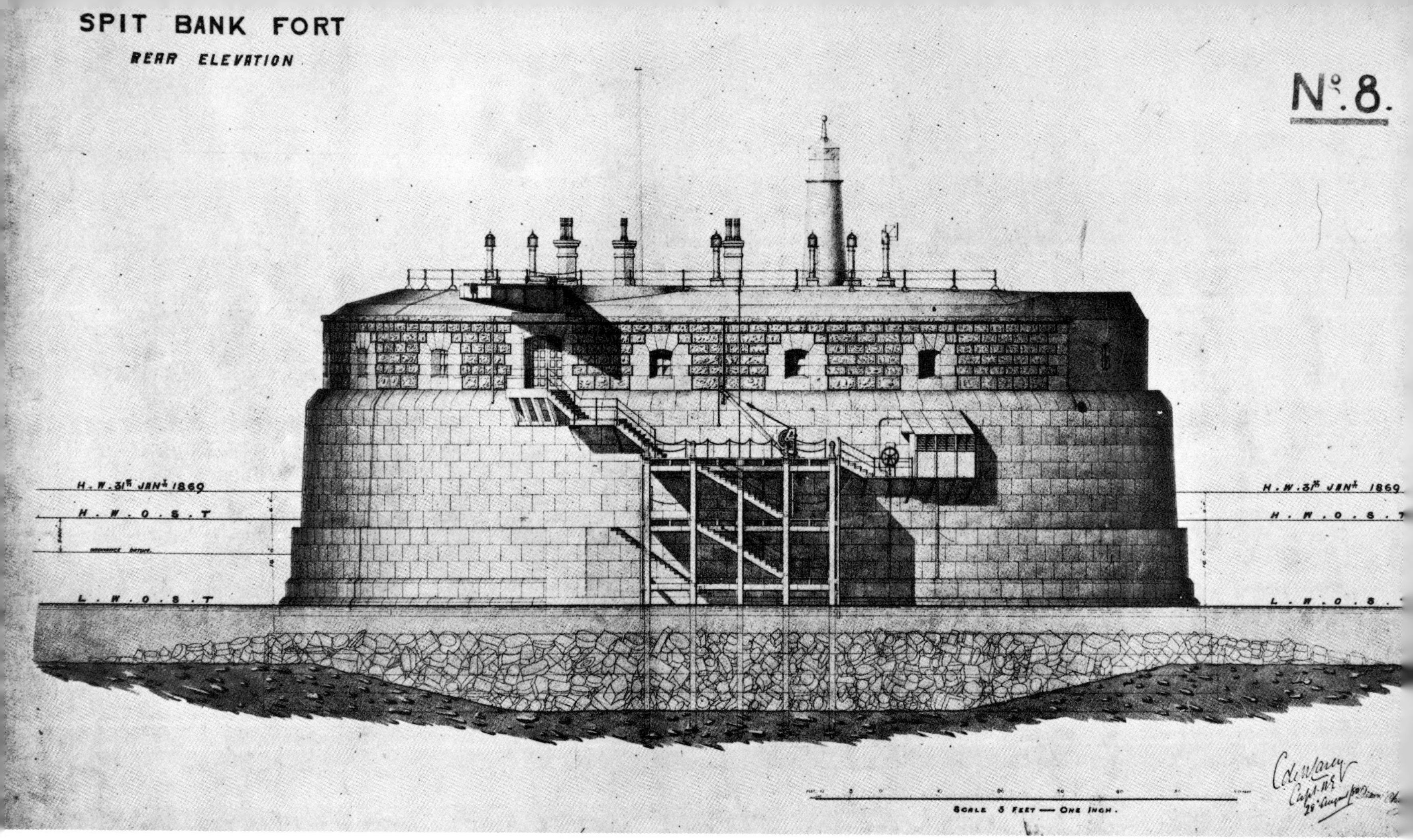

SPIT BANK FORT
REAR ELEVATION

No. 8.

10. One of the five circular sea forts planned as part of the new Portsmouth defences following Palmerston's Commission. Foundation problems led to two forts being relocated and only two forts were completed as planned, each armed with 49 cannon in granite casemates, and 5 cannon in revolving cupolas.

11. Casemated coastal battery at Milford Haven: constructed following the Royal Commission of 1858.

12, 13. Centre Bastion battery and machine gun emplacement (12), and Fletcher battery (13), both part of the Thames and Medway fortifications which were completed just before the war. By that time coastal batteries were no longer casemated.

14

15

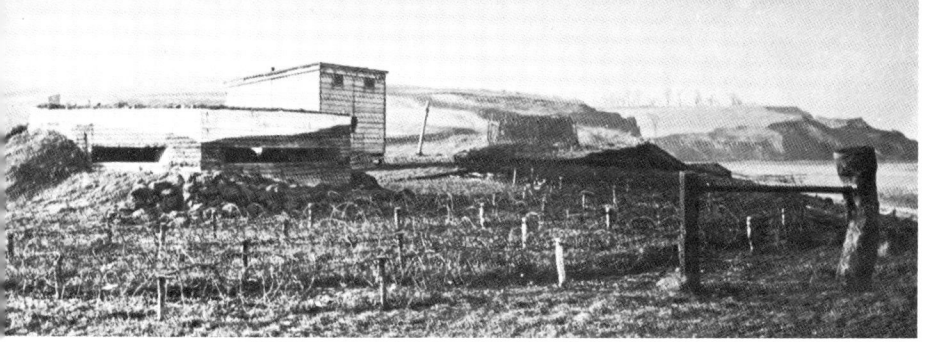

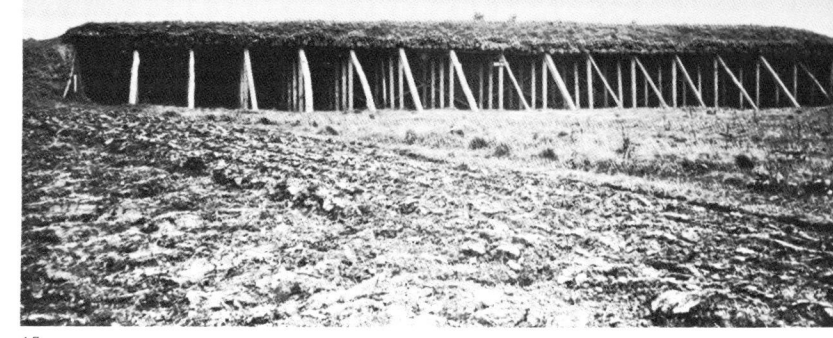

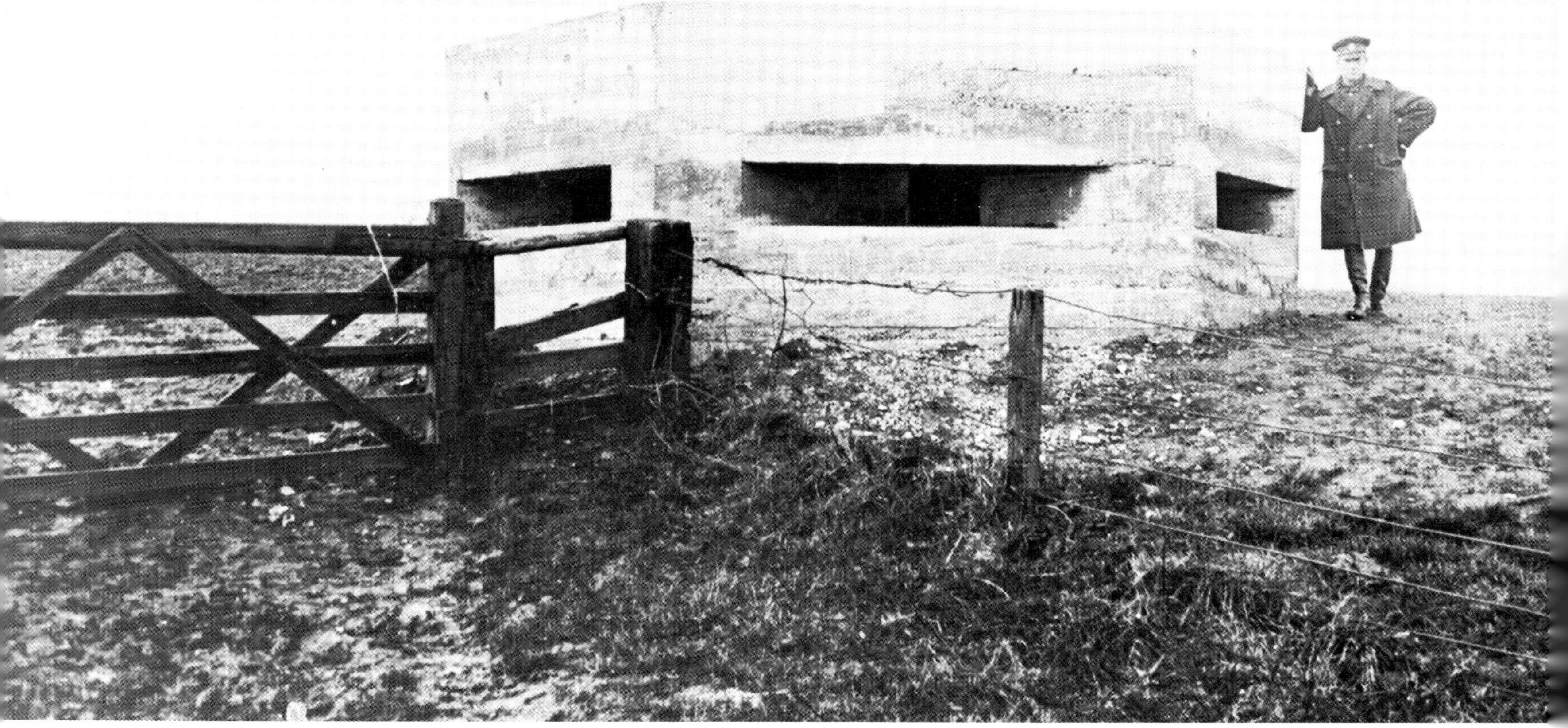

16

14, 15, 16. Standard concrete pill boxes and earthwork redoubt at the mouth of the Thames Estuary during World War I.

tions had to be protected from above as well as from the flank, and the casemate came into use. In 1860 the Royal Commission found that the new rifled guns had decisively proved their power and, though the casemate was still the approved protection for the coastal-defence gun, the new invention soon forced a change. Several objections had already been raised against the casemate: once a shell entered the embrasure it did more damage than it would have done in an open rampart due to the constricted space; it was very conspicuous and formed a well-defined target for a bombarding warship; it prevented smoke from escaping, thus seriously interfering with the functions of the gun crew; it greatly restricted the field of fire requiring the employment of a large number of weapons to cover a comparatively small area. Now, in addition, it could not keep out the shells arriving horizontally from the new and extremely accurate rifled guns (27).

The French had already adopted open emplacements for their coastal defence, arguing that the smaller target compensated for the greater vulnerability. In 1880 this type of construction, 'en barbette'—a gun platform behind a rampart and parapet, became the accepted type of gun emplacement. The British were fortunate in that the construction of defences for their principal ports had been postponed. During this period rifled artillery had proved its power and most of their artillery emplacements were therefore also designed on the open principle. Experiments with disappearing cupolas were started at Portland but these never found general use (28). The 'en barbette' artillery emplacement was to be the type generally used in both British and German (29) coastal defences until the Second World War—when protection against aerial attack became of primary importance.

The development of the aeroplane in a military context was probably the most significant aspect of the 'wider scope of defence construction' during 1914–18. In the first months of the war the first air attacks were confined to short range military targets. The Allied General Staff were dubious of the value of the aeroplane and Foch's view that 'for the Army the aeroplane is worthless' was typical (30). During 1915–16 the raids on Britain by 'Zeppelins' forced a change of attitude. For the first time the people of England felt invaded, so defenceless were they against this new form of attack. Only by the winter of 1916–17 were the defences sufficient to stop the raids, but in the summer of 1917 raids by the less vulnerable Gotha aeroplanes began. In London people took shelter in the tubes and in basements and, although little action was taken in the actual construction of shelters, the need was born. The suggestion of what effect this was to have on defensive structures was already apparent on the Western Front.

On the Belgian coast, bases had been set up for the Flanders submarine flotillas in November 1914. These bases were within air range of the Allied lines and the problem of protection against air attack became a pressing one. The submarines were particularly vulnerable to attack when in harbour, where they were both stationary and above water. A variety of shelters were constructed, most little more than a reinforced canopy cantilevered from the harbour wall, but at Bruges a large submarine shelter was built with 9 metre by 76 metre pens for eight submarines. It was this shelter which was the forerunner of the numerous German pens of the Second World War.

It was the forerunner in two ways. Firstly in the roof construction: the closely spaced pre-cast reinforced beams, with a 750 mm covering slab of reinforced concrete, were the basis of the construction using pre-stressed trusses and in situ concrete developed and used in pens 20 years later. Secondly in appearance: the Bruges pen, described in the 1920 edition of the RIBA Journal as having the 'greatness of a classic temple' (31), possessed the same monumental or neo-classic appearance which later featured so strongly in many of the pens built in France by Hitler. The significance of the Bruges pens was much more general than this however. Strategic bombing had been born and more diverse forms of fortification were to be evolved. The pens were the forerunner of not only submarine pens but also underground factories, flak towers, and both civilian and military shelters.

Similarly the pens marked the birth of a new type of war, a war which had no boundaries. The construction of bomb-proof shelters was to be the urgent concern of the civilian population as well as the front-line soldier. In concluding a history of the raids on London in 1915–18, A. Rawlinson wrote in 1923:

"In conclusion, I feel it incumbent upon me to sound a note of warning, addressed in particular to the inhabitants of our great cities. If it is thought that, because the raids on London during the war caused the death of only 500 or 600 citizens and the serious injury of over 1200 others, in addition to structural damage of over £2M, these figures bore any relation to the actual risk incurred from the air raids, a very erroneous

18

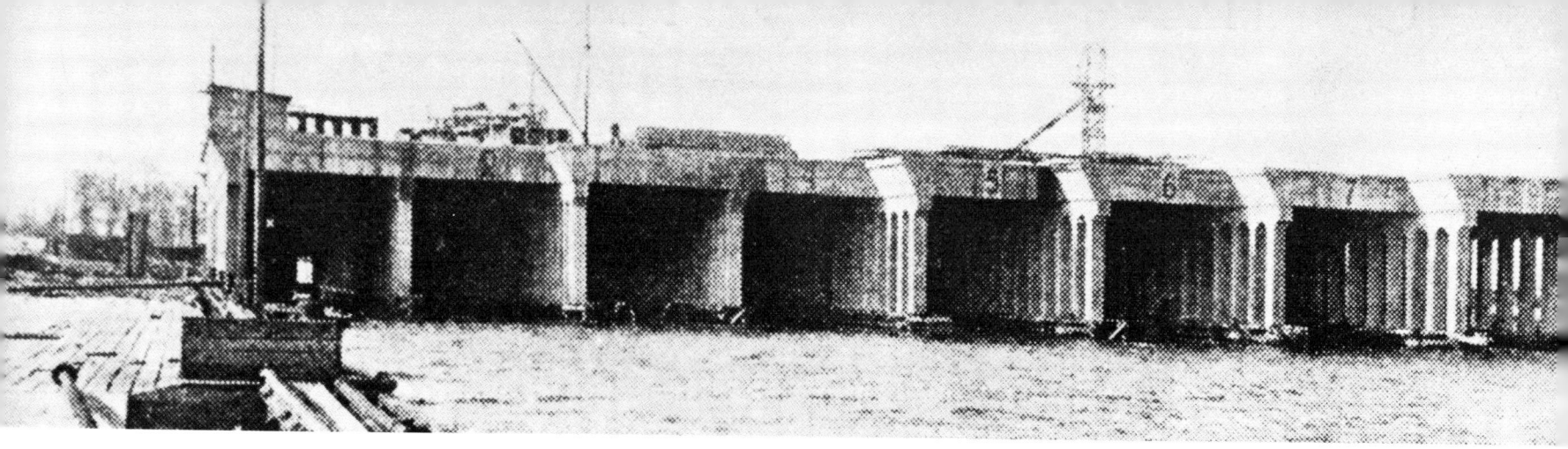

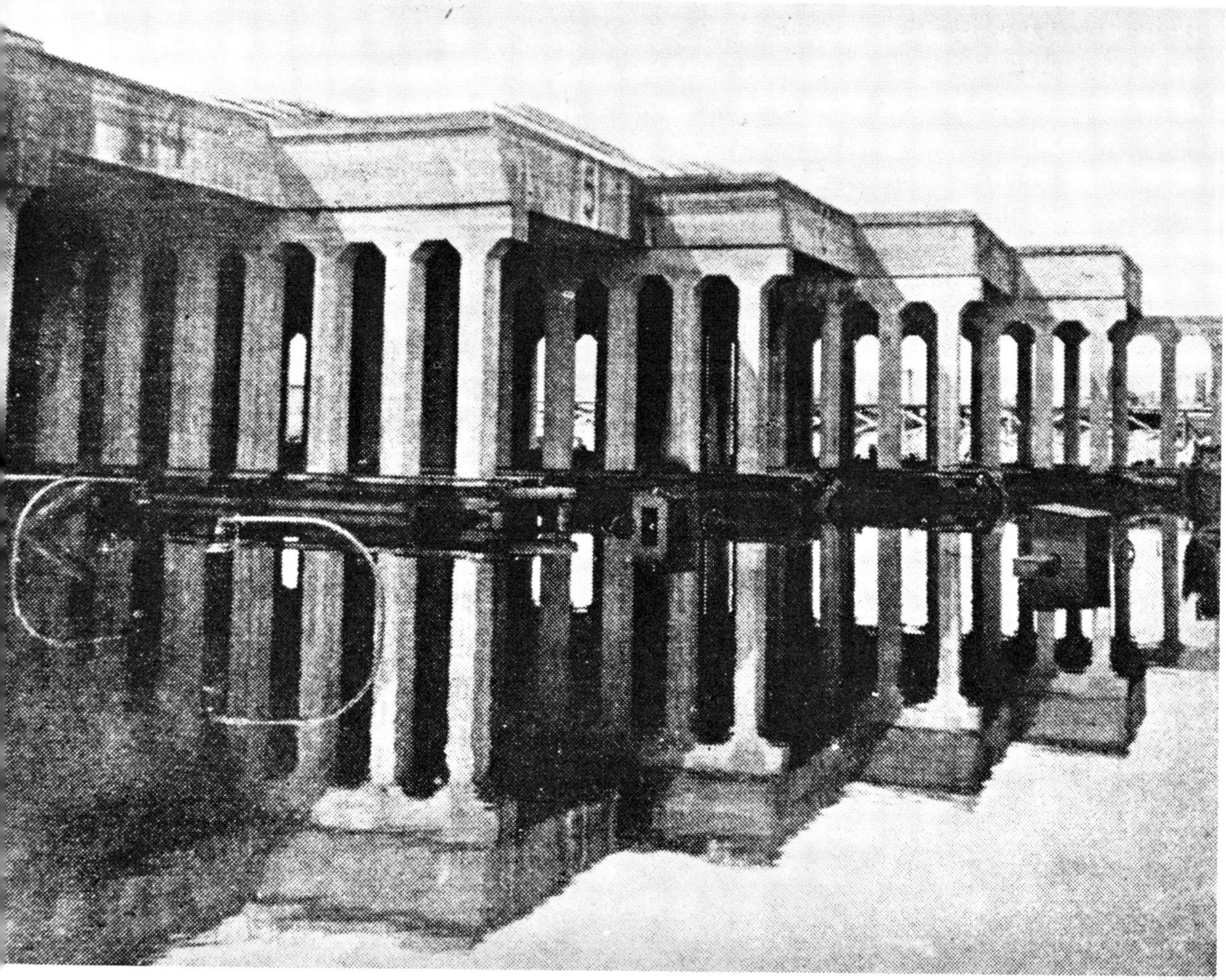

17, 18. Classical features of the German submarine pen at Bruges, 1918.

impression will have been obtained. We have every reason to be thankful that our losses were confined within those limits . . . any *future* air attacks must bear still less comparison to the above figures, for owing to the increase in numbers and efficiency of aircraft, the dangers to which the inhabitants of our great cities would be exposed if a modern air attack in force reached them are such as to *utterly defy description."* (32). (Rawlinson's italics).

4 SUPPORT STRUCTURES

Nissen huts, mobile Zeppelin sheds, Richthofen's Circus.

Events

1914 Aug. 4 British ultimatum expires: she declares war on Germany. German troops have already crossed Belgian border.
Aug. 5 British 'Works Directorate' mobilized. First 'Scales of Accommodation' for British camps based on War Office plans using tents and huts. First BEF camps built at Boulogne, Rouen and Amiens.
Aug. 11–17 BEF cross Channel.
Oct. 1 German retreat to the Aisne: re-opening of British base at Rouen.

1915 Jan. Lloyd George, Kitchener and Churchill realize deadlock. Armies grow in size. Winter shows need for better accommodation.
Spring Aylwin and Armstrong Huts are being sent from Britain. Mobile German air force group established for the attacks on Verdun: the 'Travelling Circus'.
Sept. 1 New 'Scale of Accommodation' issued which increase need for hutting.

1916 Deadlock continues. Large increase in use of portable huts, initially to overcome labour shortage. The Nissen Hut is designed. Coloured labour is being brought to France by the British for construction work.
Sept. 15 First 'premature' use of tanks by BEF on the Somme.

1917 Deadlock continues, French contractors no longer able to take large contracts due to shortage of labour and materials. British force is now 1.2 million; German is 2.5 million.
March–April German withdrawl to Siegfried Line: Allied support structures have to advance.
Nov. Revised 'Scales of Accommodation' published for BEF: increase hutting requirements.

1918 March–July German spring offensive: capture of parts of Allied support system which have been allowed to become static.
Aug.–Nov. Allied offensive: renewal of need for mobile support structures. Winter hutting programme for BEF now to provide hutted sleeping accommodation for *all* men in permanent camps.

◀ *1. German hutting on the Western Front tended to be more traditional in construction and less mobile than its Allied equivalent.*

● Clausewitz once expressed the view that: 'Superiority in numbers becomes every day more decisive' (1). Despite the advent of mechanized warfare, the military commanders on the Western Front were of the same opinion. It was in the light of this philosophy that they calmly endured the heavy losses such as those suffered by the French during Joffre's 'nibbling' in Champagne during early 1915. There in February and March they lost 50 000 men to gain 500 metres; in the April attack on the St. Mihiel salient they sacrificed another 60 000 (2). Yet the armies grew. Germany, who had a total active force of 2 million in 1914, had 2.5 million actively engaged on the Western Front by 1917. Britain, who had sent an expeditionary force of 160 000 men to France in 1914, was supporting 1.2 million combatants by 1917 (3). In addition to these soldiers, Britain supported a work force made up of Egyptians, Africans, Indians, and Chinese, as well as British contractors and workers, other service personnel and several thousand prisoners of war.

Both sides had to transport the materials, shelter, equipment, food, and every other necessity to keep these large numbers of men alive. The provision of such 'support structures', for both men and equipment on the Western Front, became an important and difficult operation. Failure in this sphere could be no less dangerous than a failure to provide bomb-proof and front-line defence structures. An extreme case was when the French General Staff failed to provide suitable accommodation for their troops at the front and this became a contributory cause of the French mutiny of 1917. Materials became as important as men. It was not without justification that the Germans named the Battle of the Somme the 'Materialienschlacht'—The Battle of Materials.

The scale of the deadlock was unprecedented and the scale of 'support structures' required equally vast. In terms of construction the variety of building types called for was extreme: hospitals, field bakeries, remount depots, veterinary hospitals, post offices, offices, billets, to name only a few (4). The answer turned out to lie in the development of prefabricated structures. This had long term effects of comparable significance, from a military point of view, to those of the idea of defence in depth; from a civil point of view they were later to be even more important. Although both the Germans and the Allies were faced with the task of supporting huge military communities on the Western Front, in the case of Britain the problem was increased by the further obstacle of the English Channel. All her goods had to be shipped out, a factor

2

3

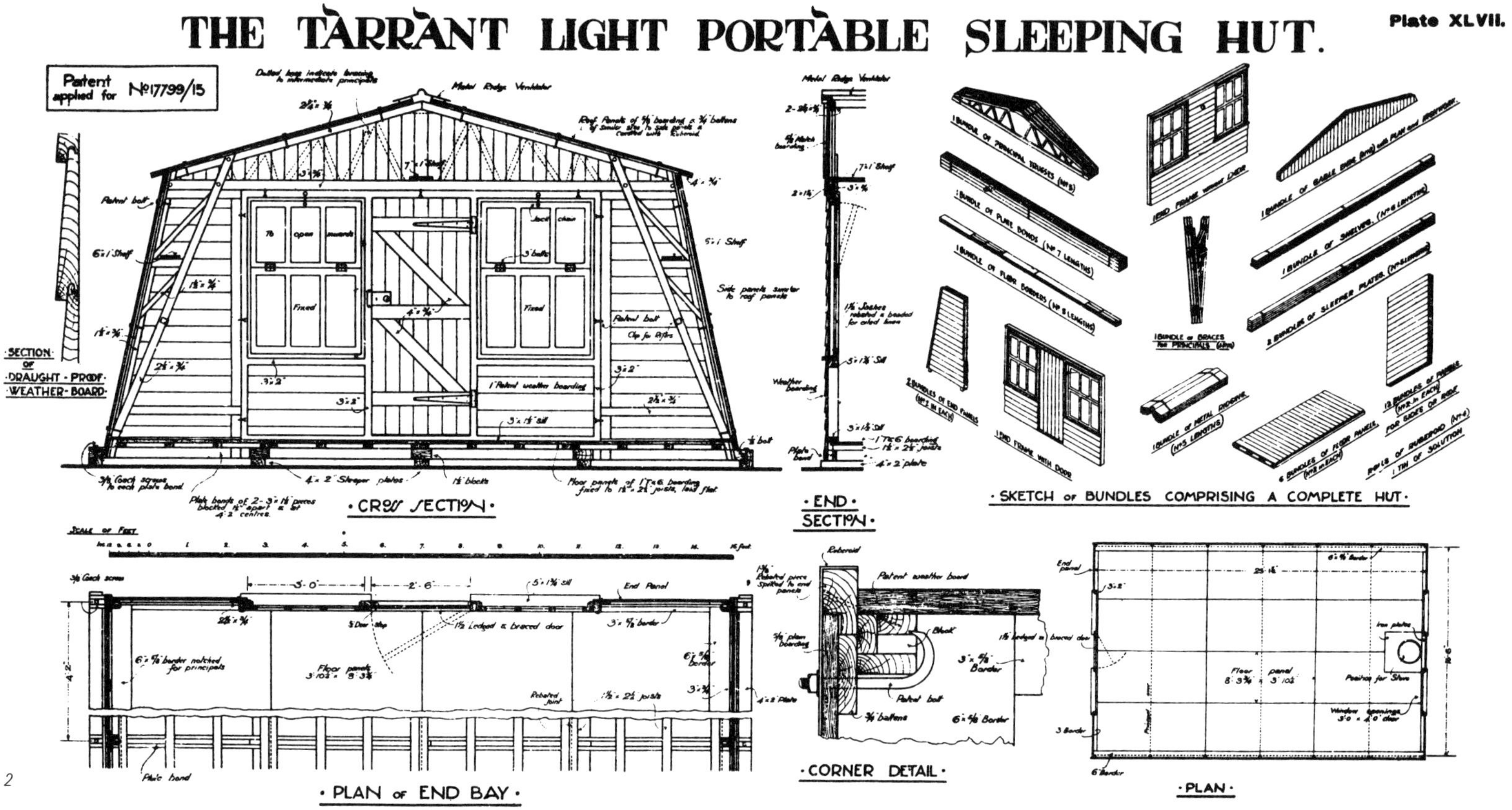

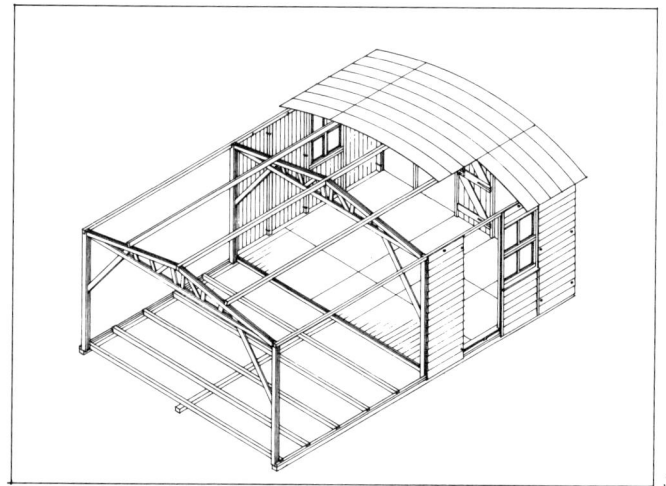

2. The Tarrant Light Portable Sleeping Hut.
3. The Tarrant Portable Hut Mark II.

which became increasingly important as the British army grew and French resources diminished. Consequently it was Britain who prompted some of the major developments in prefabricated hutting.

The official British preparations for the accommodation of forces abroad started on August 5th, one day after the declaration of war, with the mobilization of the Works Directorate H.Q. (5). The types of buildings which would be needed were investigated and the positions for the bases chosen. Generally it was decided at this stage to use in France the existing War Office designs, which had been planned for housing troops in England. But while all ancillary services were to be in huts, the sleeping accommodation was only to be tented.

Even those modest plans required a considerable amount of labour. The lack of available army labour brought about the use of civilian contractors (6). Thus M. Chouard, a large French Public Works contractor, was responsible for the camp at Rouen and two English contractors, Harbrow and Tarrant, were made jointly responsible for the camp at Le Havre. At other sites such as Marseilles and La Valentine the work was split between military and contracted labour. For the supervision of these civilian contractors a number of professional men, engineers, architects and surveyors were selected by the War Office as 'Temporary Inspectors of Works' (7). The growing pressure on the Works Directorate and the Royal Engineers gave rise to an increasing use of such personnel from the civilian sphere within their ranks. The RIBA Journal recorded in 1917:

"In a letter to Mr. Lloyd George, when War Minister, the President (of the RIBA) drew attention to the fact that many highly trained young architects were being wasted in the ranks of ordinary line regiments, and suggested that they should be appointed to cadet Corps with a view to being granted commissions in the Royal Engineers. Numerous architects have since obtained commissions in this way. As a result of a personal protest to the War Office against the claims of architects being ignored for such appointments as supervisors of building for Army purposes in France the President was asked to nominate five architects for such posts. This was done, and it is understood that their work has been highly satisfactory." (8).

The initial projects were carried out according to War Office designs, but the heavy use of timber soon necessitated local modifications. Recognizing this fault the War Office tended only to specify the extent of accommodation while the actual building designs were increasingly decided on a local level. The freedom this offered local designers, both in the Royal Engineers and the contracting firms, was to result in a large variety of designs when the production of portable huts was increased in 1916.

Huts were soon to provide an increasing percentage of the accommodation as the use of tents diminished in the permanent camps and more comfortable conditions were provided for the troops during the cold winters. The first winter hutting programme started in 1914 and only provided huts for ancillary services. In 1915 however, it was decided that nurses, and camps at particularly exposed sites, should be provided with sleeping huts. This decision was expressed in the new 'Scale of Accommodation' issued by the Works Directorate in August 1915. In 1916 the provision of hutted sleeping accommodation was extended to include permanent staff and prisoners of war. In France at this time there was a severe labour shortage and to cope with the increasing building activity coloured labour was imported. On June 30th, 1917, a 'Scale of Accommodation for Coloured Labour' was issued which provided some degree of hutting for them also. Soon after, in November 1917, a revision of the 'Scale of Accommodation' of 1915, was issued. While the former had divided into Summer Scales, Tented and Hutted, and Winter Scales, the new scales were only divided into two, i.e. Temporary and Permanent. According to the Royal Engineers Official History:

"The Temporary Scale was to be used for camps that were definitely known to be temporary and where it was necessary to get the men on the ground at a very short notice without time being afforded to provide more permanent accommodation. The Permanent Scale only differed from the original "Hutted Winter Scale" in that huts were given in lieu of tents or Aylwin huts, that the floor area allowed for dining huts was taken out of that for living huts calculated at the rate of 40 ft. sup. floor area per man, and a few minor additions, such as the provision of barber's shops etc." (9).

Even so, many of the permanent camps still relied heavily on tents in 1918 (10). When the winter hutting programme was prepared in that year, however, it was decided that all sleeping accommodation should be hutted whether occupied by permanent parties or reinforcements. By this time the Nissen hut had become the most widely used 'mobile' hut on the Western Front but it was by no means the only or indeed the first

4, 5, 6, 7, 8, 9. Erection of a Weblee interlocking hut.

hut of this type. As early as 1914 the French firm Gilet Frères of New Orleans had delivered 60 large demountable huts of French design for British use. In 1915 portable huts had been constructed in England and shipped out in order to help out the French labour and materials shortage. It was this shortage which, backed by a desire for mobility and speedy construction, prompted the wider use of prefabrication. But these huts, the Aylwin and Armstrong huts, timber structures, prefabricated in 3 metre sections, were both heavy and awkward to construct and transport.

The Aylwin hut was made in two sizes and arrived carefully packed in wooden crates. The huts were made in the form of a 'lean-to', of light wood framing covered at sides and roof with canvas, with additional roof covering of corrugated iron. Though they proved popular during the summer and autumn of 1915, they were not found strong enough for winter conditions and were discontinued after the summer of 1916. The Armstrong hut, on the other hand, was used in large numbers during the war. The hut was provided in two sizes, of which the smaller was the one generally used. It was designed at the War Office by Major Armstrong of the Royal Engineers. Like the Aylwin hut it was a timber frame structure with canvas cover. The strong construction incorporated iron stays holding the wall and roof panels together, but, although it stood up to weather conditions well, it also proved extremely cold for the occupants (11).

During early 1916 a variety of factors brought about a sudden increase in the manufacture of portable huts in France: the plans for the Summer offensive, the large amount of work still not completed on the winter hutting programme which had been planned in 1915, the lack of French labour, such as carpenters. Thus, and as a result of the freedom already given to local designers in 1914, contractors such as Tarrant and Somerville, in addition to local Work Officers at various stations, were soon competing with each other in the design of suitable mobile structures.

While previous huts had either been of the light canvas variety or, in the case of more permanent structures, constructed with timber framing on site, these huts combined the mobility of the Aylwin and Armstrong huts with the rigidity, warmth and general weather protection of those built by traditional methods on site. The Tarrant firm supplied three types: the 'Tarrant Sleeping Hut', the 'Tarrant Portable Hut', and the 'Tarrant Déchet Hut'. The wall panels of these huts were made by nailing a double layer of boarding together, the outer layer vertical and inner horizontal (12). By incorporating lattice roof trusses short lengths of timber could be utilized. For assembly special hook bolts, patented under the name of Tarrant 'Grip', were used in addition to spring-clips which folded over purlins and wall panels (13).

Among those more significant designed by members of the Royal Engineers were the 'Forest Huts' and the 'Liddell Huts'. The Forest Huts were designed by Captain R. G. Brocklehurst to provide living quarters in the forest camps and a large number were made at short notice by French contractors. The Liddell Portable Hut was designed by Major Guy Liddell. This hut was framed on the same principle as the portable huts originally sent from England. Self-contained panels of three-quarter inch boarding, or corrugated iron, wall and roof units were connected by wrought iron hinges, which enabled them to be folded together during transport and quickly erected. Though several variations were made, the principle remained the same.

All these designs as well as several later designs, such as the Weblee Hut designed in England for cross-channel shipment in 1918, were overshadowed by a design which was soon to dominate the scene in the British Royal Engineers: the Nissen Bow Hut, and the Nissen Hospital Hut. In 1917 the Architects' and Builders' Journal commented:

"Billets in existing buildings, and hutments improvised of timber, were for long the accepted means (of housing the soldiers from the front line where he again can live above ground), but on the Western Front a new type of building has lately made its appearance. This is the Nissen (Bow) hut, devised by a Canadian Engineer officer—another reversion to type, to the type of the beaver's hut or the Esquimo's." (14).
To Filson Young, a war correspondent with the British Army, writing in the Daily Mail, the appearance of these huts was even more astonishing:

"At about the same time as the tanks made their memorable début on the battle field, another creature, almost equally primeval of aspect, began to appear in the conquered areas. No one ever saw it on the roads; it just appeared. Overnight you would see a blank space of ground, in the morning it would be occupied by an immense creature of the tortoise species, settled down solidly and permanently on the earth, and emitting green smoke from a right angled stem at one end, where its mouth might be, as though it was smoking a morning pipe.

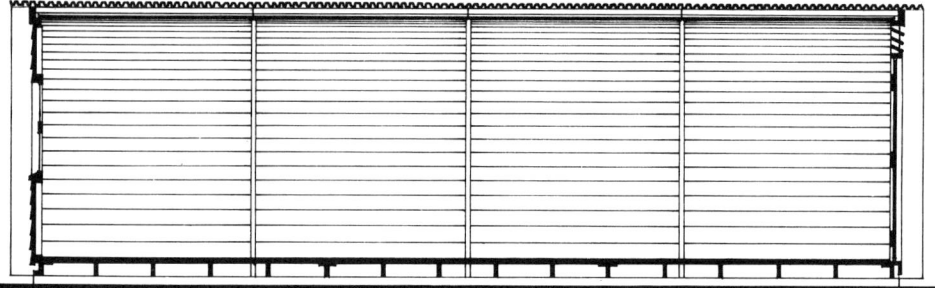
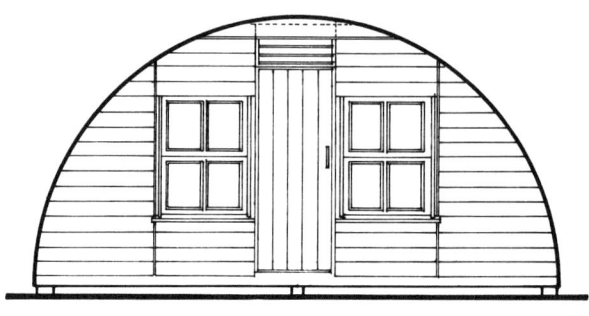

10, 11. Longitudinal section and end elevation of Captain Nissen's 'Portable Bow Hut' (10), and the hut in use on the Western Front (11). Flat sheets of corrugated iron had been used for military hutting as early as the Crimea War.

Plate LIII.

MATCHBOARD LINED HUT.

Note: The outer Corr Iron Sheets must overhang Gables by 7½" at each End of Hut

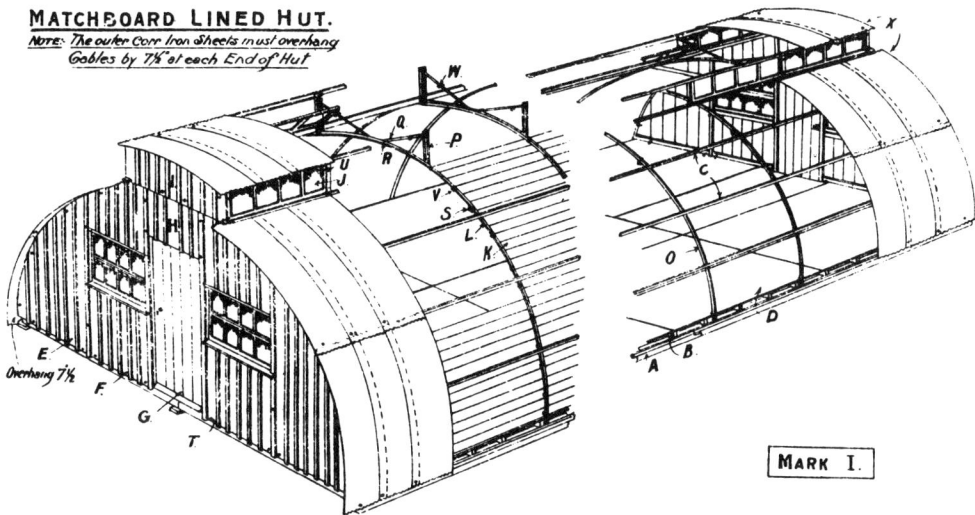

MARK I.

INSTRUCTIONS FOR ERECTING LARGE "NISSEN" (HOSPITAL) HUTS (60 ft. by 20 ft.).

(WITH MATCHBOARD LINING.)

MARK I.

1. Lay outer bearers. Unless the ground is flat it will be quicker to level up by inserting packing of wood or brick under the bearers, **rather than to level the ground**. All that will be necessary in this respect is to fill in any hollows which may allow water to collect under the hut.
2. Lay and level up centre bearers.
3. Lay floor joists and *skew nail* at ends to bearers.
4. Each bow should be examined before erection for inaccuracies of manufacture. Unless these are remedied it may be impossible to get in the lining. Bolt up rib sections with lantern light angles and braces. Raise and bolt to sills.
5. Lay floor panels on joists, notching where necessary round bows. The edges of the floor board of panels at each end of the hut must project 3" beyond the batten to obtain correct fitting.
6. Fix purlins to ribs, taking care that the shank of the bolt is tight up against the web of the T iron and that the washer is in position between the purlin and the bow. Hammer the hook end of the bolt into the underside of the purlin to prevent it slipping when tightening up the nut. Cut off the bolt above nut. Any purlins bored inaccurately should be rectified. Strut end bows temporarily with floor joists from other huts or any timber handy.
7. Attach clips to bows. If the holes are inaccurately bored, slightly bend the nail for securing the clip in order to pass it through the bow.
8. Put lining strips in under clips if necessary, placing the strips to allow room for the half-inch lining. It is important that the lining strips should extend to the end of the bows and butt on the bearers; they are supplied of proper length for this purpose.
9. Erect end panels. Fix the two window panels accurately first, equally distant from the centre, and alter the outer panels if they do not fit. The doors to be adjusted.
10. Fix the curved corrugated iron sheets on each side. All sheets lap two corrugations laterally, and the outer sheets overlap the end panels with three corrugations. The bottom sheets are fixed first. Two bolts pass through each of these sheets where they overlap laterally, and hook on to the bottom purlins. The gable end bolts only pass through one thickness of corrugated iron.

 The shanks of the hook bolts should be in contact with the upper faces of the purlins. With purlins of the correct scantling, the upper edges of the bottom sheets will also be flush with the upper faces of the purlins.
11. Fix lantern-light window frames between vertical angles by slipping them in from the top. Fix top purlins and flashing battens. Extra outside window stops must be fitted in the window frame outside the sash along the sill and vertical sides. These must be laid in paint. The outside facing strips must be well bolted up.

 A tin angle-flashing must be nailed on around each corner of the lantern light section.
12. Fix lining for raised light, on top of window frames, and secure by 1" by 1" strips nailed to purlin.
13. Fix top sheets with same overlap over end panels as given for side sheets. The hook bolts should be arranged alternately on each side of purlin.
14. Fill up the corrugations of the iron under the window sills and iron of top roof section and purlin with neat cement mixed into a paste with cotton waste teased out. This paste to be put in from the inside of the building after there is no more need for men to be on the roof.
15. Insert lining between clips and head of T iron of bows. Finish off at windows.

SCHEDULE OF PARTS AND FITTINGS FOR MATCHBOARD LINED HUT—MARK I.

Woodwork.

	Description.	No. Required.
A.	Bottom Bearers (in 4 parts)	3
B.	Floor Joists (in halves)	36
C.	Purlins for Bows and Raised Light (in 4 parts)	10
D.	Floor Panels 8' 7¼" × 4' 1¾"	35
E.	End Panels, Right and Left	4
F.	Window Panels, Right and Left	4
G.	Doors	2
H.	Centre Panels	2
I.	End Panels for Raised Light	2
J.	Window Frames for Raised Light	24
K.	Matchboard Lining for Body of Hut. Actual amount 15·76 squares. Supply, made up in 24 bundles	16 sqrs.
L.	Lining Strips for K. 240 / Spare 10	250

Ironwork.

	Description.	No. Required.
O.	Ribs in 3 sectors	13
P.	Angle Iron Frames for Raised Light	26
Q.	Braces for Angle Iron Frames	26
R.	Nuts and Bolts for Braces	26 ⎫ 52
S.	Nuts and Bolts for Ribs	26 ⎭
T.	Hook-bolts, Nuts and Washers for— Purlins	104
	Ends	60
	Feet of Ribs	26
Packed in one case.	Bearer Joints	18 505
	Corrugated Iron	288
	Spare	9
U.	Nuts and Bolts for Raised Light Frames	78 ⎫ 80
	Spare	2 ⎭
	Washer Plates for Purlins	104
	Sheet Iron Clips for Lining	290
	Wire Nails, 5"	15 lbs.
	Wire Nails, 1½"	1 lb.
W.	Window Stops	26
X.	Corrugated Iron Sheets, 7' × 2' 3"	175

12. Erection of the Nissen Hospital Hut.

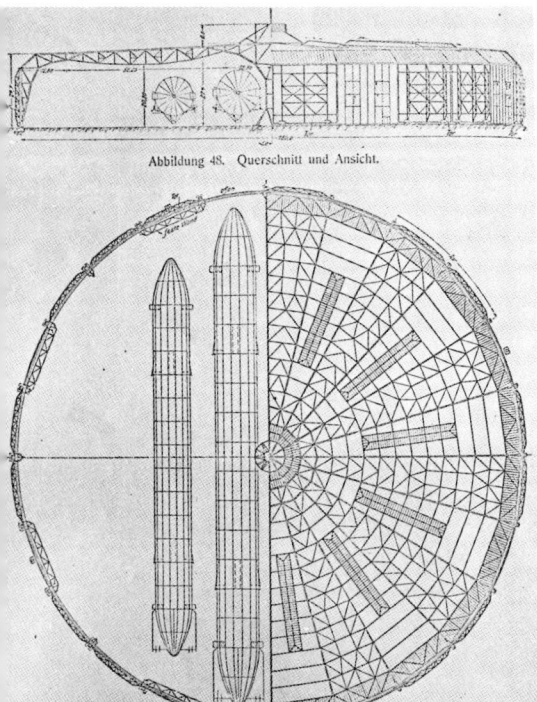

13. The large Naval airship base at Nordholz, where a variety of hangers were built. The double revolving shed (ringed) is seen head on. In the foreground are two twin sheds.
14. The circular airship hangar as proposed by Ernst Meier.
15. The problem of moving dirigibles in and out of their hangars; this was an impossible task in a cross-wind.
16. Several of the naval sheds were double sheds: view of the interior of the Alrun shed at Ahlhorn.

And when such a pioneer found that the situation was good and the land habitable it would apparently pass the word; for by twos and threes, and tens and hundreds, its fellow-monsters would appear, so that in a week or two you would find a valley covered with them that had been nothing but pulverised earth before.... The name of this creature is the Nissen Hut. It is the solution of one of the many problems that every war presents. The problem here was to devise a cheap, portable dwelling-place wherein men could live warm and dry; cheap enough to be purchasable by tens of thousands; portable enough to be carried on any road; big enough to house two dozen men; simple enough to be erected by anybody and on any ground; and weather proof enough to give adequate protection from summer heat and winter cold. All these conditions are fulfilled by the Nissen hut...." (15).

The hut was based on a 2057 mm grid, using a steel 'T' iron frame. With its simple shape—a semi-circular roof, a floor, and two ends—only a small number of different components were needed. The roof in each grid section consisted of three corrugated iron panels, all interchangeable, with a 150 mm overlap. The floor was formed by laying floor panels 1219 mm wide and 2057 mm inches long, which were also interchangeable. The two end pieces were complete, one with a door and two small windows, the other with one central window, both having louvred ventilation. Heating was provided by a Canadian drum stove and the chimney was taken at right angles through the wall above the central window. The interior lining was originally provided by tongued and grooved boarding, but later designs used corrugated iron instead.

The whole structure rested simply on the longitudinal sleeper joists and, as all connections were made by a hook bolt, the only tool which was needed for the erection of a hut was a spanner. The whole hut, when constructed, measured 8.2 m by 4.9 m and weighed a ton. It could be carried on an army lorry and each component was light enough to be lifted by two men. Due to its simplicity of detail and erection, a complete hut could be completed by four men in four hours. The hut provided sleeping accommodation for twenty-four men on the floor but, by rolling up the beds, fifty-two men could be seated. The hut thus acted as day quarters as well as sleeping accommodation.

Though several modifications were made later to the 'Nissen Bow Hut', which soon were to be known simply as the 'Nissen hut', none materially affected the initial structural or design concept. The Nissen Hospital Hut was modelled on the same idea and the only difference was the size, 18.3 m by 6.1 m, and a special window section. This central lantern, necessitated by the size, ran continuously along the apex of the arch and provided lighting and ventilation along the whole length of the hut. In addition to the initial use as a hospital ward, it was used with great success as a dining and recreation room (16).

By early 1917 it was estimated that at least 20 000 Nissen huts were already in use, providing accommodation for half a million men. With the large project for winter hutting in 1917–18 and 1918–19 the numbers continued to grow, 5000 such huts being ordered through the War Office in 1918 alone. The Nissen hut thus represented the first real mass production of complete buildings as opposed to the mass production of components, an important stage in the history of both civil and military architecture. It might even be said to have represented a first prophecy of Buckminster Fuller's Dymaxion theory: 'maximum gain of advantage from minimal energy input'. The simplicity of the individual components, the consequent ease of production, no nailing and hand fitting over several small components as in the earlier types, and the quantity actually produced put the Nissen in comparison with car production today. This speed, not only in erection, but also in component fabrication, was to be remembered during the next war when similar problems arose. It was to become not only one of the forbearers of prefabrication and mass production of buildings, but its structural shape and construction were to reappear on a larger scale in the blister hangars of the Second World War. Even more amazing perhaps is the fact that this construction proved so ideal for army camp sleeping accommodation during 1917–18 that in 1943–44, when steel became available, it was generally chosen in preference to contemporary designs despite intensive developments in this field.

The British Nissen Hut and similar designs showed a remarkable amount of flexibility in a war which was characterized by the vast stagnation of the trench front. In the Allied hutting programme this much less hidebound approach was to some extent a result of problems caused by labour and material shortages and by the inconveniences of the English Channel. Factors such as speed of construction were important but there was not the conscious strategic use of the mobility of support structures which was to be seen in the German air forces.

In the stalemate on the Western Front, although it became impossible

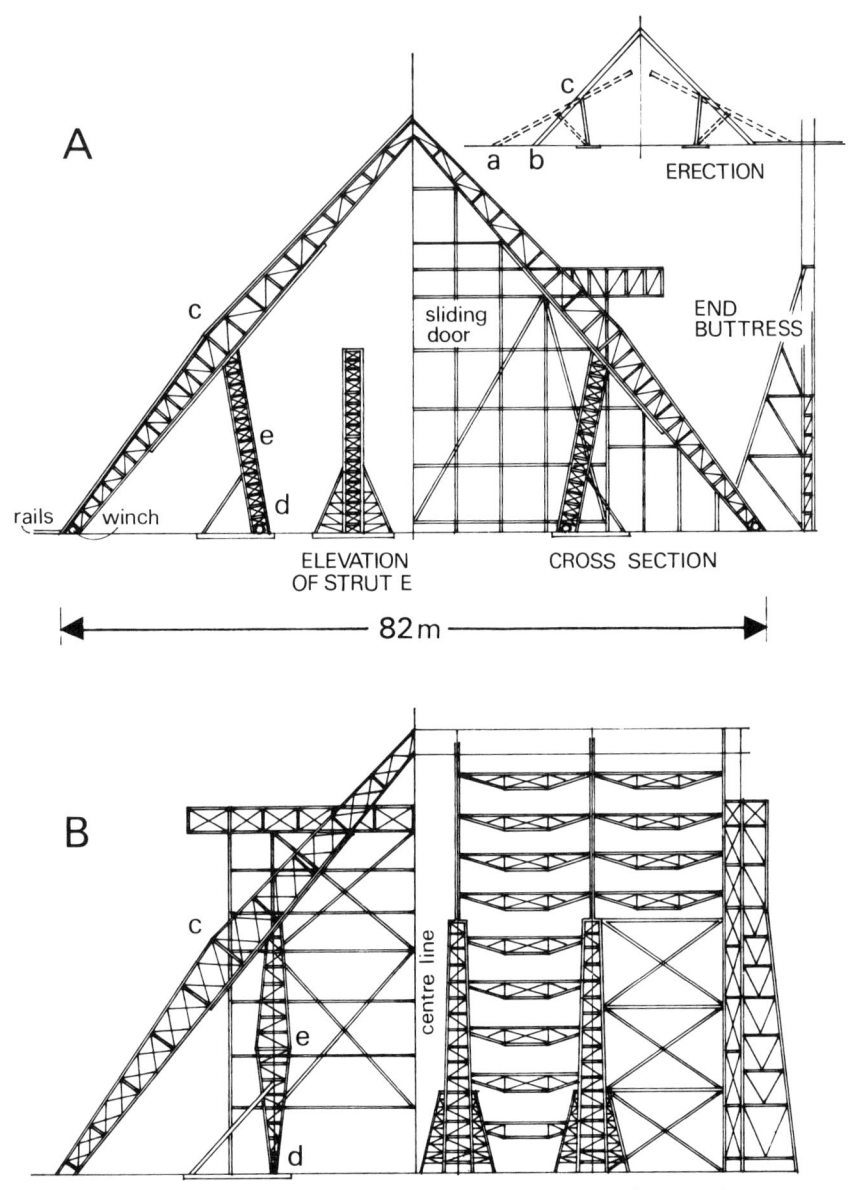

17. Two of the demountable Zeppelin sheds near Namur, A being the heavier and probably earlier type, B being almost entirely of light angle sections. The lattice beams were assembled on the ground resting on rails, large concrete blocks were constructed at d and struts e with their fixed hinged foot laid out under c. When a main beam was pushed along the ground from a to b, strut e slid along the rails on the underside of c until the beam was positioned.

to make any advance, a whole new philosophy of lateral movement was forced into existence. The Germans had the initial advantage in this, having captured the main lateral railways. This allowed them to mass troops at a given point on the front in a matter of hours. The game developed into a question of who could create a force of offensive strength before an equivalent defensive one was created in opposition. A British Intelligence Report stated:

"Taking the German front in France and Belgium as a whole, there are very few units available . . . as a central reserve to the whole front. . . . The Germans rely on obtaining the necessary troops to meet an attack by 'milking' the other portions of their front which are not attacked. Their very carefully organised system of lateral railway communication enables them to get these troops to the threatened area often within 56, and almost invariably within 80 hours." (17).

The philosophy of lateral troop movement thus invariably defeated itself. It completely ignored the fact that the more the forces concentrated the easier it became for the enemy to counter concentrate. But, although nobody ever really won the game by the strategy of 'concentrating a superior force at the decisive spot', it did create or maintain the problem of mobility. Thus methods of communication and transport were developed beyond what would normally have been available in a more fluid situation.

Despite this mobility the German army tended towards rigid accommodation for their troops, using traditional forms of construction which employed timber much more liberally than in the Allied equivalent. Prefabricated and mobile hutting did not develop to the same extent. Likewise the tent remained the dominant form of shelter but in the air war the German use of lateral movement resulted in the development of huge mobile support structures, far exceeding Allied mobile structures in size if not in their scale of application.

The provision of any form of mobile structure for the German airships (more generally known as 'Zeppelins') posed complex design problems. Before 1914 the German army had already constructed a series of fixed aerodromes for airships along the western frontier (18). Since these ships could not be left moored unprotected new designs for hangars were essential if the craft were to be used whatever the wind direction. The ideal solution was the revolving hangar (19). As the whole hangar could be turned 360 degrees, the difficult operation of bringing a ship out of its hangar in a cross-wind was avoided. Such a hangar had been built at Nordholz by the navy but, as it had taken over two years to put up, this type of construction was out of the question along the front. None of the German commanders believed the war would last that long. Another design called the 'swimming hangar', (20), which floated parallel to the wind direction, had been previously used before the war for Zeppelins on one of the German lakes, but it obviously needed a large area of water. The circular hangar proposed by Ernst Meier where the ships could be turned with the wind direction inside the hangar was considered undesirable as, housing several highly explosive craft under one roof, it was too vulnerable in war conditions. A further problem was the constant design development of the ships from 1914 until 1917 when they were abandoned by the army. G. P. Neumann writes that 'during the four years of war, beneath the pressure of dire necessity, more technical improvements were effected in airship design than would otherwise have been accomplished in ten years.' (21). As their size was more than doubled, to increase speed, lift and altitude, their hangars had not only to be quickly erected, but also flexible.

The best solution was achieved by the navy who, with their large number of dirigibles, constructed groups of hangars so that at least one of them could take off no matter what the wind direction was. The army, who had initially operated the more manoeuvreable smaller ships, and was consequently less worried about crosswinds, decided not to follow this system as their theatre of operations was too large for such concentration. Aerodromes with only one hangar, similar to those built along the border before 1914, were thus built on the Western Front and in Russia and Bulgaria. The essential difference between these new hangars and those of pre-war design was, however, their mobility (22). Whereas pre-war designs, such as those built near Cologne with 360 mm brickwork walls between truss girders, had all been permanent or semi-permanent now both the navy and army built several of their new hangars in demountable or portable construction. At Namur navy base, demountable hangars were constructed by an 'ingenious' method, as H. F. Murrell wrote in the RIBA Journal in 1920. He goes on to state:

"It might be imagined that the engineer who originated this type sought inspiration from his umbrella." (23).

The main lattice beams were hinged at their centre to a secondary

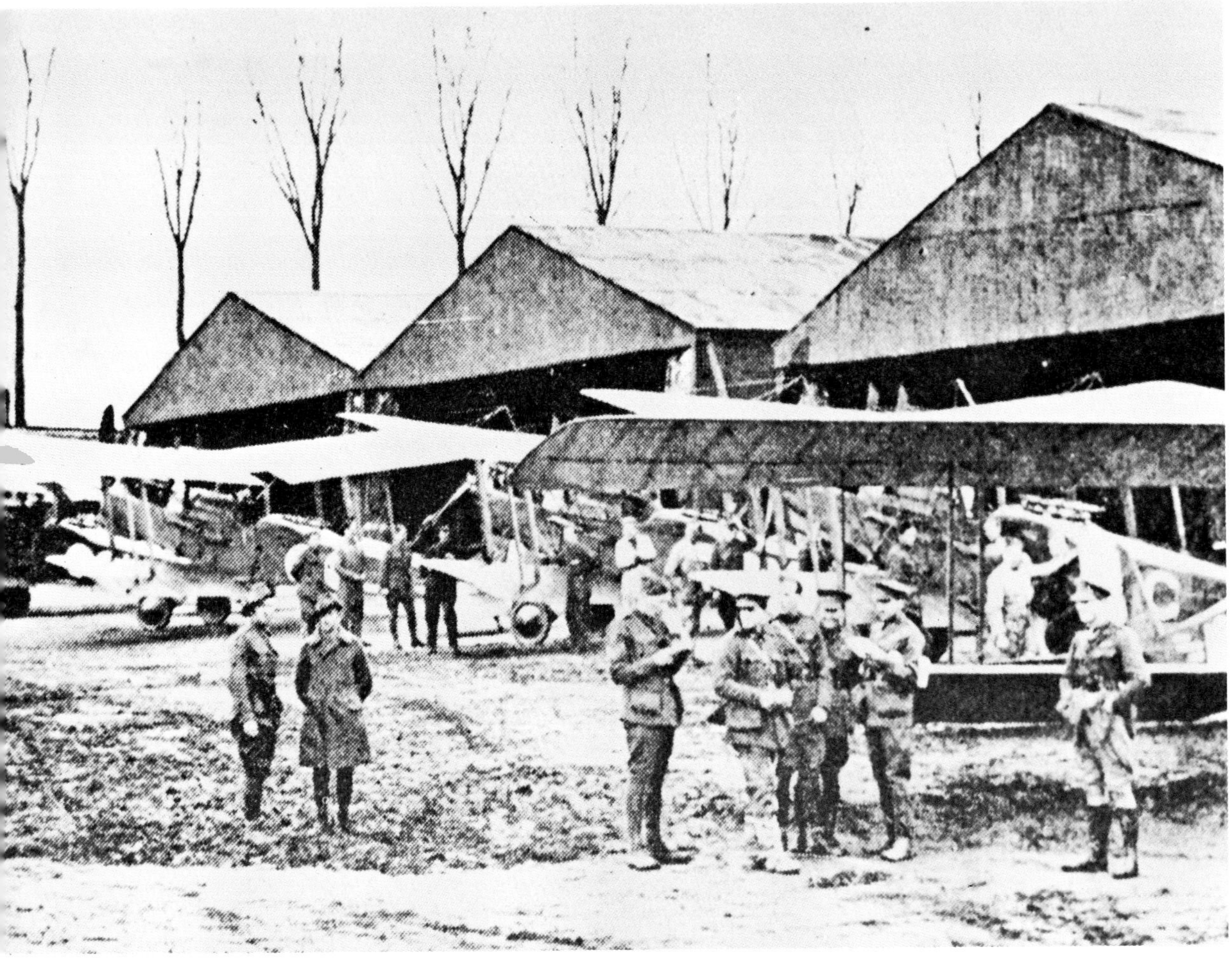

18

19

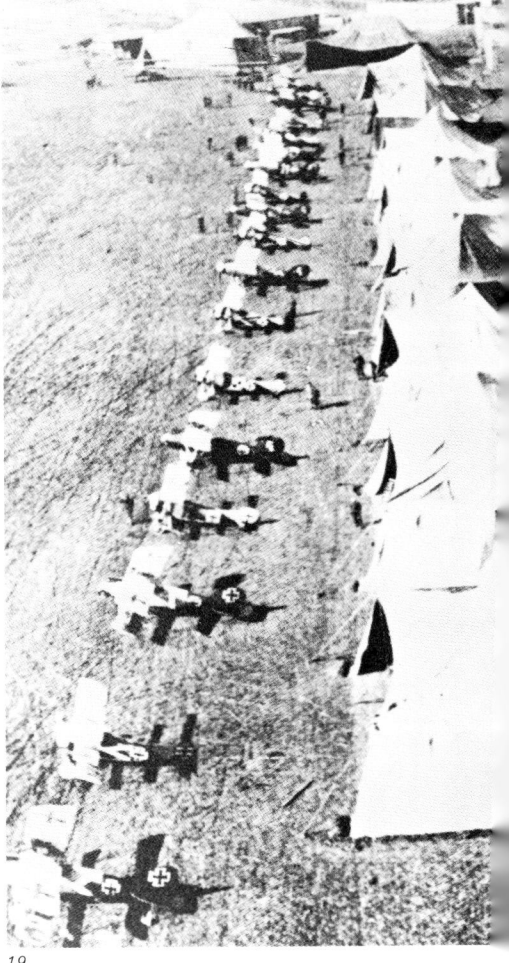

18. Allied air base: more permanent in construction than its mobile German counterpart.

19. The mobile base of the German J.G.2.

lattice beam which was also hinged to the ground. When drawing the point of the main beam, further from the ground hinge, along the ground towards this hinge on the secondary beam, the other end of the main beam rose in the air above the revolving secondary beam. The operation was carried out by a winch. When the two raising ends of both main beams rose they met to form a three pin arch. Thus the whole construction could virtually be finished on the ground before being raised.

The army, even after 1914, built constructions such as the one at Brussels which were only partially demountable in character. Here simple steel three pin arches provided the bases for steel purlins with wooden rafters covered by boarding and bituminous felt. Simultaneously, however, they provided hangars with even greater mobility than those of the navy. 'Constructed of canvas', as G. P. Neumann records (24), they were 'easily transportable'. Considering that the normal airship hangar would usually be over 30 metres high, 30 metres wide, and as much as 150 metres long, one can no longer talk of tents in the traditional sense. But the problem of the crosswinds still remained, and when the size of airships increased this became such a constant source of trouble that the army started working on a few revolving sheds in the West where the front had become static. They were started too late, and when the airships were abandoned in 1917, on account of the wide use being made of the aeroplane, they were still not completed.

The abandonment of the airship did not however mean the abandonment of mobility in the field of air force support structures. Though on a much smaller structural scale, this had started quite early. In 1915 when the Germans were preparing for the attack on Verdun, Boelcke, the German 'ace' pilot, was sent to Metz to join a unit containing both scouts and bombers. This unit was later known by the Allies as the 'travelling circus'. Here all personnel were accommodated in a train containing workshop, spares and offices. Due to the well developed railway network, the unit was able to move at short notice from one sector to another. Later this tactic of mobility was so far developed, that in early 1917 the Germans were able to regain air superiority, using these methods of lateral concentration. Thus, instead of the smaller groups of German aeroplanes, known as 'Jagdstaffeln' several such groups now formed what was known as a Jagdgeschwader. The first such group, the J.G.1, was led by Baron von Richthofen, who had formerly been a scout pilot with the Boelcke group. Known as 'Richthofen's Circus', this group was able to outnumber the Allied planes at a given point on the front by the use of mobility and concentration. While the Allies spread their energy evenly along the front this was feasible. In the Second Battle of the Somme, therefore, the German planes were in complete local numerical superiority throughout even though outnumbered by British planes by three to one on the Western Front as a whole. The Royal Flying Corps saw either none, or too many, German machines (25).

At the end of the war J.G.1 was taken over by Oberleutnant Hermann Göring, later to become Air Minister in Nazi Germany. Possibly because of his influence mobile hangars appeared with the German Luftwaffe again in 1939–45. Indeed the structural principle of mobile hangars, such as those designed for the Luftwaffe by Hünnebeck during World War II were very similar in design. A British report on German war-time engineering stated:

"Essentially it (the hangar) is a three pin arched truss construction, the whole hangar being erected complete with sheeting 'upright but on the ground'. It is then raised into position by drawing the two springing pins together and tying them; the two halves of the hangar revolving about the centre pin as they rise into the air." (26).

Even the huge demountable hangars built for the American air force in the Second World War, by the Butler Manufacturing Company, were utilizing similar principles to those used in the transportable Zeppelin sheds of the First World War: three pin steel arch with canvas opening. The Engineering News Record wrote in 1944:

"For use in the war zones a light weight steel, demountable hangar claimed to possess the 'mobility of a circus tent' was developed. . . . Using three hinged, steel arch truss of 130 feet span tied at the base as the principal framing . . . a cleverly designed hinge bolt reducing erection time to a minimum. Normally the hangar is enclosed by a flame proof, light weight canvas, suspended from the arches. . . ." (27).

Thus the structural mobility developed on the Western Front was to have long term effects reaching into the Second World War, not only in the use of mass produced hutting, but in the use of huge demountable hangars.

Out of the peculiar stalemate on the Western Front, fluid only following Ludendorff's Spring Offensive of 1918, mobility had reaffirmed itself. Not only did it emerge in the field of hutting and support structures,

20, 21. The Hünnebeck Hangar of World War II: its three pin arch structure was fully demountable (21). The hangar was assembled flat on the ground with the two springing pins resting on rails (22). When the pins were drawn together the centre of the hangar buckled upwards to form an arch.

22. One of the lightweight transportable hangars made by the Butler Manufacturing Company for the USAAF during World War II.

but in the elastic 'defence in depth', the tank and the aeroplane. Ironically the first major post-war military construction programme took the form of huge and costly fixed defences which took insufficient account of this new mobility. The French Maginot Line was to be a strategic catastrophe. More sophisticated than either Hitler's West Wall or Atlantic Wall, the Maginot Line must today rank as Europe's most expensive and complex architectural or military folly.

PART TWO 1918–40

5 MAGINOT LINE

Design and Construction of French Defence

Events

1919 Preliminary studies by General Staff.

1920 Higher War Council informed.

1922 May 22 Commission appointed for further studies under Joffre (later Guillaumat).

1925 Final conclusions reached and on Dec. 5th Higher War Council adopts discontinuous system of fortified regions. Commission for further study appointed under Painlevé.

1926 Nov. 6 Commission presents report.
Dec. 17 Meetings and Discussions of War Council begin.

1927 Sept. 30 Decree appointing an Organization for Fortified Regions.
Dec. 29 Programme of siting, basic design, and order of urgency approved.

1928 Feb. 17 First three sections out to tender.

1929 Maginot becomes Minister of War.

1930 January Maginot gets vote of credit despite criticism.
June 30 Evacuation of Rhineland, extensions proposed by Maginot but refused by Pétain.

1940 May 14 German Panzers cross the Meuse.
June 17 Guderian reaches Swiss Frontier at Pontarlier, outflanking the complete Maginot Line and a large part of the French forces.
June 28 Maginot Line surrendered to the Germans intact.

The First World War came to an end with the Treaty of Versailles. Its conditions were considered harsh by some of the more far-sighted Allied leaders, and bitterly resented by the Germans. Their generals, having got over the virtual collapse of their army in 1918, now maintained—ominously—that they had not been defeated in the field. It was hardly surprising, therefore, that the French General Staff were soon undertaking preliminary defence studies.

One of the first conclusions, drawn by both military staff and generals, was that the next war should not be fought in France. The invasion of the Ruhr in 1923, allegedly to put pressure on Germany to pay her war debts, should be seen against this background. When the decision to withdraw these troops in 1930 was made, not only from here but from the entire Rhineland, it increased the need for the army to find a means of securing the 'liberated' industrial zones of Alsace-Lorraine from attack and to block the classic invasion routes of the Moselle and Sarre valleys. They found valuable backing for their defence plans in André Maginot.

Maginot, born in 1877, had abandoned his position as an under secretary of state for war to join the army as a private during World War I. Having been seriously wounded at Verdun he resumed his political career later in 1916, first as a deputy, then as Minister of Colonies and Pensions, and from November 1929 as Minister of War. During the war he had experienced the value of solid protection against German attack, both for the soldier and the nation. After all, the Germans had not been able to make a major break-through of the trenches or the Verdun fortifications. Similarly, a basic distrust of Germans had been born, which the events of the post-war years did nothing to dispel and which was increased by Locarno and the demilitarization of the Rhine. When the decision was taken to advance the date of evacuation from 1935 to 1930 he described it in a speech in January 1927, as a 'veritable crime against the country' (1). He began to see that the only answer to the 'growing German menace', was the construction of fixed and impregnable defences.

At the same time the army was holding a series of meetings discussing the broad principles of defence proposed by a 'Frontier Defence Commission' together with detailed studies of the terrain. Both Joffre and Pétain had been involved in this work but had come up with different solutions.

1. One of the Maginot Line's artillery blocks near Bitche as it remains today.

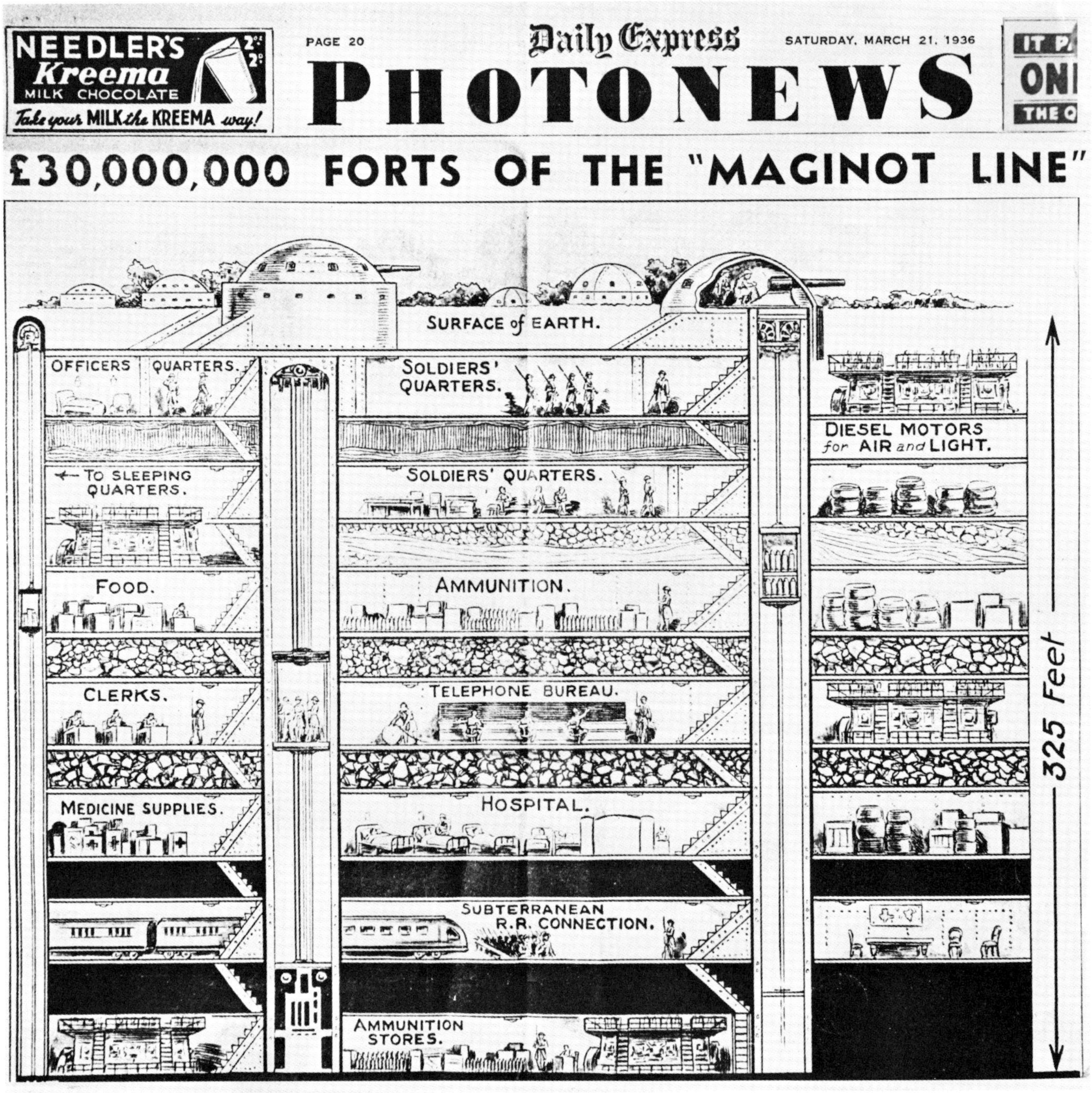

PAGE 20 — Daily Express — SATURDAY, MARCH 21, 1936

PHOTONEWS

£30,000,000 FORTS OF THE "MAGINOT LINE"

2. Diagrammatic section through a Maginot Line fort as it was popularly imagined in 1936.

Joffre believed, as did most of the generals at that time, that the German defeat in 1918 had been a military defeat caused by the French offensive and he failed to recognize the effect of the blockade and the trench deadlock. Unlike the French General Staff in 1914, he did not however believe that massive offensives were in themselves a sufficient defence, and he was prepared to strengthen the 'offensive defence' with fortifications. Looking at Verdun with its fortified area on a circular system, which had withstood the massive German onslaught in 1916, and Fort Douaumont, which had sustained the bombardment of 120 000 shells (largely French) there with relatively little damage, he proposed a similar system of fortifications now. Large fortified areas should form a discontinuous fortified zone along the German border. The fortified areas would provide the bases for troop concentration and counter-attack. Should the Germans march between each area, they would open their flanks to strategic attacks, pincer movements and the destruction of their communications. His ideas were not, in fact, much different to those of General Séré de Rivières, who had created the French fortress zone following the French defeat in 1870.

Pétain had been in command of the French defences by Reims, the only sector where the final 1918 German offensive had been successfully beaten back. Unlike Joffre he had realised some of the lessons of the new German methods of linear 'defence in depth' and had implemented them there. Although detailed information on these German methods was being published by the French Second Bureau from June 1917 (2), the French army in 1917–18 had not completely adopted them. Officers who doubted their value received support from Foch, the Chief of Staff. Similarly, in the British army copies of translated German documents were being circulated, but the British took more notice of the constructional aspects of defence in depth than in its theory. Pétain was therefore the only Allied leader who grasped and tried to implement these German ideas during the First World War and, when it came to the post-war defence of France, Pétain naturally pushed forward these theories again. His success with them and of course his own experience at Verdun enforced his opinion that fixed fortifications were of value; but he had also seen Verdun, with its lines of communications almost encircled by German troops, as the terrible mincing machine of human lives which it became. Instead of fortified areas on a discontinuous circular pattern, he insisted on a continuous linear one, the pattern which he had used with success in the trenches near Reims. His successes both there and at Verdun, his concern for the safety of his troops which had made him able to quell the French Mutiny in 1917, had both given him a reputation. Apart from that his policy of defensive rather than offensive was a more popular vision in the 20's and when the committee came to choose between the two plans, they chose Pétain's system of continuous linear defence. Joffre resigned.

Believing that fixed defences would also allow for a reduction in standing manpower, as well as additional time for mobilization, large sums of money were allocated to the project. The programme of siting and basic design of fortifications as well as the order of urgency set out by the Organizing Commission for Fortified Regions, were approved on December 29th, 1927. On February 17th, 1928, the first three of the twenty sectors were put out to tender under War Minister Paul Painlevé. These were intended to enable a more detailed feasibility study to be made. However, Maginot, the greatest believer in fixed fortifications in France, became War Minister in November 1929, and probably took little notice of such studies. As the Line's political parent, Maginot gave it his name and defended it against criticism, which came mainly from the press, from Briand for political reasons, and from the Socialists who refused to vote such enormous sums for military purposes. Despite this and with the assistance of both external events and Premier Tardieu he managed to get a vote of credit allowing full scale work to begin in January 1930, five months before the evacuation of the Rhineland.

The original military concept was still Pétain's, however, and he together with General Weygand formed the main military backing. Pétain's ideas had now moved closer to those of Painlevé, Minister of War previous to Maginot. He realised the implications of tanks and aeroplanes, no longer acting only as protection for infantry but also as an offensive weapon on their own. Further protection was necessary and so, in addition to the already well known anti-personnel measures (mines and barbed wire), anti-tank mines and obstacles had to be incorporated. Ordinary trenches had been overrun by the Allied tanks in 1918. Large tank ditches overlooked by casemates housing armour-penetrating guns were now to be used, and the troops were all to be protected by concrete fortifications. Fixed fortifications had been built, semi-underground, in the First World War with steel and concrete

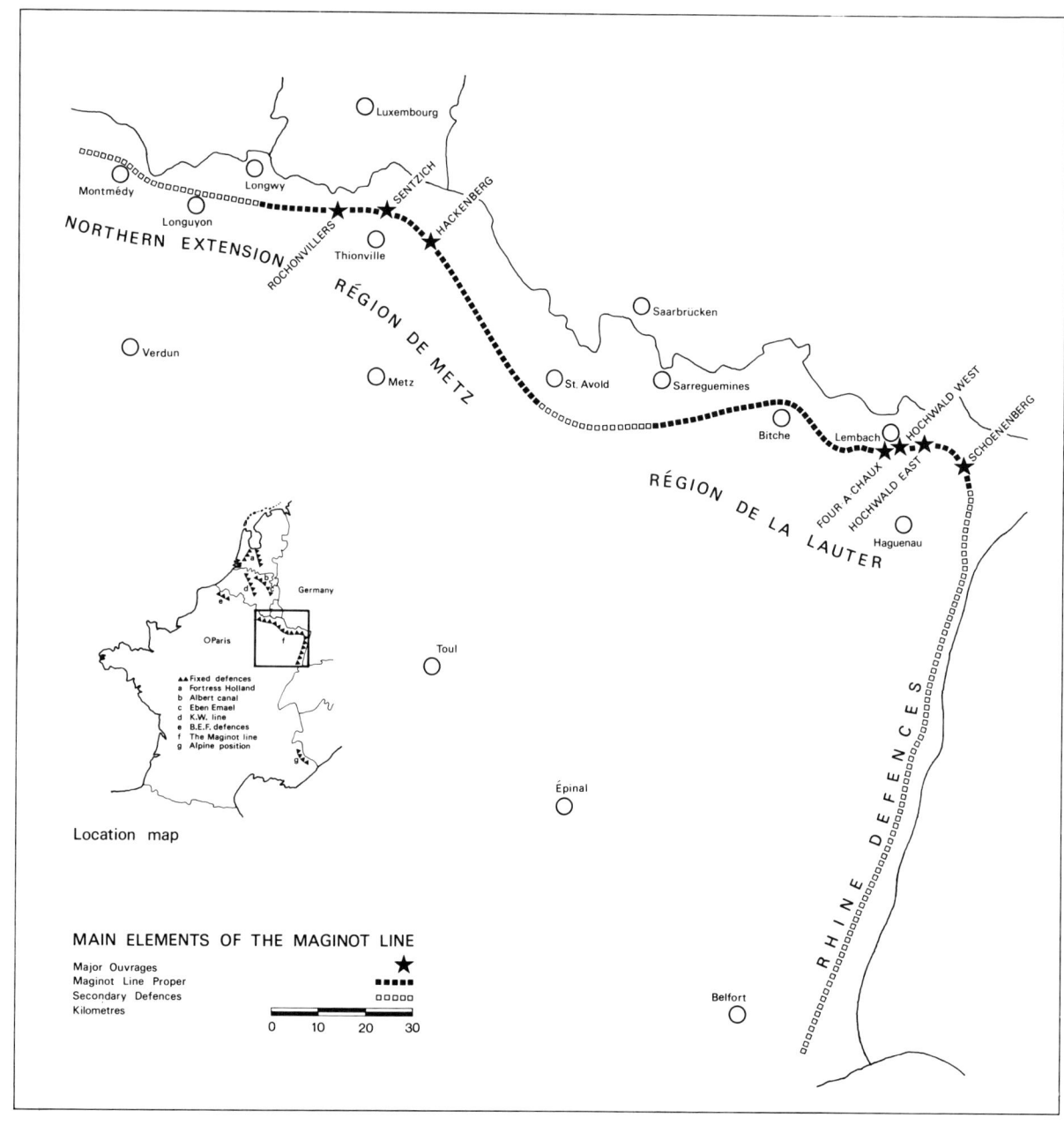

3. Map of the Maginot Line and other fortifications on which France relied for defence against Germany. In 1940 Guderian's panzers were to simply avoid them by striking through the undefended Ardennes sector.

protection against artillery. Now that aeroplanes could carry even larger explosive shells additional security was needed, which found structural expression in the deep underground nerve centres of the Maginot Line. The gun positions were given added protection against air bombardments. Continuous line fortifications were thus abandoned in favour of a grid of fixed fortifications but although each one essentially represented the old circular system or ring fortress, their integrated total still formed a system of defence whose dominant axis was linear (3).

Because of the way that the design evolved, it cannot be said that the final solution was purely the creation of Pétain and Painlevé. Its emergence was slow and was probably influenced by the opinions of many others. The detailed design and work was the responsibility of Generals Weygand, Debeney (Chief of Army Staff) and hundreds of unknown engineers but, like the main overall concept, it was probably greatly influenced by the 'personal interest' of other high-ranking officers. It is impossible, in looking at the design of the Maginot Line, to pick out a single name as architect of the entire project but if a single name must be picked out it is the name of a battle not of a man. For Verdun continued to be the legend which inspired such great effort, as James Eastwood wrote in 1939:

"... the concrete of Verdun stood up well to the enormous strains imposed upon it, and when in post-war years military technicians had leisure to digest the lessons learnt at such cost, Verdun was remembered and it was realised that fortresses were not a thing of the past." (4).

Ironically the legend of the Verdun fortress was misleading. After the fall of Liège in 1914 many of its guns had been withdrawn and put at the disposal of field commanders by a decree of summer 1915. What remained was a shell, even if it was a strongly constructed one. Instead of an all round defence a single trench position was taken up beyond the forts and the casemates were used simply as shelters for the troops. Verdun was a battlefield rather than a fortress, yet it did show the strength of fortifications, if not in the way the legend would have it. Liddell Hart reinforces this point in his book 'History of the World War 1914–18':

"Forts Douaumont and Vaux fell into German hands, and when they were recaptured in October the French found that months of tremendous bombardment had made scarcely an impression. The underground cover remained intact, not one field gun turret was destroyed, and hardly any of the casemates rendered unoccupied. It was a grim jest of fate that the French should have thrown away their shield for a target, through a hasty assumption that fortresses were valueless." (5).

Verdun was a massive killing ground for both sides. It was defended by an unprecedented expenditure of blood rather than fortifications, but somehow the 'defence of Verdun', became synonymous with 'fortress' and a legend grew, on the strength of which the French were now preparing to fortify an entire border at an estimated cost of two million pounds per kilometre.

The final design of the Maginot Line was both deeper and more complex than Verdun had ever been. In detail design the Line was in some ways a development from the ring fortresses of 1914 but while the Verdun forts were designed to resist artillery and men, the Maginot Line was designed to resist the new machines of war. Whereas the large Verdun forts had resisted the enemy fire mainly by the thickness of their concrete, the main nerve centres of the Maginot Line were deep underground, completely out of the enemy's reach. Finally, while the bastions of 1914–18 had ultimately relied heavily on trenches for their communications, all parts of the Maginot Line fortifications could be reached by a system of tunnels and lifts. Men and munitions could thus be moved into position without danger of enemy interference. The similarity with a battle fleet is obvious. Comparison of its large 'ouvrages', 'casemates', 'Avant-Postes', and fortified houses with a battle fleet consisting of capital ships, cruisers and various small craft on the perimeter, is striking. Steel furniture and the interior design of the cramped underground quarters also had a distinctly naval character. Maginot himself called his line a 'subterranean fleet' (6), perhaps affectionately. Alanbrooke was later to call it 'a battleship built on land' but in a far more critical context.

The Maginot Line proper was to cover two sectors: Région de la Lauter and Région de Metz. The first ran from the Rhine to a point just west of Bitche, and the second from Falquemont to just north of Thionville and the iron fields, including the frontier town of Longwy in the line. Pétain had the line brought back to Longuyon, however, arguing that the position would be jeopardized by dead ground across the frontier. Defences along the Rhine, defences in the Alps against Italy, defences along the Belgian border, were added to the line later as extensions but these were extremely primitive in comparison (7).

4

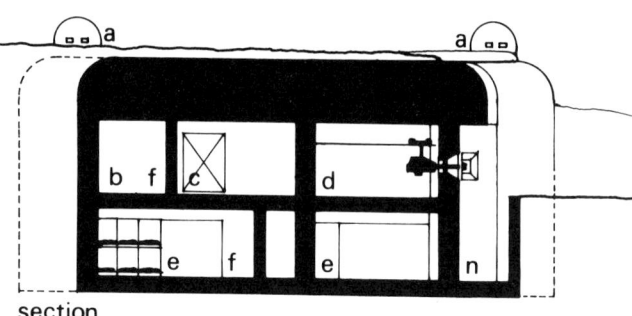

section

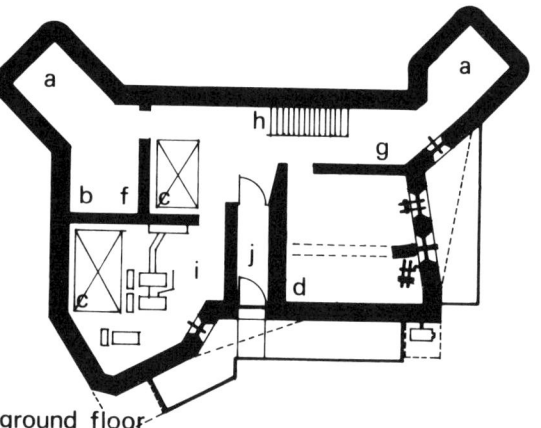

ground floor

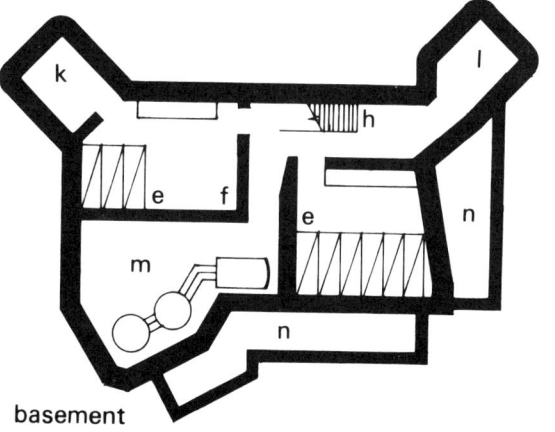

6 basement

4, 5. Recent photographs of a medium sized casemate between Sedan and Montmédy.

6. Section and floor plans of a typical casemate:
   a. observation cupolas
   b. periscope room
   c. water tank
   d. gunroom
   e. dormitories
   f. telephone exchange
   g. cooking area
   h. stairs
   i. engine room
   j. entrance
   k. toilets
   l. wireless room
   m. ventilation chamber and stores
   n. moat

7, 8. Recent photographs of a casemate on the Maginot Line near Rohr Bach: the fluent forms of such casemates reflected both French national character and her considerable experience in the use of reinforced concrete.

9

10

11

9, 10. Casemate flanking an anti-tank ditch, part of the Hackenberg defences, shortly after construction (9) and as it exists today (10).

11. Avant-poste.

12. Diagram of the detailed layout of the Maginot Line defences: the main fortifications were situated 6–10 kilometres from the frontier.

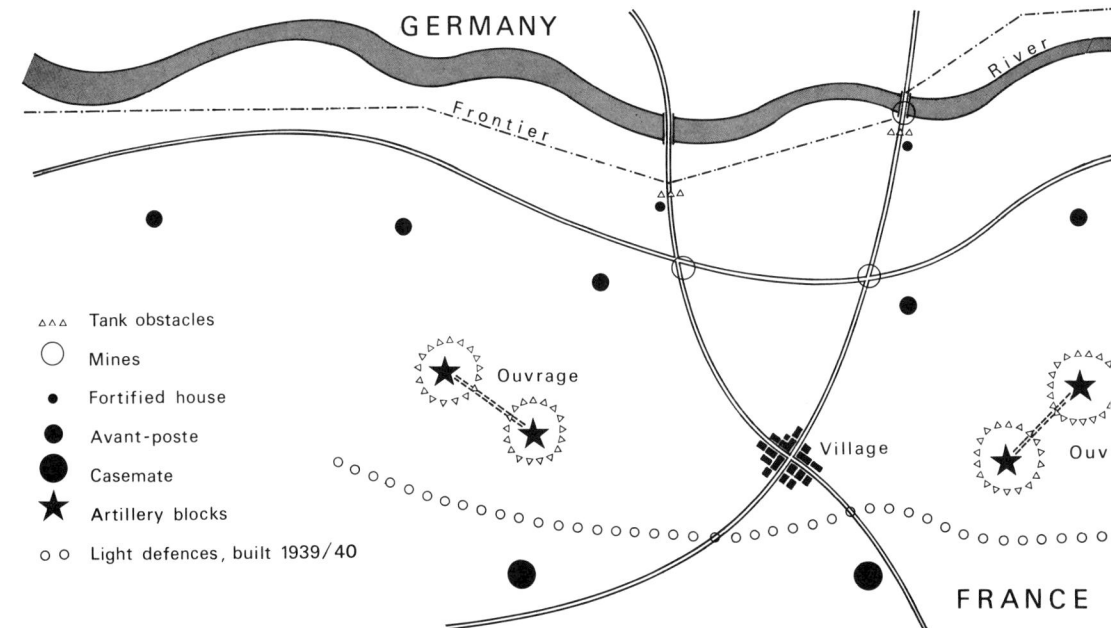

The actual siting of the individual fortifications on the line was governed by topographical and strategic factors. Plains were dominated as at Rochonvillers, and towns such as Lembach, situated in valleys, could be protected. Furthermore, each Fort and 'Avant-Poste' was situated so that not only could its field of fire cover the ground between itself and its neighbour, it could also cover the neighbouring fortifications. Each fortification was thus designed to take direct hits from its neighbour without damage and a combination of concrete, steel and earth, known as 'method of protection No. 3', was tested with short range howitzer fire for this reason. Although these tests were successful the engineers were still not satisfied and Robert Leurquin states that 'it was then decided to give the casing a thickness capable of resisting three projectiles falling at the same point of impact.' According to James Eastwood, 'in the course of construction the thickness of the cover was again tripled' (8), and the final result was a minimum roof cover of 3.5 metres which was additionally covered by heavy layers of earth. The potential value of such heavy protection emerged in the circumstances under which the Belgian fort Eben Emael was captured in 1940. Using special troops the Germans were able to overpower the garrison by lowering incendiary explosives down gun-turrets, ventilation shafts and similar openings from the roof, where they themselves were out of the field of fire (9). In the Maginot Line the neighbouring fort could sweep the fort with its fire, killing the enemy without danger to the men inside.

Each position had to have natural cover, sites for observation posts, minimum dead ground in the field of fire, maximum area of fire, locations for anti-tank and personnel obstacles, and the potential to build hard surfaced roads up to each point of entry. If possible the fortifications were built into the hillside. In flat country artificial earth mounds were constructed and the forts covered with turf.

Starting at the border, the river crossings were equipped with anti-tank obstacles and minefields, in addition to fortified houses. After a succession of minefields and anti-tank obstacles at cross-roads and other likely spots for enemy manoeuvre, the 'Avant-Poste' was the first fortification intended to seriously delay and give warning of enemy attack. These had a force of 25 to 35 men manning them. They were well equipped with armament, munitions and stores, and had quarters for officers and men. Protected by heavy concrete, they were expected to offer quite substantial resistance in addition to warning the fortifications further back by direct telephone communication.

The main fortifications were a line consisting of casemates and reinforced with ouvrages. The casemates were always of two storeys with firing chambers on the top floor and munitions, stores and quarters on the lower. In front an anti-personnel ditch was constructed, hindering the enemy from reaching the gunports. Above, a series of armoured turrets could cover the immediate surroundings with machine-gun fire. Each casemate had its own diesel-generator which provided electricity for lighting and amunition lifts as well as for ventilation. Two such casemates a little apart but connected by an underground passage, usually both situated in the same hill, were known as 'coupled casemates'.

At intervals of 5 to 8 km the line of casemates was reinforced by an ouvrage, the massive underground forts for which the Maginot Line is famous. Surrounded by an outer tank barrier of erect rail lengths bedded in concrete, followed by an inner personnel barrier of barbed wire and steel posts, the forts themselves were of three main types, small, medium and large. The medium and large were only different in size, the large holding up to 1000 or 1200 men, the medium some 500. The small forts were also fully self-contained units housing approximately 200 men, but consisted of only infantry blocks, while the medium and large had both infantry and artillery blocks.

Like the coupled casemates, one could also talk about 'coupled ouvrages'. Each fort was divided into two similar groups of fortifications, usually 1.5 km, but sometimes up to 2.5 km apart. The fighting units were subdivided into infantry and artillery blocks, situated well apart for flexibility when under attack. The former were equipped with 37 mm and 47 mm anti-tank guns. Using their telescopic sights, observation for these guns was from the firing position. This was later found to be a weakness, as the Germans, after their study of the similar Czech fortifications (10), discovered that their best marksmen were able to blind a fortress through the destruction of these telescopes. The artillery blocks held the larger guns such as the 75 mm retractable gun in a steel turret. With 360 degrees field of fire and calibrated and ranged to extreme accuracy for its time, it was a formidable weapon. Other large guns such as the 81 mm, 135 mm and the 340 mm were housed in huge concrete casemates with supporting machine-gun turrets above and around,

13 14

3, 14. One of the large casemated artillery blocks: part of the ouvrage at Rochonvillers.

5. This diagram of an underground ouvrage was published in the late 1930's. Although the casemates and other details are incorrectly shown, the diagram gives an impression of the basic design.

6. Two of the main types of cupola used on the Line. The flat artillery cupola on the left was of the retracting type, the cupola on the right was fixed and gave machine gun protection and observation.

17, 18, 19, 20. The tunnels and galleries which connected the various parts of an ouvrage.

similar to the casemate fortifications.

Each artillery block was a complex affair in itself. Below the gun came the magazines, and then the quarters for the men manning the gun. Deep below this came the Block Command Post with magazines, quarters, command post and telephone operators. The fort commander would get his information from observation posts in cupolas of special steel with a 3 metre external and 0.9 metre internal diameter. These would usually be sited somewhat away from the main gun position in pairs and in addition to serving as observation posts they served as the main outlets for the ventilation system, creating rather draughty conditions for the observer. His information would be used in conjunction with maps and panoramic photographs of the surrounding countryside, by the Block Commandant directing his fire from his command post. The actual manipulation of the guns was done by special crews from a control room directly below the firing-room, using an ingenious system of hydraulics, mechanization and automation. Each block was then connected with the main headquarters by an underground gallery leading from an airlock. Geographically the ouvrage H.Q. was situated midway between the two main groups of fortifications. In addition to the quarters and office of the fortress commandant, it housed the main operation and communication nerve centre of the fort with the main command intelligence room, telephone switchboard and calculators, kitchens, infirmary and barracks. All were deeply underground and connected by tunnels to the H.Q.

Contemporary impressions of these quarters are quite fascinating in themselves:

"... there were reports that the main section of the Maginot Line had been equipped with cinemas, barber shops (complete with white-coated attendants) even cafes with splashes of local colour from North Africa to make the Moroccan troops feel at home." (11).

The actual conditions were far from being as comfortable as that quote, from a book published in 1939, suggested. According to more recent authors reality was much more squalid, with cramped living quarters:

"... holding nine bunks in three layers, the topmost men having the utmost difficulty in squeezing themselves under their blankets without hitting their heads on the roof." (12).

Dripping concrete walls and glaring light bulbs created a cavelike environment which was completely cut off from the outside world. Men were soon reported to be suffering from 'concretitis', psychological depression, ear trouble and other signs of stress emerging from their unconventional surroundings. Light bulbs were whitewashed to reduce glare, lack of sunshine was compensated for by exposing the men to ultra violet light, but attempts at drying the concrete electrically were not entirely successful. The equipment could all be protected by water-tight fittings and compartments, but not so the men. The problem was solved by sleeping them under canvas and later by the erection of small collapsible villages, above ground, on similar principles to the mobile camps of the previous war. When danger threatened, back they went to their underground quarters. Plainly this Wellsian underground world was no environmental utopia, and the underground urban complexes which have been proposed by post-war idealists are based on very doubtful precedent.

From the centre of the fort, long communication tunnels, with several changes of direction and fortified points, led to the final airlocks and main entrances. There were two of these, one for personnel and one for supplies. They were usually situated far to the rear of the main fort and strongly fortified. A small electric train, named the Metro, carried supplies from the entrance to all the different parts of the fort, through tunnels or galleries 2.5 m wide and 3.0 m high. This created a problem as the slope of the tunnel floors was dictated by the train and this in turn dictated the slope of the drainage. It necessitated the use of septic tanks, sterilized by alkalis; since deodorization could not always be effectively treated at this time, the resultant stench can only be imagined. The system also required continual maintenance, as one historian records:

"... made all the more necessary by the reluctance of the garrison troops (a reluctance common to all armies) to realise that lavatories were not proper places in which to dispose of old razor-blades, cast-off clothing and other such practically indestructible objects." (13).

In the same way as the commandant's office formed the brain of the underground monster, the plant rooms formed the heart. Occurrence of serious fault here under attack would not only break down the functions of the one fort, but as all the fortifications depended on each other it could cause progressive failure over a large area. It was here, in the attempts made at environmental control, that the design of the forts lay not only close to, but in some aspects ahead of contemporary civilian counterparts. There was general central heating and the 'Metro'

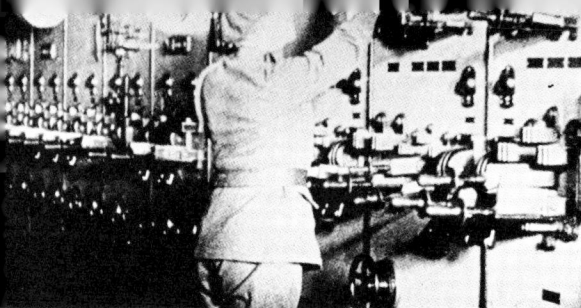
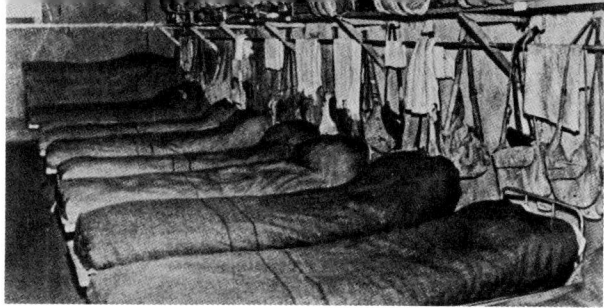

21. Some of the complex plant installed for lighting and ventilation.

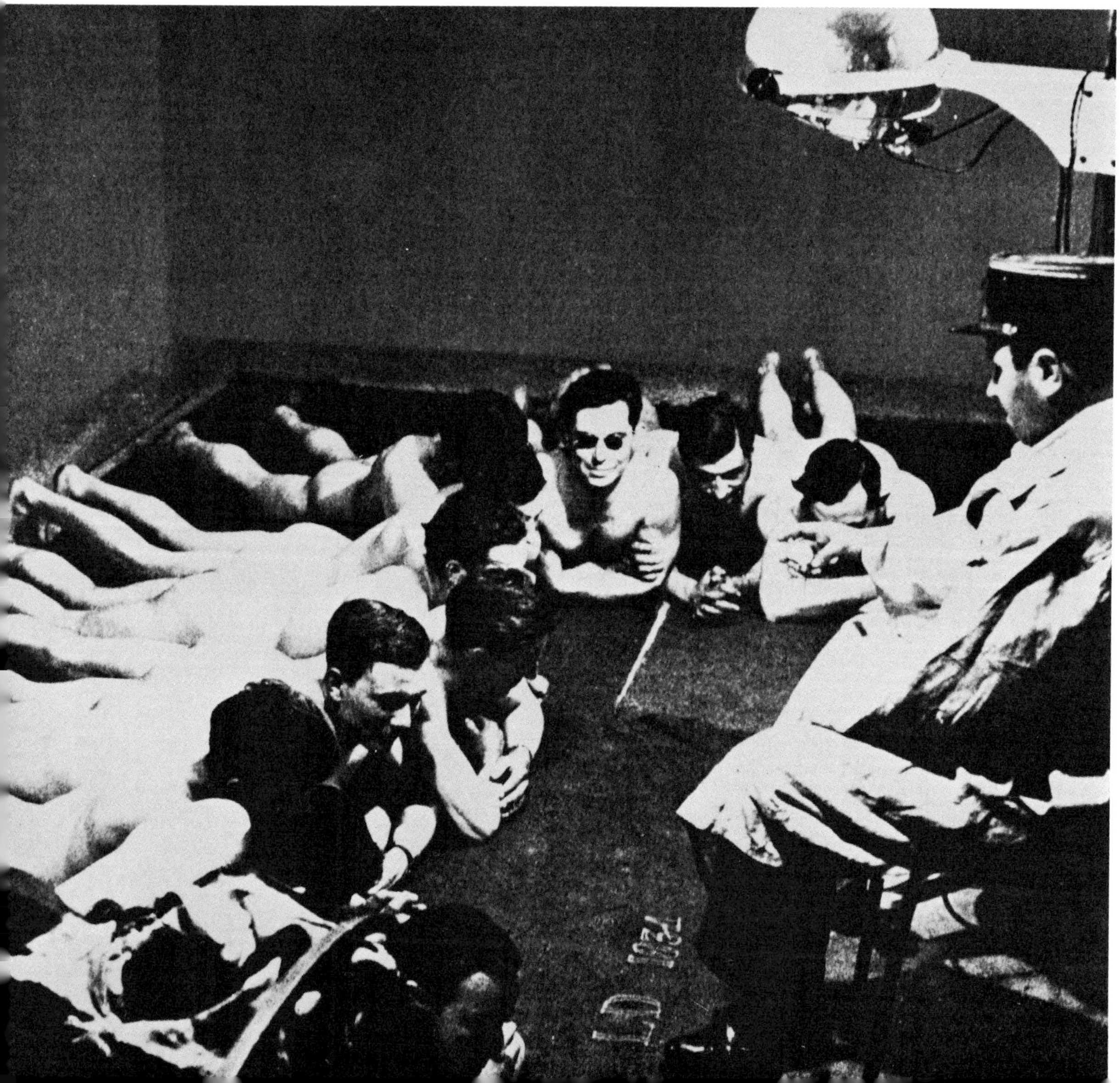

22. View of the garrison's quarters: accommodation was seldom as spacious as this and cramped conditions were to be a major criticism.

23. Sun lamp session for the troops: such treatment was intended to balance their artificial environment.

for all horizontal transport of personnel and goods. Numerous electrical lifts for vertical transport of munitions were installed. Supply of hot and cold water to all parts of the fort was supplied from steel lined reservoirs. With the memory of the use of gas in the 1914–18 war still in mind, a complete ventilation system using positive air pressure, huge filters, and a series of airlocks at all entrances was perfected. No gas known at that time would have been any threat to the occupants of the fort. At the same time this ventilation system dealt with fumes from the fire rooms.

A huge amount of A.C. current was necessary for all these services. During peace the forts got their electrical supply from the national grid, but should this be cut they could take over themselves. Each ouvrage had four diesel generators, each developing 250 horsepower and one of which was kept in reserve. These were water-cooled, each power station having six water tanks of nearly 50 000 gallons capacity for this purpose. Like the tanks gravity-feeding the rest of the fort with drinking water, they were replenished by pumping from deep sunk wells. A further six reservoirs held fuel, and there were two reservoirs for lubricating oil. Automatic self-closing fire-doors could isolate the powerhouse from the rest of the fort in seconds if there should be an appreciable rise in temperature. Should the fort's own power supply fail, deep buried cables from neighbouring forts could provide adequate supplies.

With the addition of large stores of food and munition the forts were fully self-contained and could hold out for almost indefinite periods of siege. One of them, with its vast tunnel communications and underground facilities is described by Vivian Rowe in his book 'The Great Wall of France':

"Somewhere deep underneath the turrets, totally invisible, is the great extent of the fortress itself, a troglodyte city of extraordinary dimensions." (14).

Had it been erected for any but this highly military purpose with all the attached secrecy, without doubt the Maginot Line would have been acclaimed as the greatest subterranean architectural and environmental experiment or achievement of our century. But even with the atmosphere of secrecy and spy-mania which existed in Europe during the late '30s, European military commentators wove a somewhat fantastic, and at times inconsistent, picture of these fortifications. 'The Times' correspondent described the fortifications as completely invisible from land and air. At the same time he stated that on his tour from Dunkirk to the Riviera, he was always within sight of huge fortifications. James Eastwood described the line in 1939:

"... there is little possibility, if the lines are strong enough, of successful flank attack ... the Maginot Line runs from coast to coast." (15).

Even the military writer Liddell Hart overstated the extent of the line and in his book 'Europe in Arms' in 1937 he described the line as extending 'to Dunkirk in the north, and to the Jura in the south, along the Swiss frontier' (16).

The only people who were not fooled apparently were the Germans. Even in May 1937, before they got the additional information through the 'rape' of Czechoslovakia, a map appeared in 'Wehrtechnische Monatshefte'. Though the Rhine defences were not shown differently from the Maginot Line proper, the two sectors Metz and Lauter were surprisingly accurately drawn. Most of this information was probably gathered through fairly innocent labourers from the German side of the Rhine, for a surprising aspect of the construction of the Maginot Line was the large use of foreign labour. Although France had her own unemployment problems at this time the Line's construction force consisted of only 50 per cent French labour, the rest being German, Polish, Hungarian, Russian and Czechoslovakian. Maginot exposed a particularly extreme example of this practice when it came to his notice that the job of constructing an important military bridge near Metz had been given to a German firm from Essen, using German labour and materials. It was rumoured at the time that the small percentage of French labour working on the line reflected the fact that the work was so heavy that only foreigners, desperate for work of any sort, were willing to do it. It seems more probable however that this situation was in reality a result, not of the government giving contracts directly to foreign firms, but of the French firms under government contract being forced by the pressure of work to resort to sub-contracting in order to meet their completion dates. When tenders were invited the prices quoted by the German firms were invariably lower than those of their French counterparts (17). Anybody with a will to penetrate this type of French security would surely not have found too much difficulty. Yet descriptions such as impregnable, modern, invisible, indestructible, from coast to coast, etc., were to become the comfort of France and England in the years leading up to the war. Eastwood wrote in the year war was declared:

"Barbed wire awaits the men, tank-traps the tanks. The soldiers of the

24, 25. Recent photograph of a personnel entrance to an ouvrage near Bitche (24). These were positioned far to the rear of a fort where there was also a separate entrance for the the delivery of supplies (25). All entrances were heavily defended and additional defensive posts were positioned inside the tunnels.

Maginot Line call the tank-traps 'Asparagus beds'. These uncomfortable beds consist of rows of steel spikes, set at varying heights. The tank approaches and sticks on the spikes. The anti-tank guns then come into action and destroy the tank at point-blank range. Should the tank escape this ordeal there is another row of 'Asparagus', this time fitted with high explosive. Needless to say where steel tanks cannot penetrate human bodies stand little chance. . . ." (18).

Everybody was not equally impressed and General de Gaulle and Paul Reynaud, representing the ideas of mobility as opposed to fortifications, were the main critics. After the First World War the idea of mobility developed in isolated pockets, but the majority of the World's General Staff ignored these ideas of armoured formation, at least in carrying them to their logical conclusion. In 1918 a Chief General Staff Officer to the Tank Corps, Colonel J. F. C. Fuller, wrote 'Plan 1919' which conceived a blitzkrieg principle of fast armoured warfare. These ideas were the inspiration of Captain B. H. Liddell Hart's similar concepts and were eagerly taken up by Generals Guderian and Lutz in Germany. Undeceived by the 'stab in the back' myth, at least in terms of military reality, the German General Staff had arrived at the opinion that the tanks had been the decisive factor in the Allied victory of 1918. But the 'blitzkrieg' concept which they evolved did not find favour with British, French, or American General Staff until after 1939. The French General Staff were perhaps even more tragically conservative than their allies. General Debeney, the Chief of the French General Staff, opposed de Gaulle strongly and claimed that defence in modern warfare was more important than attack. Egon Eis in his book 'Forts of Folly' perhaps overstates the naiveté of pre-war military France:

"As late as 1935 Debeney still wrote that mechanised units would no longer be able to manoeuvre as freely as they had in the last war, and Pétain added his voice to warn against exaggerated faith in air forces. . . . These rationalisations were eagerly snapped up by the French Press, for France, having spent vast sums on its bastions, could ill afford additional expenditure on tanks and planes. . . . Thus, what small advantages the Maginot Line might have offered were cancelled out by the weakening of France's army. . . ." (19).

After visiting the Line at Welshtenburg in January 1940, Lord Alanbrooke wrote in his diary one of the most balanced judgements of the Maginot Line at this time:

"The fort reminded me of a battleship built on land, a masterpiece . . . the whole conception . . . is a stroke of genius. And yet . . . little feeling of security, and I consider that the French would have done better to invest the money in shape of mobile defences . . . than to sink all this money into the ground . . . garrison of over 1000 men, over seven kilometres of passages, four vast diesel engines, electric baths, automatic gun controls, and all round an astonishing engineering feat. But I am not convinced that it is a marvellous military accomplishment." (20).

But if only de Gaulle in France and Alanbrooke in Britain, along with a small group of fellow critics, doubted the value of the Line, Hitler and his followers in Germany had no doubts that the line was completely valueless:

"I shall manoeuvre France right out of her Maginot Line without losing a single soldier." (21).

Which is, granting him the right to make slight political overstatements, exactly what he did. For Guderian's panzers outflanked the Line in little more than a month in the summer of 1940, by a rapid advance through the Ardennes forest. The latter had been thought 'impenetrable', a narrow front, and attempts to include it in the Maginot Line proper had been strongly opposed by Pétain in 1934. Despite the fact that individual forts proved to be tactically strong during the battle, the Maginot Line thus proved to be a strategic and national disaster.

6 THE AUTOBAHN AND WEST WALL

The birth of the Todt Organization, the highway system in Germany and the military implications of the West Wall.

Events

1922 Fritz Todt joins the Nazi Party.

1923 Minister of Economics proposes an 'Autobahn' system.

1930 Fritz Todt publishes paper on 'Reich Autobahn'.

1933 Hitler forms government and Todt becomes head of the new 'Reichsautobahnen'.
Jan. Gen. von Blomberg appointed Minister of Defence.

1935 German rearmament programme announced.

1936 April Occupation of Rhineland and first fortifications begun in Saar.

1938 Feb. 4 Hitler takes over from Blomberg and creates OKW.
May 28 Todt becomes involved in the building of the West Wall.
Aug. Labour shortage as work is to be increased.
Sept. Large numbers of men moved in from labour camps.
Nov. Adam is replaced.
Dec. Original plans for 'Autobahn' finished.

1939 Floods show up defects of fortifications, along the Rhine.
July SS moved in to tighten control on OT corruption.
Sept. Most of the work finished by time of invasion of Poland.

1940 Work on fortifications stops.

1940–44 West Wall robbed of armament to supply the Eastern Front and the Atlantic Wall.

1944 Sept. Germans retreat to the West Wall.

1944–45 Strengthening of West Wall takes place. New fortifications built which lengthen it to the coast.

1945 March 25 Last resistance of West Wall broken.

◀ *1. Autobahn bridge designed by architect Paul Bonatz as illustrated in the National Socialist's 'Neue Deutsche Baukunst' in 1941.*

● The construction of expensive fixed fortifications in Europe during the 1930's was not confined to France alone. Although in Germany Guderian and Lutz evolved concepts of fast mechanised warfare the German passion for the concrete bunker, much in evidence on the Western Front in 1915–18, once more emerged as a characteristic under Hitler. The Third Reich was to construct bomb-proof shelter and fortifications of every conceivable type during 1936–45, far exceeding the Maginot Line in the sheer mass of concrete such works consumed. The first major part of the programme to evolve was the West Wall, which was started in 1936 facing France's Maginot Line. Its strategic origins can be traced fairly directly to German experience in 1916–18 but in certain aspects the West Wall owed some of its existence to the German Autobahn project.

There are in fact several reasons why the motorway system built by the Nazi government is relevant to a study of military architecture. Firstly the provision of an efficient network of roads cannot be divorced from the need for mobility which, the German General Staff stressed, would be a feature of the next war. Secondly, Fritz Todt, its planner, and many of the private firms engaged in building it, moved directly onto the West Wall construction. An understanding of the Autobahn project therefore explains some of the construction and organization methods employed on the fortifications of 1938–39.

The first ideas for an Autobahn system had been presented by the German Minister of Economics in 1923. The then weakly governed Weimar Republic had however neither the strength nor the facilities to start such a massive programme, and the plans were forgotten. In 1930 the idea was revived in a paper by Dr. Todt, entitled 'Proposals and Financial Plans for the Employment of One Million Men.' Todt, a doctor of engineering, had served in the First World War, first as a Lieutenant of the Reserves, and later in the air force where he was wounded in air combat in 1918. After receiving the Iron Cross and the Order of the House of Hohenzollern at the end of the war, he entered the construction firm of Sager and Woerner in Munich, where he had previously studied. The firm specialized in road and tunnel construction and Todt, becoming its manager, must have gained valuable experience and influence there. In 1922 he joined the Nazi Party and soon became one of Hitler's close friends. Later he was one of the founders of the Nationalsozialistischer Bund Deutscher Technik. Using the SS training

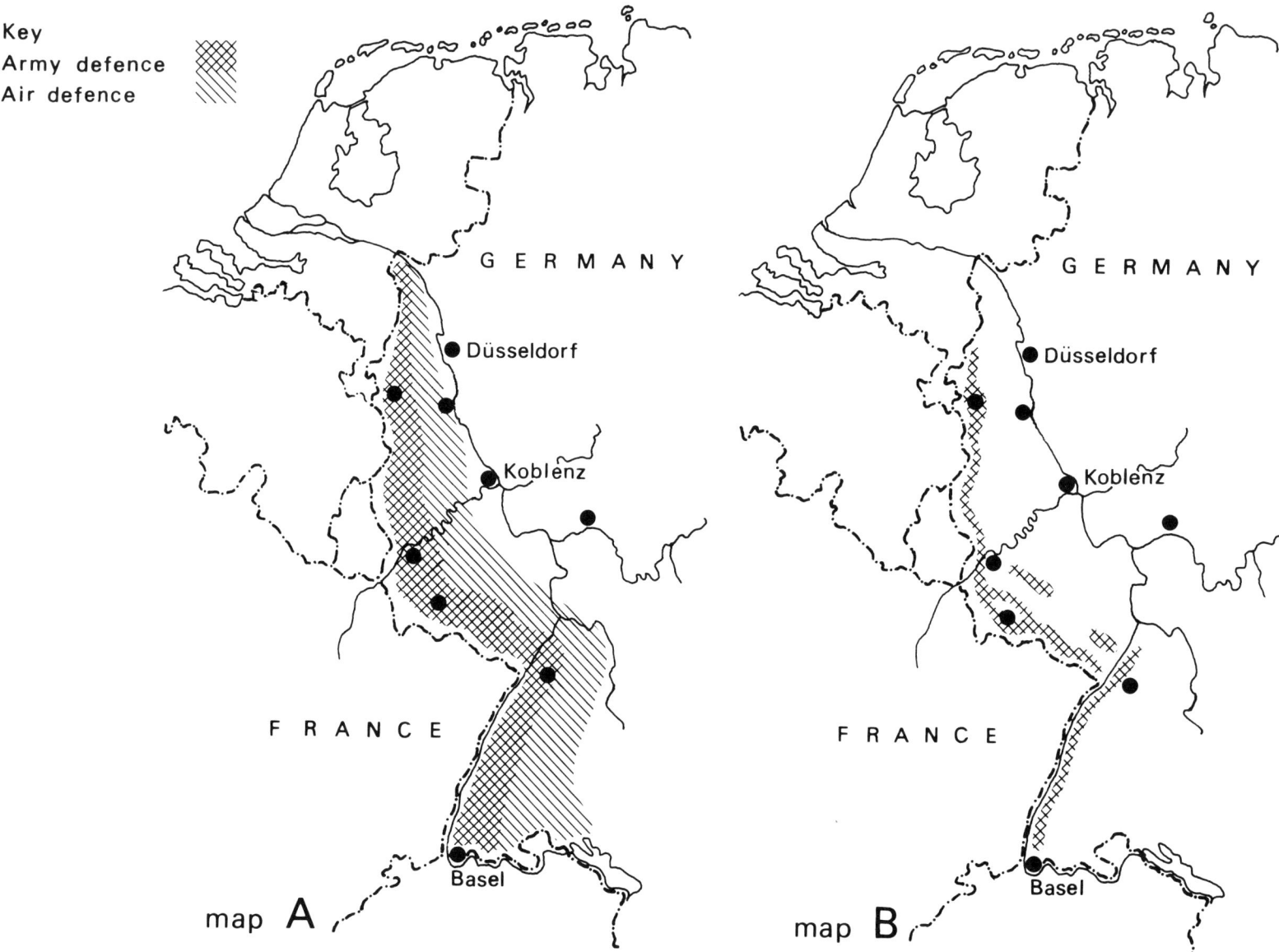

2. *The West Wall as political propaganda: Map A shows the West Wall defences as published by the Nazis in the late 1930's. Map B shows the Wall as found by the Allies in 1945.*

school at Plassenburg, this league paid special concern to the lack of independence in the German production industry and the general unemployment problems of the time. Through suggesting the opening of new fields of industry they hoped to solve both problems. Todt, with his previous experience, naturally concentrated on the highway system and the essence of his 1930 paper was that his plans would give Germany the best and largest motorway network in the world as well as offer a solution to the unemployment problem (1).

When his friend Hitler won the election in 1933, a state-owned corporation, the 'Reichsautobahnen', was established by decree on June 28th of that year. Todt became its administrative leader as 'Inspecktor für das deutsche Strassenwesen'. The corporation was made a subsidiary to the 'Reichsbahn', the German railways, to avoid unnecessary competition. The ultimate decision on planning and strategic objectives was given to the German Supreme Command, led by Blomberg since January, and Todt himself was directly responsible for implementation.

Though the building of the Autobahn was theoretically to relieve unemployment in Germany, French suspicions were soon aroused. The French General Serrigny, making an analysis of the highway system, came to the conclusion that there would, even with the immense innovations taking place in motor transport, only be one of two reasons for the project—either the German love for the monumental or their need for a communication network for warfare. Of the two, he personally favoured the latter. Germany at that time, however, maintained that the massive size and heavy construction was to compensate for the lack of a system of waterways such as that existing in France. The true reason why the project went ahead was probably not a single one, however, but a combination of different considerations: Hitler's fascination with grandeur ('best motorway system in the world'), Goebbels' interest in any material with a 'propaganda potential', the army's interest in transport generally and more specifically for the new armoured columns. All helped to bring about the acceptance of the plans but there is little doubt that the strength of construction and layout were based on military strategy rather than civilian needs.

After the 1914–18 war the army had not seen their defeat as either fair or ultimate. Officers and men claimed to have been 'stabbed in the back' by the politicians and so when General von Seeckt set about reducing the German army to the size which the French and the Versailles Treaty allowed for keeping internal order, his ingenious idea of creating a nucleus or skeleton, to which the bulk could be quickly added, was greeted with enthusiasm by the German military staff. Seeckt also introduced the theory that a small nucleus could strike and beat a large army if it consisted of well trained men and its mobility and mechanical strength were superior. Seeckt himself had resigned from the army in 1926 but it is more than likely that the later strategies of the General Staff, hand picked by him in this period, were a development from this. When German motorways built in the 30's were built strong enough to take the axle weight of today's traffic, it seems more probable that the actual design criteria was the German tank rather than any forecast of future transport tonnage. Similarly the layout, which one would expect to form a triangle between Berlin, the Ruhr and the Saar, instead forms a grid with little interest paid to these industrial and political centres, while the ability to offer mobility for large amounts of troops east-west and along the border emerges clearly.

Once these criteria were fulfilled, however, the German propaganda machine took over. Bridges, for instance, were given a great amount of aesthetic consideration for this reason. A visit by a committee of British architects and engineers in 1946 led to the conclusion that:

"Possibly British engineers have underestimated the appearance values in the search for economy. The same cannot be said of the Germans; indeed the probability is that political factors and propaganda values have outweighed those of economy." (2).

With its multitude of long span bridges in masonry and 'rough board' shuttered concrete, the motorway system became not only what British engineers called 'the greatest post-war German engineering development' but also became an important asset in the new German national socialist cultural movement. A recent article by J. B. Weber in the Architectural Association Quarterly goes even further than this:

"The most impressive achievement of the Third Reich, without any doubt, was the Reichsautobahnen . . . especially the realization of the importance of the relationship of Autobahn and landscape." (3).

For the building of the bridges the 'Reichsautobahnen' would prepare a survey including preliminary design for selective tender. Each contractor would then, in addition to his tender on the official design, submit his own alternative design (4). As the work progressed the firms used

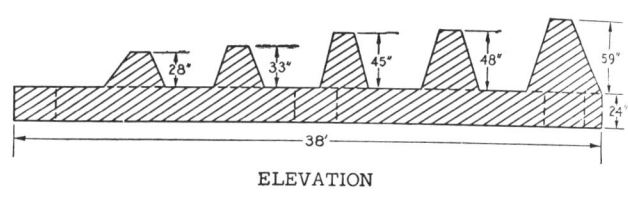
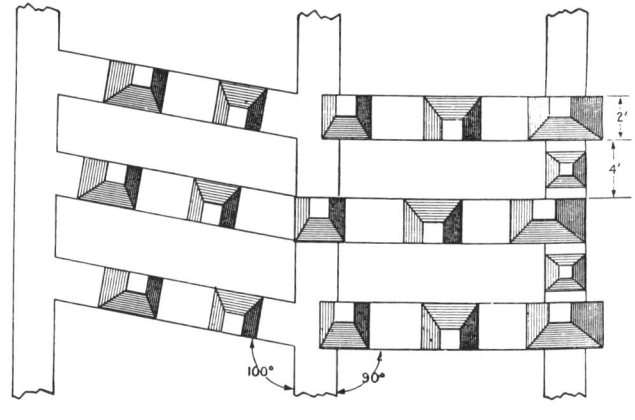

3. 'Dragon's teeth': row of four. Two such rows were sometimes used to form a row of eight.

4. Gate through the teeth, the concrete slots holding steel 'H' beams when under attack.

5. Diagram of teeth construction: row of five.

for the selective tendering became increasingly involved with the corporation. Thus by 1938 corporation manpower, the contractors that were used, and the corporation engineers and management personnel under the ultimate direction of Fritz Todt, formed a huge, if still not tightly knit, building organization. It was with this organization to back him up that Todt, when asked by Hitler in 1938, was able to take over the construction of the West Wall and virtually completed it in just over a year.

The strategy behind the West Wall (5) was far more complex than has often been suggested. Critics, such as Egon Eis, suggest that the whole line was a bluff, an attempt to be 'one up on the Joneses':

"France had her line, and so Hitler had to have one too." (6).

Others have suggested that the West Wall was built in order to free the army in the west during expansion in the east, while contemporary German writers maintained that the Wall was necessary for the protection of Germany from her aggressive neighbours. In reality the reasons for the project were not only a mixture of all these factors, but also the outcome of continuous political evolution. Having been reduced to a skeleton force and forbidden to build fortifications in the West in a zone extending 50 kilometres east of the Rhine, the army had been unable to intervene with the French occupation of the Ruhr in 1932. Before 1933 the weak government had refused point blank to support even modest ideas of defence, keeping strictly to the terms of the Versailles Treaty. Hitler, however, gave the army a free hand to carry out defence construction and in 1934 strong defences were subsequently built in the east, across the plains between the Oder and the Warthe. Nothing was as yet done in the west but in May 1936, the year following the announcement of the rearmament programme, the Rhineland was re-occupied and the army immediately 'dug-in' to secure their new position. Within a month fortifications were started in order to protect the vulnerable Saar, following the principles for such defence which had already been developed by the General Staff in 1933.

These first fortifications were extremely modest compared with the scale the West Wall was later to take on. They were purely defensive and purely military in purpose. With Blomberg, a believer in the luring defence, as Minister of Defence since January 1933 and as Commander in Chief of the Armed Forces and Reich Minister of War since 1935, and with independent men such as von Fritsch as Commander in Chief of the Army and Beck as Chief of General Staff, Hitler's influence over the army and its projects was minimal. The initial building project was thus not a result of Hitler's ideas on fortifications but of Hitler giving the army a free hand. Only for the materials was the Army dependent on civilian authorities, as Fritz Todt was in control of all German building materials for his Autobahn project. This independent situation was not to last for long, however. Early in 1938 Hitler seized control of the armed forces, replacing Blomberg, Fritsch and Beck. His decree of February 4th began:

"From now on I take over personally the command of the whole armed forces. . . ." (7).

and he created the organization OKW, Oberkommando der Wehrmacht, to which the army, navy and air force were subordinated, taking over Blomberg's office as Commander in Chief himself. At OKW his new men Keitel and Jodl were soon to become Hitler's office boys. Brauchitsch and Halder at OKH, Oberkommando des Heeres (Army High Command), were more independent but had to carry out OKW orders. Now in strong control at OKW, 'The Greatest Fortress Builder of all Time', as Hitler was later to call himself, soon changed the West Wall plans of 1936.

Instead of the short line along the Saar river a giant system of fortifications was planned, stretching 560 km from Switzerland to the point where the Rhine enters the Netherlands. To complete it Hitler called in Todt and his gigantic building organization to assist the army engineers in May, after Czechoslovakia had taken a defiant attitude towards German indications of aggression (8). Though many officers within the General Staff undoubtedly welcomed these fortifications, their purpose was no longer a purely military one. While the Maginot Line was surrounded by secrecy, the West Wall was widely publicized. Articles, books, photographs and films were issued by the Nazi propaganda machine under Goebbels. The picture painted was of a defensive line immensely strong, both deeper and longer than the French Maginot Line. Two consecutive lines, the Army Zone 20–30 km deep, and the Air Defence Zone or 'Luftzone West' 30–50 km deep, were to stretch continuously the whole 560 km. The strength of these zones was described by J. Pöchlinger in 'Das Buch vom West Wall', published in Berlin during 1940:

"Next to the Army Zone lays the Air Defence Zone West. Here are

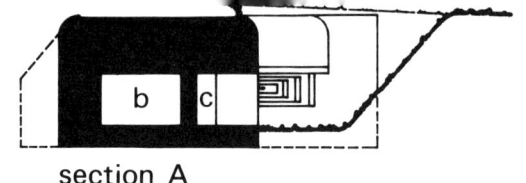
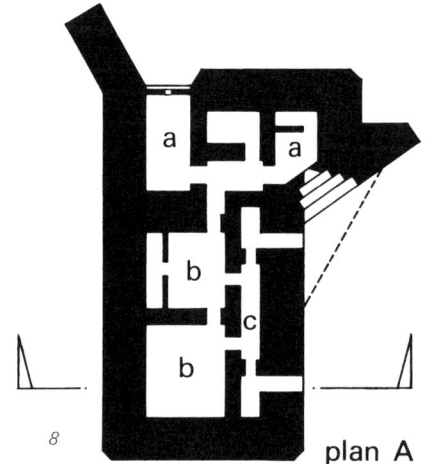
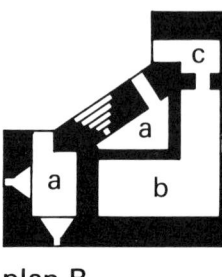
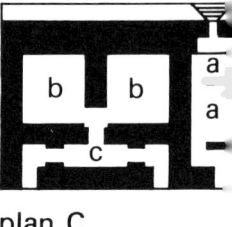

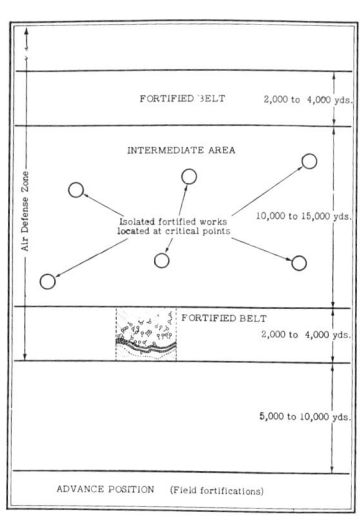

6. Aerial view showing the cluster pattern of installations, typical of both the continuous line behind the tank obstacles, the first fortified belt, or individual clusters further back.

7. Close up of a small installation.

8. Three typical decentralized installations on the West Wall:
   a. gunroom, b. troop quarters with bunks, cooking facilities and telephone,
   c. gasproof entrance.

9. Diagrammatic section through defences.

similarly deep staggered rows of Bunkers for light and heavy Flak that are arranged after the same system as the bunkers in the Army Zone, and build a wall of fire several kilometres wide in the sky." (9).

As Hitler took pains to learn all the statistics of the Maginot Line, it may well be that he saw the West Wall partly as a status symbol but the 'Wall' was nevertheless to serve a useful purpose. Through its exaggerated propaganda the Wall was to deter the Allies from fulfilling their obligations in the east and give the German people a sense of security. Some critics maintain, however, that there was no need for a West Wall, considering the superiority of the Germany army and the defensive attitude of France but it should be remembered that the theory of mobility was not proved a successful one in the field until September 1st 1939. Prior to this the army was in the middle of rearmament and no doubt welcomed the extra political and military protection given them by the West Wall. This consideration of political/military propaganda is expressed well by the military writer Liddell Hart in his book 'Strategy':

"If the Allies declared war in fulfillment of their obligations they would automatically forfeit the advantage of defence.... If they merely tapped on the Siegfried Line (West Wall) they would manifest their impotence, and forfeit prestige. If they pressed the attack, they would only pile up their losses and weaken their own chances of subsequent resistance when Hitler was free to turn westwards." (10).

The ultimate value of this propaganda was proved both in 1939, when Pétain started an attack at the Saar without any serious attempt at penetration, and in 1944, when Eisenhower decided to spread out and await reinforcements before attacking. In the view of many experts the latter resulted in extending the war unnecessarily into 1945.

The actual West Wall as planned and built by the army and Todt, though brilliant in its design, had nothing of the strength painted by Goebbels. The huge difference between the propaganda wall and the real West Wall is best described by the fact that General von Rundstedt simply 'laughed' when he saw it. While the propaganda wall would have been able to throw back an attack by its defences, the real wall was only a tool to be used by a mobile German army. The rigidity of the propaganda wall was unsuited to German ideas on mobile warfare and the magnitude inferred could not have been completed in the time available. Though the West Wall was often compared with the Maginot Line, they therefore had little in common. While the French implemented a Verdun type of static defence against mobility, the Germans under such men as Seeckt, Blomberg, Guderian, and Lutz had developed their defence in depth to incorporate the mobility of modern warfare. Robert O'Neill makes a particularly apt comparison in his recent book 'The German Army and the Nazi Party':

"The Maginot Line used the 'hard thin skin' approach, which was valueless once an initial penetration had been achieved. The German fortifications were built on a 'Milky Way' principle, affording depth with economy of construction, and excellent mutual support for their defending positions." (11).

Increasingly stronger lines of defence were intended to break up and slow down the enemy momentum while the army, concentrated behind the line, out of enemy reach, prepared to deliver a crushing counter-offensive. In this way it was planned to use tanks and troops in a mobile battle within the deep fortified area where all advantages would be with the defenders.

Following the principles evolved by Fritz von Lossberg at Arras, in 1917, the advanced position was only weakly held by field fortifications such as discontinuous trenches, fox holes, barbed wire entanglements, observation posts and mines. The first line of serious resistance, the 'First Fortified Belt' (12), came 5000 to 10 000 metres behind this advanced position. Initially two such fortified belts had been envisaged stretching all along the border. Even at an early stage, however, it had been decided that need for this was overruled by both the time factor and economic considerations. Instead a double 'Belt' was only to be built where natural fortifications were weak or in areas of high strategic value. With this in mind the German maps published during 1939–40 are misleading — probably deliberately so.

Starting at Switzerland, it was decided that where the Rhine formed a natural tank obstacle along the French border, only a single 'Belt' was required. From the Rhine through the Saar region to Trier the fortifications were at their strongest. By Saarbrücken, in the centre of this region, the 'First Fortified Belt' split to form a 30 km long double 'Belt', the two belts being 11 km apart, and 16 km behind this a third 'Belt' was built directly behind Saarbrücken where natural obstacles were weak. North of Trier, however, only one 'Belt' was used where the Ardennes formed a great natural obstacle. Further north the defence then strengthened once more. The single line split into two 'Belts' about

10, 11. Construction of one of the few installations equipped with a cupola.

12. Larger installation of the 'Decentralized' type.

8 km apart with Aachen in the middle, in order to protect the southern approaches to the Ruhr. This section was strongest north of Aachen. From halfway between Geilenkirchen and Aachen the two 'Belts' again merged. Though strong and occasionally backed by a single cluster of pill-boxes it soon faded into a thin single line and almost disappeared north of Geilenkirchen where the Maas could be used as a forward obstacle.

The 'Forward Belts' and the 'Rear Fortified Belt', where it existed, consisted mainly of pill-boxes in cluster pattern varying in depth from 800 metres to 5 km, giving mutual anti-tank and machine gun fire. To the front of this a continuous passive anti-tank obstacle was formed either by natural (rivers, lakes, etc.) or artificial obstacles, followed by barbed wire and mines. Several types of artificial tank obstacles were used, both steel and concrete, but the most common were of the 'dragon's tooth' variety. These were truncated concrete pyramids laid in rows four or five (13) pyramids deep and increasing in height from 800 mm in the front of the rows to 1400 mm in the back, or 1750 mm at the back when in rows five deep. Cast on a grid of interlocking concrete beams projecting 600 mm above the ground, this line of continuous tank obstacles was probably the most outstanding visual feature of the West Wall. The line of dragon's teeth stretched continuously through towns and forests, over hills and marshes, even bridging small rivers, and gave an impression of strength and continuity which must have played a major part in convincing the Allied intelligence of the strength of the Wall.

The pill-boxes and small casemates for anti-tank guns were situated so as to give a continuous field of fire covering the obstacles in front as well as a deep barrage fire of mutual support within the cluster pattern. A series of loop-holes provided fire in all directions, though mainly forward and flanking, but the capture of one installation would only bring the enemy in contact with a host of new positions. Each position was carefully sited with consideration given to fire effect, cover and concealment. Fire effect was given priority over cover, but it was attempted wherever possible to merge positions with the surrounding terrain and to keep size to a minimum so as to reduce the available target area.

Each installation was further treated as a self sufficient unit and manned by approximately 15 men. They therefore contained facilities in addition to the 'Kampfraüme' (firing rooms) such as ammunition stores and tool rooms, food stores and living quarters. Here fold-up bunks, washing facilities and messing equipment would be found but, although electric lighting and telephone communications were provided by underground cable, the units lacked the environmental refinement of the Maginot Line. For emergency use candles and paraffin lamps were provided. Toilet facilities were provided by bucket latrines and a cast-iron stove provided the combined means of heating and cooking. Some original research was undertaken in the case of the stove. It had to be capable of being hermetically sealed within seconds for camouflage reasons, a tricky affair as a pressure of three atmospheres would build up in the chimney. For protection against gas attack the troops were provided with gas masks and chloride lime but, in addition to these provisions, manually operated ventilation equipment, gas-proof doors and steel embrasure shutters were provided in the larger installations.

Steel cupolas were sometimes incorporated as can be seen from the illustrations of the Rhine fortifications but generally their use was restricted due to their high cost and the inherent transport problem. However, where they were provided, they were never of the rotating or retractable but only of the fixed type, a continuous field of fire being provided by the six or four ports for automatic weapons around their perimeter. These could be closed with steel shutters and observation carried out through a centrally placed periscope.

The pill-boxes and small casemates, known as 'decentralized works', were further strengthened by field works such as trenches and foxholes. The infantry forces for these positions were provided with large semi-underground shelters, known as 'closed works', generally placed to their rear. These huge concrete shelters had no emplacements for guns and were only protected by rifle embrasures near their entrance. Their primary function was to provide shelter for the infantry during air and artillery bombardment before the crucial counter offensive, to house relief units and to store ammunition. Lacking fire power, the shelters were protected by additional obstacles and field works.

In military terms, however, the main feature of the West Wall was not to be found in its building construction, but in the mobility it provided. According to a translated extract by the US Intelligence of the German field manual 89 'Die Ständige Front' (the Stabilized Front):

"The Germans teach that defensive combat on a stabilized (fortified) front is conducted generally according to the same principles as the

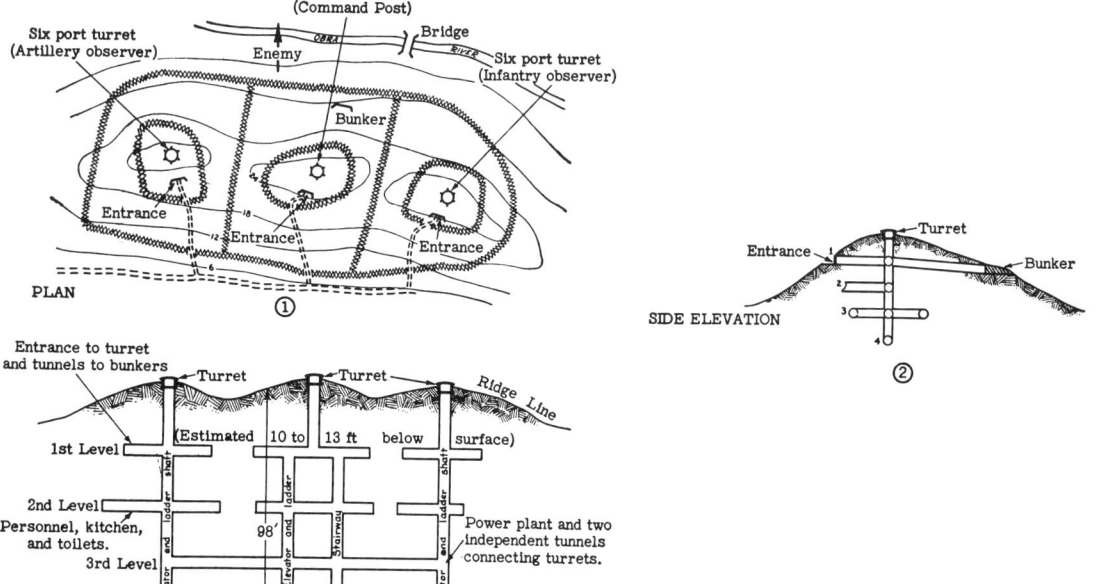

13. Artillery block in the Czech fortifications.
14. Large works built east of Berlin as a miniature Maginot Line on the model of the captured Czech fortifications.

defence in a war of movement, the chief consideration being tactical organization of the defensive fire and forces." (14).

To enable such mobility the Wall was permanently manned by only a skeleton force, although during attack it was to be manned by a full infantry division per 8 km of front. Equipment such as anti-tank guns and machine guns were to be brought in with the mobile defenders while the skeleton force looked after the arms and other equipment which could not be moved, although fixed equipment was kept to a minimum. Again, for reasons of mobility, the heavy artillery was provided with open emplacements and not casemates but of those small casemates that did exist for anti-tank guns only a few could take any larger calibres than the 37 mm generally used in 1939. In 1944 this was to be a serious fault for by then even the 75 mm guns, which could be mounted in some of the casemates, were inadequate to cope with the Allied tanks.

Originally the 'Advanced Position' and the 'First Fortified Belt' were to have been the Army Zone and this was then to have been followed by a separate Air Defence Zone of similar strength. As the First Belt in many cases was the only belt, however, this strict division was not always possible and though most of the flak could still be situated behind the First Belt, much of it had to be moved forward into the original Army Zone. Where further fortifications were added then an intermediate area was constructed, 10 000 to 15 000 metres deep. This was followed by a 'Second Fortified Belt', identical to the First but generally weaker, and by additional air defence, flak towers, barrage balloons, warning devices and military airports for pursuit aircraft. Flak batteries and searchlights were under the command of the Luftwaffe and mixed with the other fortifications throughout the West Wall defences.

The propaganda machine published several photographs of the flak towers on the West Wall (15) but it put considerably greater emphasis on the so-called underground forts—Germany's answer to the Maginot Line. In these forts large underground galleries contained power stations, kitchens, quarters, toilets, ammunition and food stores as well as mechanical transport such as trains and lifts. According to the official historians, however, no such forts seem to have been encountered on the West Wall by American or British forces in 1945 and the only fort seen by an impartial observer which fits this description was situated on the west bank of the Obra, east of Berlin (16). As the German propaganda photographs of such installations were published after the capture of the Czech fortifications (17) in the Sudetenland, in 1938, it is probable that they were either from these fortifications or from a little known east wall, east of Berlin by the Obra, where Hitler had ordered the building of a miniature Maginot Line, prior to the invasion of Poland, based on the Czech fortifications.

The size of the area to be fortified in the West Wall, 560 km long and between 12 and 30 km deep, made it an impossible task for the army and they had not even finished their 1936 plans when Todt was called in. The choice of Todt was quite logical. Through the Autobahn project he had gained control of all building materials and both control of and respect from the large number of building firms he employed. The original plans for the Autobahn would be completed by December 1938 and so when Todt proposed the new West Wall scheme, one which would provide employment for a third of the building industry, the reaction among the Autobahn building firms was naturally favourable. Bringing with him his personal staff, Todt set up his headquarters in Wiesbaden together with a drawing office which was to produce working drawings to army specifications. This nucleus rapidly expanded as firms moved from the Autobahn project. By 1939 some 350 000 workers were employed by Fritz Todt (18). 100 000 men from the 'Arbeitsdienst', 90 000 'Festungspioneere' (army engineers) and various other army personnel were working on the West Wall a total labour force of 500 000. These men poured over six million tons of concrete, laid three million rolls of barbed wire and constructed 22 000 individual fortified works in 18 months

Three hundred and fifty trainloads of materials plus an equivalent by road arrived daily (19), which gives some indication of the intensity which the programme reached. In this situation it would be reasonable to assume that the vast scale of the project would necessitate a large degree of standardization of design. This was not the case, however, with the exception of the dragon's teeth constructed by the Arbeitsdienst. The reason can be found partly in the army organization created by Hitler and partly in the organization led by Fritz Todt.

By the time the army had started its work in 1936 Hitler had already done his utmost to divide authority rather than centralize it. When he made himself Supreme Commander of OKW, in 1938, he continued this philosophy. General Adam, who had been in command of the West Wall,

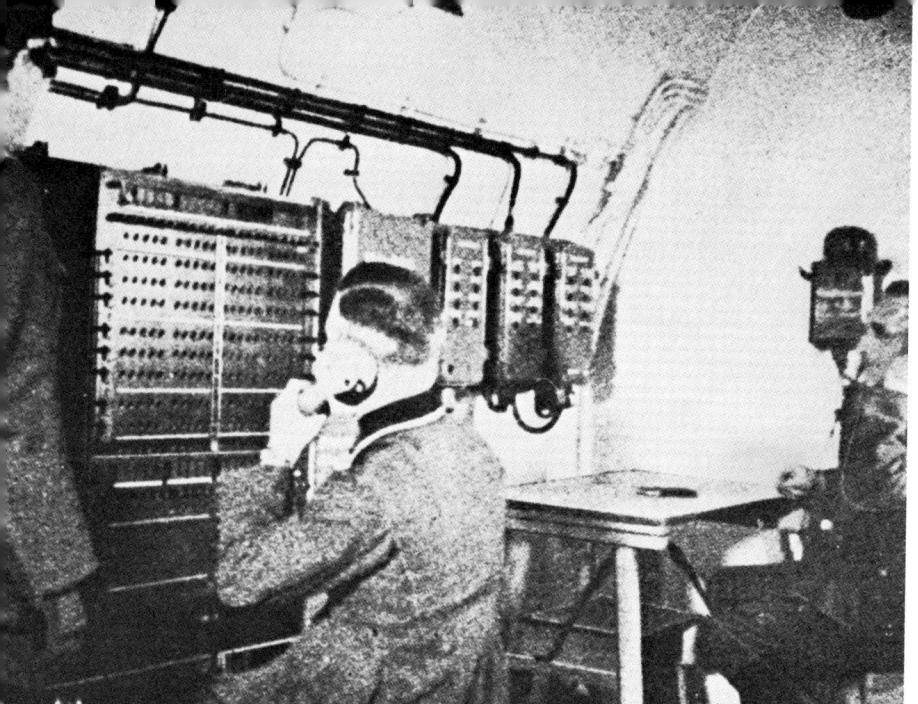

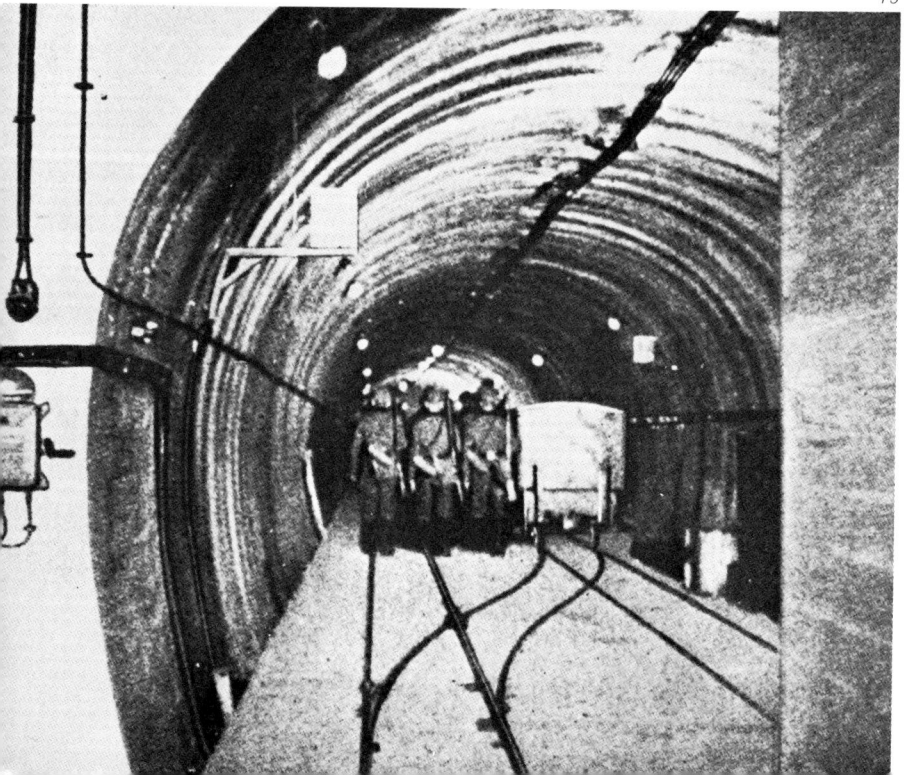

15, 16, 17. These photographs of the interiors of large closed works were published in Germany as being the West Wall in 1939. Although underground works did exist on the West Wall, their date of publication, and the sophisticated facilities illustrated do however suggest that the photographs may have been from the East Wall or captured Czech defences—published as part of West Wall propaganda.

was removed in November and at the same time Warlimont replaced Jodl as head of the National Defence section without being given any authority over the West Wall strategy. The general confusion this caused suited Hitler's intentions. He ordered the individual commanders to go ahead with the construction within their section without reference to their neighbours. Though Hitler was later to show a detailed interest in the building of fortifications on the Atlantic Wall, he was now only interested in the propaganda aspects, his architectural aspirations being concentrated on his plans for the rebuilding of Berlin.

Todt could have been expected to introduce standard design, but he had certainly not incorporated standard design in the building of the Autobahn bridges and he was not to do it now. Instead of introducing a centrally controlled administration, he continued to look on his organization as a task force which would carry out work delegated by the army. This can be seen from the proud comment he made on September 6th 1938. Hitler had just given the organization the name 'Organization Todt' (OT) at Nüremberg, when Todt exclaimed:

"We are being called the Organization Todt, without ever having organized." (20).

Another factor was the increasing amount of responsibility given to Fritz Todt. Though he continued to show interest in the large forts of the 'closed type' due to the engineering problems involved, his post as Minister of Armament soon left him little free time. Consequently nobody really tried to introduce standardization, and, though an overall plan was adhered to, one finds that out of the 28 'Decentralized' units used on a typical 650 metre stretch of the 'First Fortified Belt' there were over 20 different designs. Some basic patterns were probably used and modified to local conditions but the extent of standardization was minimal when compared with the BEF pill-boxes in France, based on only five different designs, or the British coastal fortifications of either 1914 or 1940.

The lack of design standardization meant a lack of economy in the use of building resources but an even more important fault in this respect was the way in which corruption developed. Without any central control or direction the OT firms, left under the command of local army officers and engineers, soon started to increase their profits by selling materials, which the government had issued, to local farmers or by using them for private projects. In July 1939 the SS was called in and they arrested several contractors and officers (21). Much damage had already been done, however, and strengthening of existing works had to be carried out. Moreover, the flooding of the Rhine in 1939 showed that not only was construction weak but also siting was faulty.

Not all these defects were the fault of the OT firms. Much of the problem had been caused by haste and inefficient central control. The long studies of concrete resistance to artillery bombardment undertaken in France had no German counterpart. Monolithic construction of varying quality was employed and basically in only two thicknesses: the 'Feldmässig 40 cm' standard, and the strengthened 'Feldmässig 150 cm'. As a result of the defects shown up in 1939, however, a tightening of control was established and new works were built with far better control over the quality of the concrete. Thicknesses on several of the old fortifications were increased up to 'Vorpanzer' and 'Schürze' standards. 'Vorpanzer' meant that an additional thickness of reinforced concrete was added to the front of the installation to increase their resistance to artillery bombardment, 'Schürze' was a 150 cm thick—600 cm deep concrete wall cast between the circumference of the installation and sheet piling laid parallel, which prevented shells exploding under the installations as well as protecting against soil corrosion.

Such improvements did not make the West Wall impregnable and it was propaganda which brought more rewards than concrete thickness. In Germany, R. T. Kuhne wrote in 1939:

"The overall foreign political and military importance of the West Wall is that foreign countries, enemies to us, always try to compare their own importance and military value favourably with ours through false reports. But it helps no longer, the West Wall stands and is unconquerable. . . . No enemy can approach this bulwark of steel and concrete unpunished. Even an attack with the strongest means, will, despite the largest sacrifices, soon break down under the fire of defensive armament." (22).

Foreign correspondents were suitably impressed. A Canadian newspaper called the line invincible, whilst an observer from Argentina predicted that any attack against this giant bulwark of steel and concrete would be 'smashed'.

The real defences did not have the strength envisaged by the army staff in 1936 or by contemporary critics but their many defects were compensated for by a loosely woven structure which enabled improvisa-

tion. As an English writer accurately stated in 1939:

"There are undoubtedly weaknesses in the Siegfried Line; it has been hurriedly built; the materials are not, perhaps, of the highest quality. But the sound principles on which it is built give it a great inherent strength. It is a formidable instrument in the hands of the modern German Army." (23).

The line of 1944 was not the line of 1940 however. Hitler had made no preparations until the end of August 1944 as he believed the Allies could be halted on the Somme–Marne line. Nearly all the permanent signal equipment and munitions had been removed to reinforce other theatres. The concrete shell was being used to shelter bombed out families or stores and the keys for the installations which had been locked up could not be found at OKW. Even more important still was the fact that the casemates for the 37 mm anti-tank guns of 1940 could not house the 75 mm and 88 mm anti-tank weapons necessary in 1944. The West Wall as it stood when the Germans retreated to it was thus obsolete. To bring it up to date von Rundstedt needed six weeks.

Only a miracle could give him this time but, like the French 'Miracle of the Marne' in 1914, a miracle did occur and in September 1944 the Allied advance faded away. The reason was that Eisenhower wanted to pause and regain momentum before penetrating the West Wall defences. His evaluation was a pessimistic one based on propaganda and rumours rather than accurate intelligence reports. Similarly the Allies made another faulty assumption in believing the German war machine to be incapable of strengthening these defences if allowed the time to do so. This momentary slackening gave the Germans the chance they had not dared to hope for. The OT strengthened the existing fortifications and extended the line to the coast. New troops were raised by lowering the age of conscription and by combing factories for surplus manpower.

The result was that when the Allies attempted to advance again they found a defence which was much stronger than expected. General Patton managed to take part of the defence around Aachen by November but did not breach the line. The Western Front had become static, at least temporarily, but this situation was soon to be altered by Hitler's final gamble. Scraping together all the armour available, he launched the ill-fated Ardennes breakthrough, instead of following von Rundstedt's advice of merely retaking Aachen. After the initial success of surprise the ultimate defeat of this 'desperate gamble' left the Germans without the possibility of launching any other offensive.

The West Wall had always been intended to contain an enemy offensive until the momentum was broken and a counter-offensive could be launched with planes and panzers to recapture the lost areas. The Luftwaffe had been crushed by the introduction of Mustangs during the American day bombing, and now the Ardennes offensive had broken the panzer. By fanatic defence they could slow the Allies down, but there were now no mobile reserves available to throw them back, an essential feature in the West Wall strategy. In early February Eisenhower was once again ready to launch an attack from the position the Allies had held in September 1944 (24). The American official historian writes:

"Without question, these fortifications added to the defensive potentiality of the terrain along the German border; but their disrepair and the calibre of the defending troops had vitiated much of the line's formidability. . . . It could in no sense be considered impregnable. Nevertheless, as American troops were to discover, steel and concrete can lend backbone to a defence even if the fortifications are outmoded and even if the defenders are old men and cripples." (25).

On March 23rd the Rhine was crossed and by the 25th the last organized resistance on the West Wall had been broken.

18. Seventh U.S. Army troops pour through the West Wall on March 26th, 1945.

PART THREE 1940→45

7 FORTRESS BRITAIN

Organization and construction of English defences.

Events

1922 Geddes Committee suggests reducing air force to eight squadrons but stopped by public pressure. Instead a sub-committee is set up, known as the Steel-Bartholomew Committee, which proposes first plans for British air defence.

1924 Rommer Committee proposes revised and extended air defence.

1932 Disarmament Conference fails to put restrictions on strategic bombing.

1933 Nazi election victory.
Nov. 14 First meeting of Defence Requirements Committee decides that Germany is to be considered the ultimate danger now.

1934 Rearmament begins. Navy finds that its coastal defence, with only few exceptions, had not been modernized for 30 years.

1935 Reorientation Committee proposes air defence zone against the east: Germany, rather than the south: France.

1936 Success with radar causes outer artillery zone in air defence system to be abandoned.

1937 Germans march into Austria. Air defence zones extended. Policy of British Continental intervention readopted.

1939 Invasion of Poland followed by declaration of war.
Sept. 3 BEF sent to France.

1940 Apr. 9 German invasion of Norway and Denmark.
May 10 German invasion in west starts.
May 28 Dunkirk evacuation completed.
June First installation of emergency coastal batteries. Ironside becomes C-in-C for home forces—GHQ Line started.
July 20 Alanbrooke becomes new C-in-C for home forces.
July Battle of Britain starts.

Aug. Alanbrooke reorganizes home defence.
Sept. The 'Blitz' begins. Decision to build permanent batteries at Dover in answer to German batteries at Cap Gris Nez.
Dec. 18 German directive for attack on Russia marks end of invasion plans.

1941 Installation of emergency coastal batteries completed.
May The 'Blitz' ends.

1942 June Last of large batteries at Dover completed and number of coastal batteries at the maximum achieved during the war.

1942–43 Building of Sea Forts takes place (designed 1941). Switch to invasion preparations (D-Day).

● While the Franco-German border saw frantic defence construction in the 30's, with both the Maginot Line and West Wall, it was not until 1940 that similar activity was prompted in Britain. Following Dunkirk Churchill, both PM and Minister of Defence, delivered his now famous speech:

"We shall not flag or fail, we shall go on to the end, we shall fight in France, and we shall fight on the seas and oceans, we shall fight with growing confidence and growing strength in the air, we shall defend our island whatever the cost may be, we shall fight on the landing grounds, we shall fight in the fields and in the streets, we shall fight in the hills; we shall never surrender." (1).

Churchill's prose did not mention, however, that in the field of fixed fortifications little had been done to prepare Britain for such a battle (2). Before the war the government had felt that the cost of such defences was prohibitive and, subsequently, the navy had only modernized a few of its fortifications on the coast during the interwar period. The only defence that did exist in any magnitude was air defence but strangely enough the building of this was due to political and civilian pressure rather than any desire for it among the air staff (3). Among the civilian population of 1914–18 the raids on London, first by Zeppelins and then by Gotha bombers, had made a substantial psychological impact (4). Although the destruction and casualties caused was com-

◀ *1. Anti-tank defences being constructed on the east coast in July 1941.*

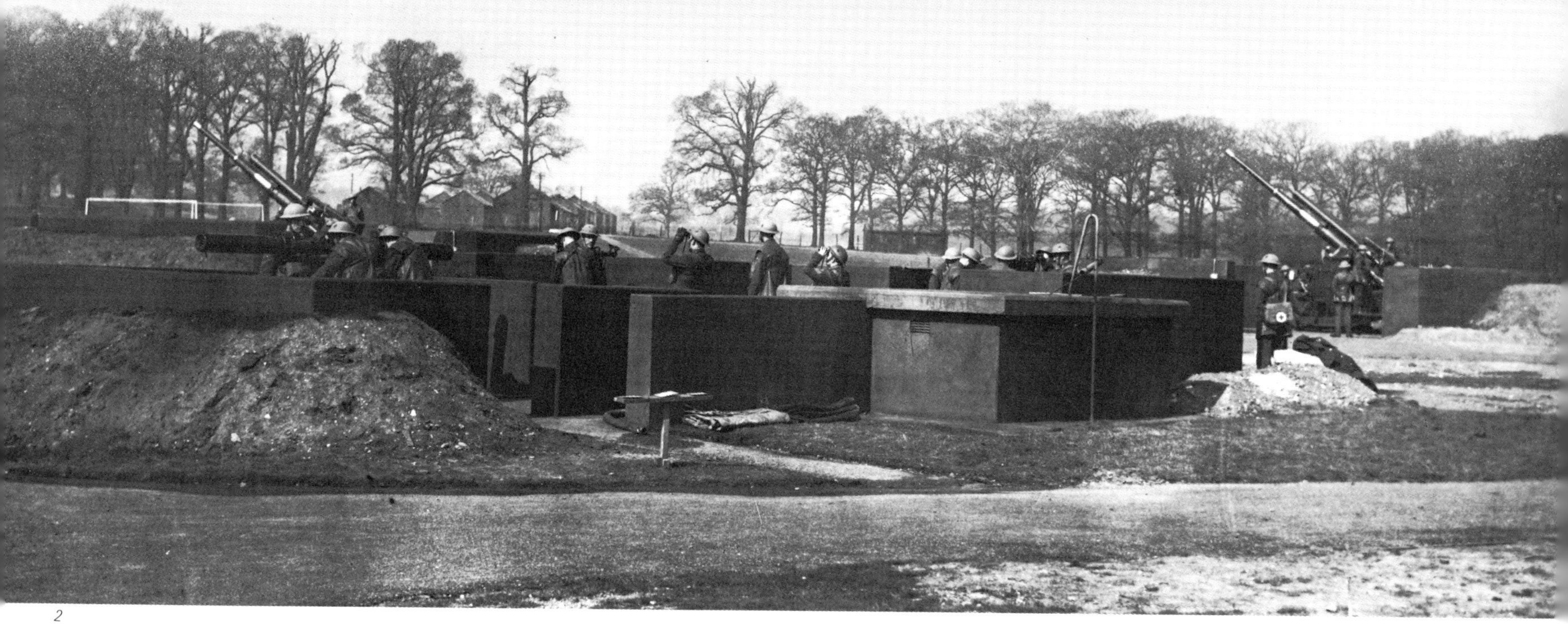

2

3

4

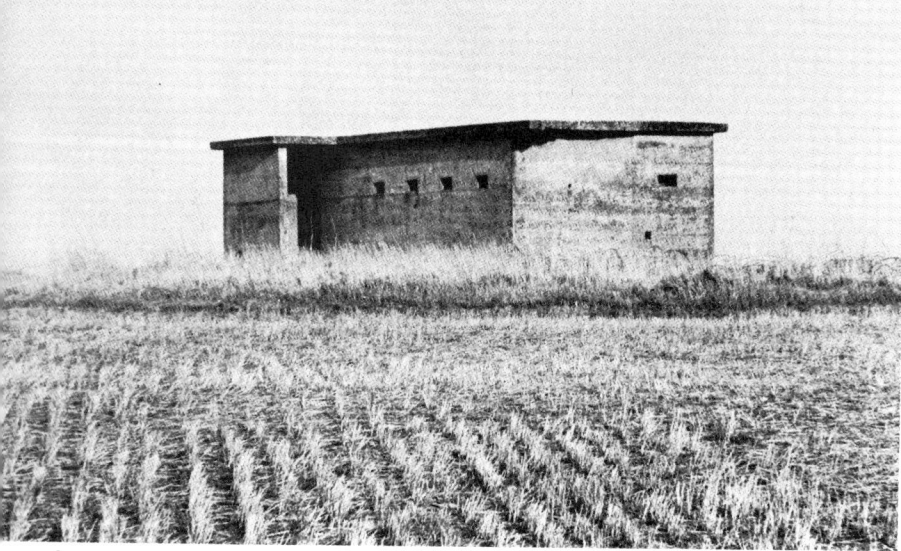

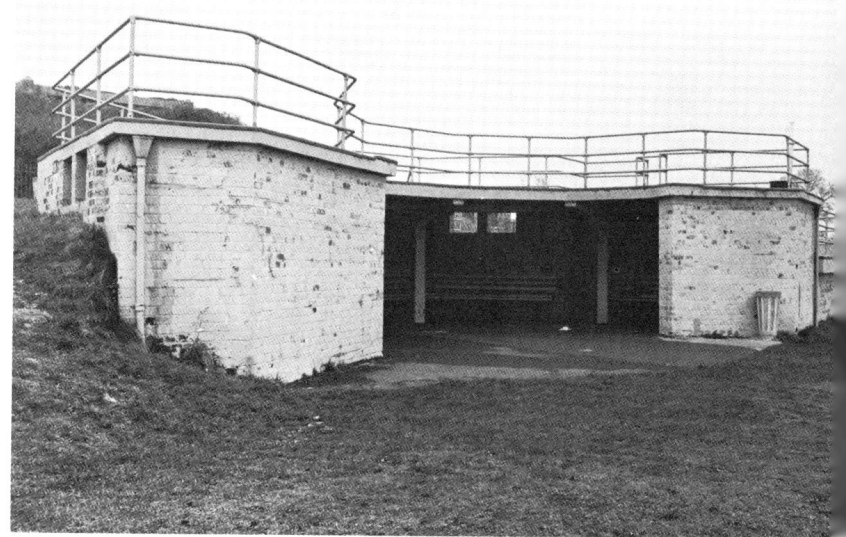

2. British anti-aircraft battery.

3, 4. British coastal artillery casemates as they remain today.

paratively small the public of 1918–39 demanded protection in the form of searchlights, artillery and fighters, against an 'invasion' by enemy bombers, which could, they argued, 'put an end to civilization as we know it' in one blow (5).

The first plans for such a defence had already been started early in the 20's when a joint army and air force committee (6) under Air Chief Marshal Trenchard appointed Colonel W. H. Bartholomew from the War Office and Air Commodore J. M. Steel from the Air Ministry to look into the matter. The defence proposed by this sub-committee, known as the Steel-Bartholomew Committee, was to run continuously from Duxford in Cambridgeshire and east of London to Salisbury Plain. Basically it contained four different sectors: an inner artillery and searchlight defence for direct combat with enemy planes, a fighter zone for air to air combat, an outer artillery zone only firing during daylight and intended to break up enemy formations rather than to engage individual planes, and finally a belt of advanced observation posts. To give the fighters and artillery sufficient time to get ready, these zones were placed well to the rear with early warning stations situated on the coast incorporating huge concrete acoustic mirrors (7). As France was considered the potential enemy at this time, both this plan and the revised plan by Rommer and Ashmore were directed against attack from the south, not the east. Only in 1935 did the Reorientation Committee (8) propose that the air defence system should include the north rather than the south west, as Germany was now to be regarded as the threat, but the basic zonal arrangement was still kept and this was not changed until 1936.

The air staff had argued that the lack of early warning would allow the fighter defence to attack only when the enemy bombers had already dropped their cargo. The huge concrete acoustic mirrors erected for warning purposes had indeed met with only limited success but the Committee for Scientific Survey of Air Defence, led by Henry Tizard, which had supported acoustic mirrors, did not give up hope and in 1935 started to experiment with radar for this purpose. With radar equipment effective warning was finally made possible and, as the fighters could now attack the enemy in his approach over the sea, the outer artillery zone was abandoned (9). Instead the fighter zone was to be the only continuous defence zone and in 1937 it was extended north beyond Newcastle. Vulnerable areas were given a local artillery and searchlight protection similar to the previous 'inner artillery' zone. Balloon barrage protection was also envisaged, but lack of time and money prevented it from being finished by the outbreak of the war.

So far, it had been argued that as the enemy could approach only by air or sea, this air defence, together with naval superiority, would provide safety against an invasion. The army was left the task of rounding up small detachments of enemy paratroopers and the coastal fortifications were considered more in the context of naval base defence than as a defence against invasion. By the time of Churchill's speech, however, events in Norway and France had cast doubt on all these suppositions. The German navy and air force had managed to carry sufficient troops to Norway under the very nose of the British navy and the bombing of naval vessels had proved a thorough disappointment. Dunkirk had similarly shown the ineffectiveness of the Luftwaffe in interfering with the naval operations of the evacuation and in France the army had seen at close quarters how the German panzers had penetrated their lines at high speed and completely outmanoeuvred the Allied forces. The need for some form of coastal defence against invasion was now seen as a desperate one. The Chiefs of Staff had come to the conclusion that 'the crux of the whole problem is the air defence of this country' but meanwhile the country must be 'organized as a fortress on totalitarian lines.' (10).

As a result, the building of 'emergency' fortifications began as a matter of urgency. The first thought was to prevent the enemy getting a foothold and so the coast was strengthened and the central reserves brought forward. With the BEF back, it was argued that the situation had been greatly improved. Up to a point this was true but although the men of the BEF were successfully evacuated from Dunkirk, their tanks, artillery, and other equipment were abandoned. Having started late in the arms race the replacement of these was a slow process. Despite this the evacuated BEF, together with the central reserves, were now moved forward to cover the positions immediately behind the beach and on the coast itself the construction of pill-boxes, obstacles and gun-emplacements went forward feverishly. The coastal defences around Dover were given first priority, as this was considered the most likely area for a German landing but the harbours which had previously provided bases for navy and convoy movements were similarly treated. Later this defence was extended, forming an almost continuous line between harbours from

5. One of the standardized six-sided concrete pill boxes on the east coast.

6. Local modification of pill box design (recent photograph).

the Humber to Dartmouth, depending on topographical and industrial factors.

During the First World War coastal batteries had been placed in open emplacements behind earth banks or concrete parapets, 'en barbette'. Casemates had been found restricting and their embrasures were vulnerable with the accuracy of rifled artillery (11). But by 1940 the low-flying bomber had become a greater hazard, and casemates were introduced once more along the English coast. The casemates, grouped in batteries, were not the only structures which had to be constructed. A battery of three 6-inch guns called for operational force of four officers and 135 men, a battery of two 4-inch guns three officers and 90 men. As a result quarters had to be built for these men in addition to observation posts, plant rooms and searchlight units. The necessary new batteries, known as 'Emergency batteries' were finished in seven instalments, a total of 153 batteries, over the last six months of 1940 (12).

In addition to these, three long-range batteries were constructed (13) near Dover as an answer to the large German batteries on the French coast at Cap Gris Nez. In May 1940 Churchill had already suggested the installation of long-range guns at Dover but all that could be provided were the two 14-inch naval guns known as Winnie and Pooh. These were insignificant when compared with the large German batteries such as the 'Batterie Lindemann' and the 'Batterie Todt', but Colonel K. W. Maurice-Jones writes in 'The History of Coastal Artillery in the British Army':

"If the enemy could engage British shipping passing through the Straits, shell coastal defences, and bombard Dover, the Coastal Gunners were equally resolved to stop German ships moving through French coastal waters between Dunkirk and Boulogne and to bring counter-battery fire to bear on the German guns." (14).

It was decided to build three new batteries for this purpose in September 1940. Fan Bay with three 6-inch guns was ready in February 1941, South Foreland with four 9-inch guns in October 1941, and Manstone with two 15-inch guns in June 1942 (15). The batteries had an extreme range of 23 000, 28 000, and 38 500 metres respectively. At the same time as the coastal batteries were being built, pill-boxes, tank traps and wire entanglements were erected. Layout of all these defences was decided on locally but for large installations Brigadier A. Burrows travelled up and down the coast equipped with a rifle and a case of maps, siting battery layouts so that a continuous field of fire was formed. For smaller defences, siting was done by the local commander in most cases.

The actual labour consisted of a mixture of experts from the Coastal Artillery School, sailors from the Royal Navy, gunners who were to man the guns when they were in position, and Royal Engineer Services with contract labour. This was organized first by Major R. Shrive and later by Major C. S. Woodsford. Most of the building expertise and 'quick-setting' concrete naturally went into the construction of the large coastal artillery emplacements, leaving the small pill-box installations and obstacle construction worse off for both man-power and materials. As a result of this the small installations were more frequently modified in design and considerable improvisation occurred. Brick, or even pebbles from the sea-shore were substituted for concrete, and as there was little control during construction of the smaller installations, design was often misunderstood, the men employed in their construction having had little if any building experience. One of the men working on the fortifications north of Hull as a joiner with a local building firm describes the situation as follows:

"The labour exchange asked for volunteers to go constructing pill-boxes and tank defences. . . . On to this job all sorts of riff-raff came. Some who thought a drawing was something you see on a lavatory wall. The 'all sorts' included trawlermen, fish-bobbers, labourers and various tradesmen. I never saw anybody of authority other than the foreman. The drawings were from the Defence Ministry and no check was made to my knowledge. . . . My guess would be that 20 per cent (of the labourers on the site) were experienced in building and none in placing reinforcing rods and shuttering. . . . As for concrete mixing, I think we all took that as a matter that any silly bugger knows how to do." (16).

On the sites in this area the administration problems were increased by the presence of trawlermen from the Humber fishing ports. To be tactful, a foreman passed his orders through their captain. This chaotic situation was hardly comparable with that of the Maginot Line sites or indeed the BEF French defences on which the War Office, Directorate of Fortifications and Works, had based their main designs for the coastal defences. Despite this the smaller works, as well as the batteries, were finished by 1941 using standard designs for both the batteries and many of the pill-boxes (17). Hexagonal concrete pill-boxes were constructed

7, 8, 9, 10, 11, 12. Camouflage of British coastal and land defences during July–October 1940. Despite ingenious camouflage many of these defences clearly have the features of the standardized pill box designs.

all along the coast and on a regional basis there is evidence that standardization also occurred. In the South West region, for example, the same square brick observation posts can be found at both Avonmouth and Plymouth. Haste had, however, been a major factor in their construction, and apart from the design modifications caused by shortage of labour and materials, faulty siting proved to be a major cause of failure.

Along the cliffs some of the smaller installations fell into the sea during construction and many fell later. Even batteries, such as the one at Hunstanton, had to be abandoned after a year when the edge of the eroding cliff had arrived dangerously near the foundations of the emplacement (18). On the Holderness coast of Yorkshire, pill-boxes were even built half way down the 'boulder clay' cliffs, regardless of the notorious local erosion rate of two to three metres a year. These were cases of poor siting due to disregard or ignorance of local geology but there were also cases where the basics of tactical siting had been ignored due to the chaotic and overworked state of the administration. A machine gunner stationed in one of the pill-boxes on the north-east coast, previously having been with the BEF in France, recalls that many of the pill-boxes were 'useless' from a machine gunner's point of view. Limited fields of fire tended to face the sea and did not give the 'enfilading' or sweeping fire along the beaches which was desirable for machine guns to be used effectively.

The siting of the installations may have been faulty but their camouflage, where it was applied, showed an element of bizarre genius. The standard designs such as the hexagonal pill-boxes were hardly recognizable by the time the camouflage experts had finished. Here the English approach was quite different to the German and French equivalent. Both on the Maginot Line and the West Wall, as well as on the Atlantic Wall, attempts at camouflage were related primarily with existing topography. Overall form was blended with the landscape but although this method made the installations difficult to locate by air it did not always work so effectively at ground level. Embrasures could not be easily camouflaged in this way and the circles of barbed wire and tank obstacles around the installations made detection even easier. In the case of Germany in particular, however, it must be questioned if any further camouflage was in fact desired. The expression of great strength given by massive concrete installations was probably a highly desirable feature for both national morale and international propaganda purposes. Photographs gave the German citizen a belief in the strength of the Reich and broke down the enemy's confidence. This 'architectural' philosophy can also be seen as an extension of the confused attempts by the National Socialists to give Germany a strong cultural and national identity. Both the French and the Germans could point to their defences in 1938 and say: 'Look . . . our country is invincible'. The English in 1940 could hardly have done the same. Fortifications were disguised as haystacks, ice cream kiosks and seaside roundabouts. Even those emplacements which did look like fortifications were sometimes dummy installations constructed out of cardboard and paint. Except for large casemates and open emplacements, whose position could not have been hidden because of the visibility of gun flashes, it appeared as if the individual camouflage teams had tried to outdo one another in fancy-dress and imagination.

The value of this type of camouflage is debatable. Though brilliant against the initial attack, once recognized for what it was the effectiveness would be reduced considerably (19). Its most important feature, however, was probably that it must have confused any assessment German intelligence made of the actual strength of the defence system and kept them guessing what percentage of British beach kiosks were in actual fact concrete emplacements. The inventiveness shown in this form of camouflage was perhaps an English characteristic and was apparent in other aspects of the defence problem. The tubular scaffolding used by builders found other uses as coastal defence booms and beach obstructions. Coastal roads were lined with perforated pipes which were connected to petrol tanks. But ingenuity of this kind could not solve the major strategic defect—the lack of depth in the defence system (20).

General Ironside, who had become C-in-C of the home forces in June 1940, soon realized that this thin crust could not hold a powerful enemy offensive. Once they had penetrated it, he argued, the Germans would be free to tear the 'guts' out of the country (21), as had happened in Belgium and France. He himself was a believer in fortifications and had seen the contributions of Verdun as valuable in the First World War. He had, however, realized that the main weakness of these fortifications was that they were permanent. The large forts built in both France and Belgium at the turn of the century had been outdated before they were finished due to the rapidity of weapon development. In a lecture given in 1927, he had therefore drawn the following conclusions about future fortifications:

"They are not to be permanent as of old, but quickly built. Rivers have become more important and all such tank obstacles as marshes, forests and the like. Fortresses must also be of an extensive area. It seems to me that the pill-box defence of an area is the way to provide delay against tanks. Such an area will be difficult to locate and reconnoitre from the air and it can be quickly organized; I think there are many areas which in future must be made tank-proof—our bases and our aerodromes. All such areas will become real stumbling blocks to the enemy seeking for great mobility and attacks against flanks and rear. They can hold the enemy up until the field army can be brought to bear." (22).

Following these theories thirteen years later, he started building a form of defence in depth. Its object was to prevent an invader from reaching London and the industrial Midlands. Forward defences and obstacles were built to slow down the enemy after breaking through the coast defences. The home forces, being kept continually short of almost everything they needed, would in this way be given additional time to concentrate for an attack against an enemy who was expected to achieve a high standard of speed and hitting power.

The plan, which was worked out in June 1940, called for a continuous rear line, known as the GHQ Line, behind which the GHQ reserve was to be organized. The line was to be formed by building anti-tank obstacles and pill-boxes, wherever possible following natural obstacles such as steep inclines and waterways. It was to run from Richmond in North Yorkshire to the Wash, and then through Cambridge to the Thames at Canvey Island; south of the Thames it was to continue through Maidstone and Basingstoke to Bristol. In front of this line a series of 'stop lines' would be erected. Five such lines were to run through the Eastern Counties to check an advance, either towards the Midlands or across the open uplands north of London, from the vulnerable beaches at Lowestoft. Three others were to cross Surrey, Kent and Sussex, delaying approach from the south towards London. The troops in front of the GHQ Line would, by exploitation of these forward lines, be able to break up and confine the enemy while mobile troops from the GHQ reserve could organize the counter-attack which was to deliver the 'killing punch'. The battle was therefore to be mobile within the defence zone but, as the enemy was expected to have a higher degree of mobility than the British, the lines of obstacles and pill-boxes on the beaches and on all roads leading in from the coast were to reduce this mobility as much as possible. Accordingly, most of the 786 field guns available in June were sited near the coast to cover the most likely landing places, while the better part of the 167 available anti-tank guns were kept back in the GHQ reserve (23). It is no coincidence that the theory behind this defence seems similar to that of Germany's West Wall. The theories of defence in depth, used by the Germans in 1917, had been published then as 'Translations of Captured German Documents' by the British General Staff (24). Ironside, who was the commander of an infantry division from 1914–18, must have studied these in great detail.

An unfortunate feature of this form of defence was one which had already been apparent in the coastal defences, namely haste. Initial planning had been undertaken in the beginning of June. Already in the third week of June 150 000 civilians, beside troops, were engaged in building these defences. Instead of central control and design, a local group, often without any military training, would be asked to fortify an area. The result was pill-boxes sited in the wrong direction, useless strongpoints and road blocks that an armoured vehicle could drive round without difficulty.

The heaviest criticism was however levelled against the basic strategy. By stretching out the divisional lines of defences, the manning of the stop lines alone would consume most of the divisional strength, leaving none for the reserves. As reinforcements from GHQ reserve would need at least 12 hours to reach them the divisional commanders felt that they could offer only very limited resistance against enemy penetration, if any at all. To add to this dilemma, there were insufficient anti-tank guns to support both the forward lines and the GHQ lines or the armour needed to give the counter-attack a good chance of success. It was thus argued that the defence still depended too much on a coastal 'crust' without providing the necessary reserves to hold it in the places where penetration could be expected. It was maintained that: 'to make no major attempt to halt the enemy until a great part of the country had been overrun was a suicidal policy' (25). When met by this criticism Ironside could, however, give the reply that at that time the trained troops and the material resources needed to make the defence

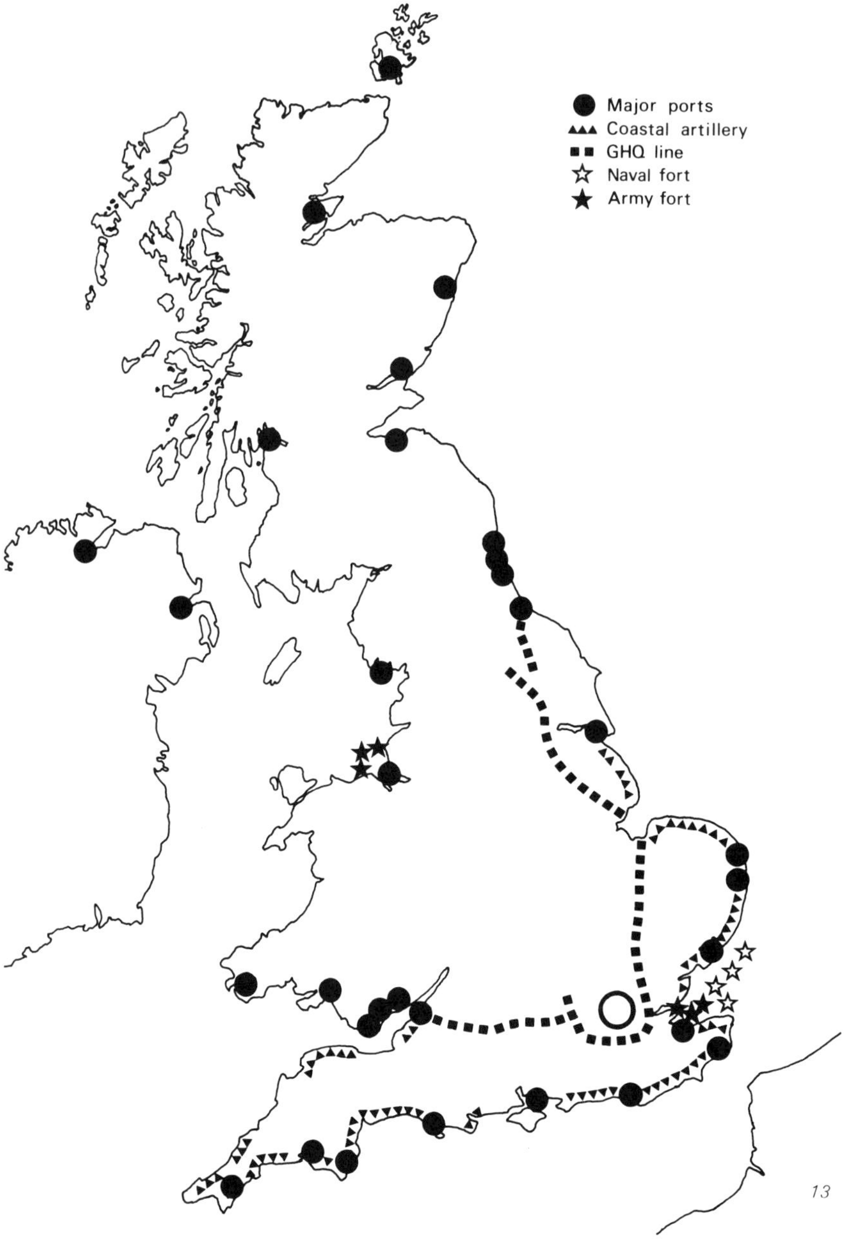

13

14

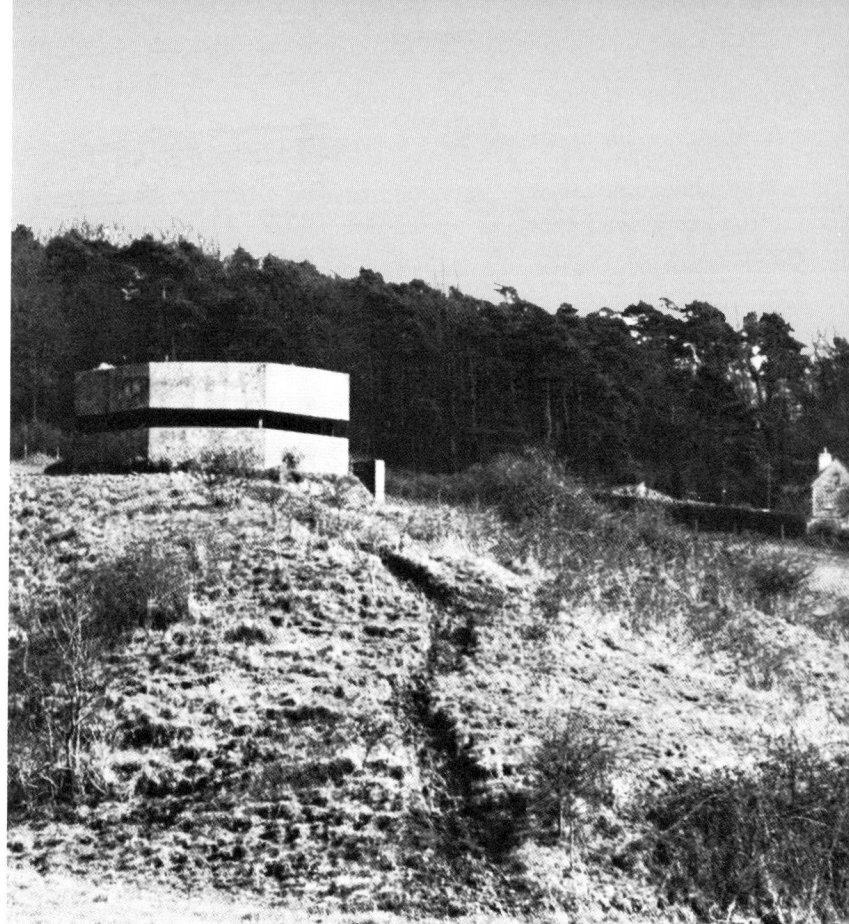

13. *Map of the British defences (the Sea Forts indicated were not built until 1942–3).*

14. *Pill box on the GHQ line between London and Bristol.*

more mobile and offensive simply did not exist.

On July 20th Alanbrooke took over the defence. He immediately set about giving the defence a more mobile character. In early August, he proclaimed his intention of stamping out the idea of Linear Defence and of limiting fortifications to nodal points. Of the road blocks designed to slow down the enemy, he wrote:

"Another form of defence which I found throughout the country and with which I was in total disagreement consisted of massive concrete road blocks at the entry and exit of most towns and many villages. I had suffered too much from these blocks in France not to realise their crippling effect on mobility. Our security must depend on the mobility of our reserves, and we were taking the very best steps to reduce this mobility.... I stopped any further construction and instructed existing ones to be removed where possible." (26).

It should be remembered that by the time Alanbrooke reorganized the defence in August, 1940, most of the 'emergency' coastal batteries and defences had been erected. Even so it was not with confidence that he set out to prepare for a German invasion.

Fortunately none of these defences were ever to be used. Hitler had demanded air superiority before launching an invasion. He did not achieve it and by spring 1941 he had turned his attention east. Thus it might be said that the British air defence had succeeded beyond all expectations, although it must be pointed out that several historians have since cast doubt on the real reasons for Britain's success. Such factors as the lack of German experience in amphibious operations made the 'Seelöwe' invasion plan an unlikely proposition. Equally, a number of high-level studies made by the Luftwaffe in the 30's which have since been published, show that the Luftwaffe would not have been able to bring about the capitulation of Britain either and this could only have been achieved by an actual occupation of the country (27). Add to this Hitler's own indecision over the invasion and the conclusion emerges that Britain was helped by events as much as she helped herself.

Although the threat of invasion was reduced, the need for air and coastal defences was not removed. Instead, defence against raids became their major role. The existing coastal and air defence had two gaps that the Germans had been exploiting, the Thames and Mersey estuaries. Flak ships had so far patrolled these waters, but it was now decided to replace them with more efficient permanent forts which were to be floated out and sunk into position, the 'Sea Forts' (28). The idea had been suggested to the Admiralty quite early in the war by G. A. Maunsell, a civil engineer. Due to the amount of research necessary before the project put forward by Maunsell could be undertaken however, they were not built until 1942–43: at a time when other coastal fortification work was being cut back.

The idea of building fortifications at sea was not new. The Admiralty had in fact already been considering the construction of permanent forts in the estuaries, the so-called 'Mystery Forts', but using traditional dam methods. The plan was dropped following Maunsell's suggestion (29). Similarly the methods suggested by Maunsell were not entirely new. Churchill had made the same suggestion over 20 years earlier. 'He has at least a hundred ideas a day,' Roosevelt once remarked, 'of which four are good' (30). In 1917 Churchill wrote:

"Concrete vessels can perhaps be made to carry a complete heavy gun turret, and these, on the admission of water to their outer chambers, would sit on the sea floor.... Other sinkable structures could be made to contain store-rooms, oil tanks, or living chambers." (31).

Again Churchill's idea was based on a certain amount of precedent. J. L. Lambot had exhibited a boat made in concrete at the 1855 Paris Exposition. Concrete barges and even ships were also constructed on a small experimental scale in both the USA and Britain during 1914–18 to overcome problems caused by the steel shortage (32).

The military objectives of the 'Sea Forts' project are set clearly by John Posford in 'The Civil Engineer in War':

1. to break up enemy aircraft formations approaching a target such as London or Liverpool from the sea;
2. to prevent the laying of mines by enemy aircraft in the navigable channels serving the great ports;
3. to prevent enemy E-boats carrying out raids on shipping and coastal targets in the estuaries;
4. to obtain early warning of the approach of hostile air or sea forces by means of radio location and direct telephonic communications with the shore; and by fulfilling 1, 2, 3, 4, to release for other duties the 'flak' ships which had hitherto patrolled the estuaries, and to provide the missing links in the chain of land defences on either side (33).

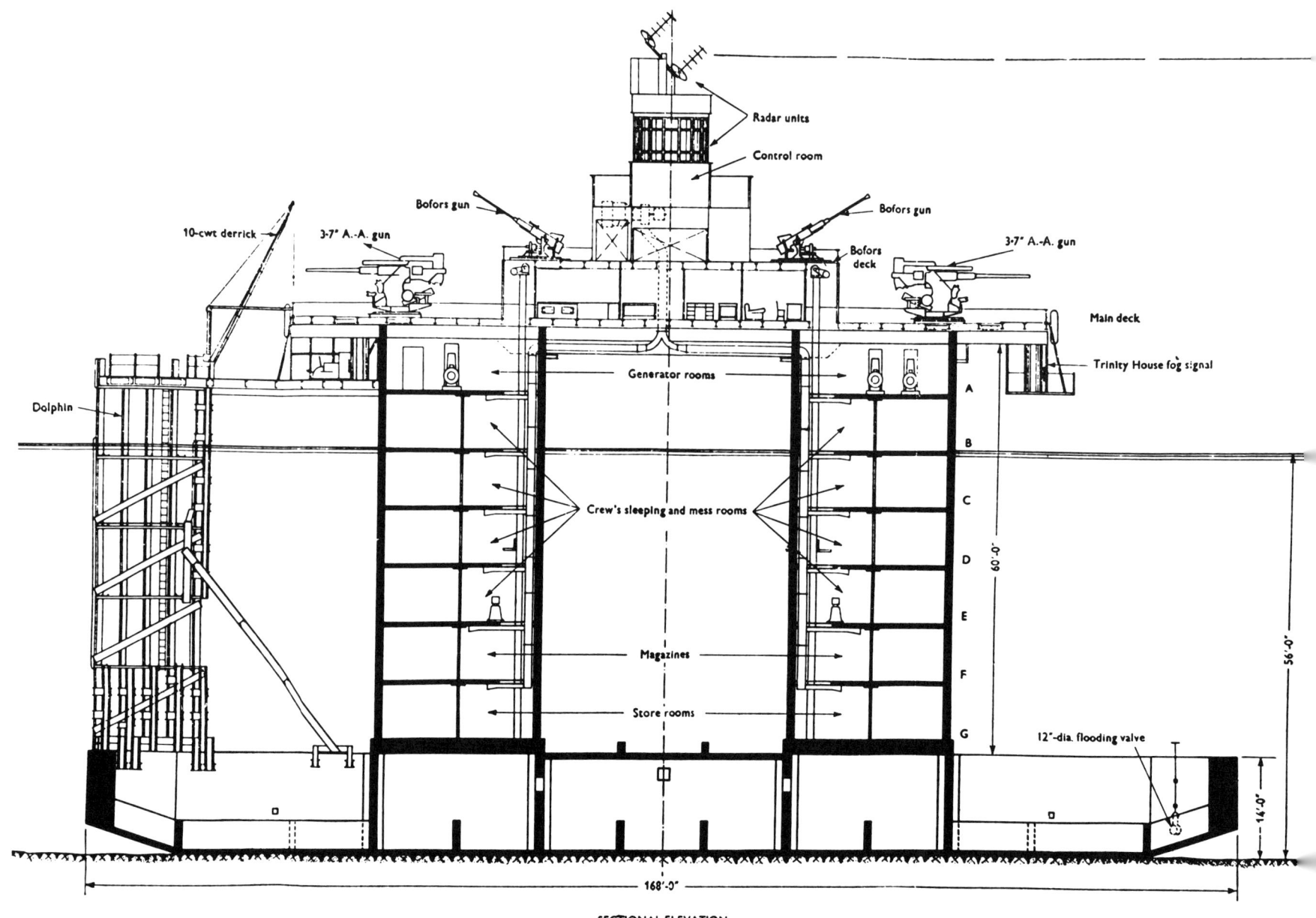

SECTIONAL ELEVATION

NAVAL FORT

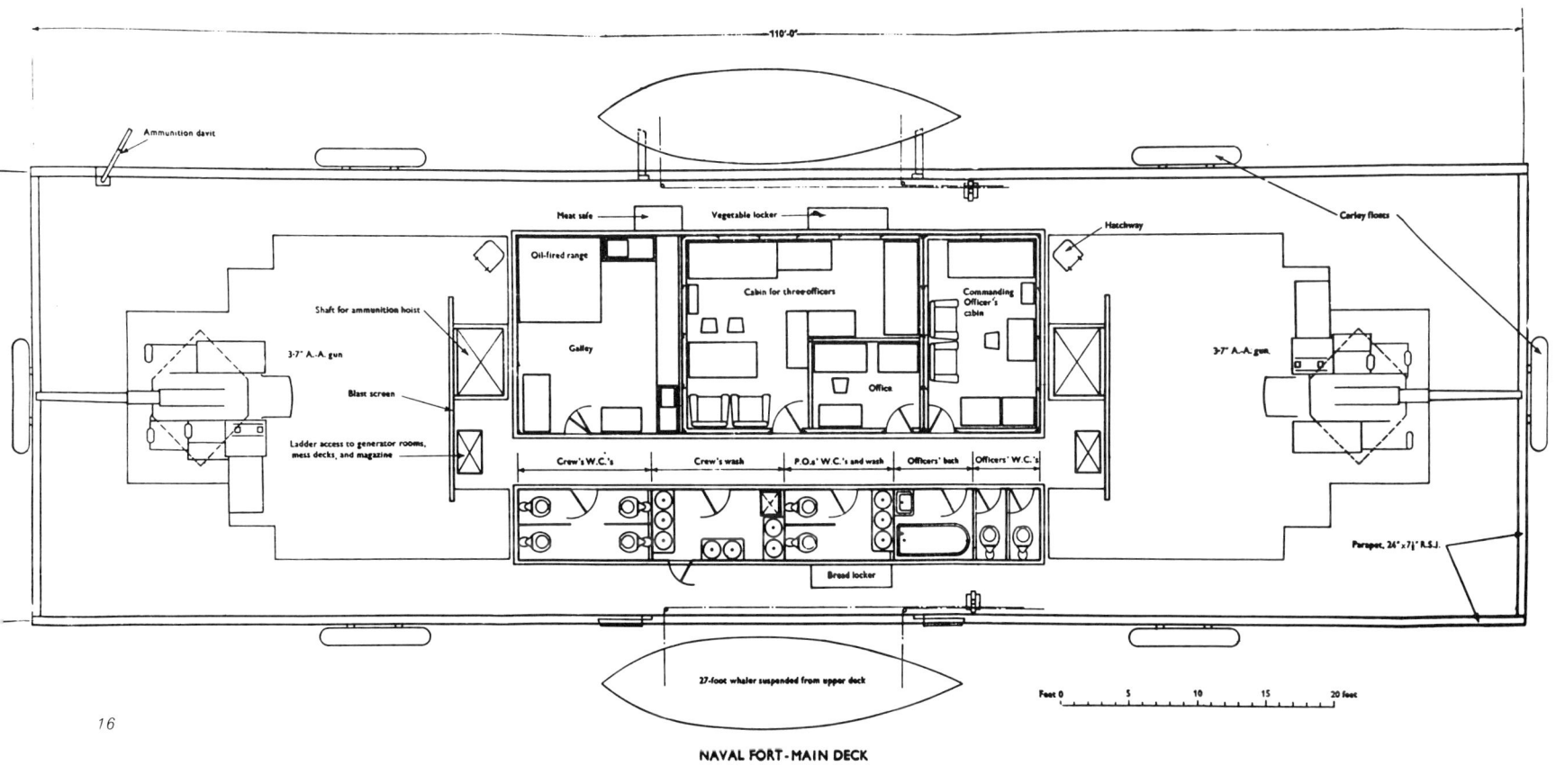

15. Section through a naval fort. The main structure was of reinforced concrete construction while the platform and superstructure were of steel.

16. Plan of a naval fort's main deck.

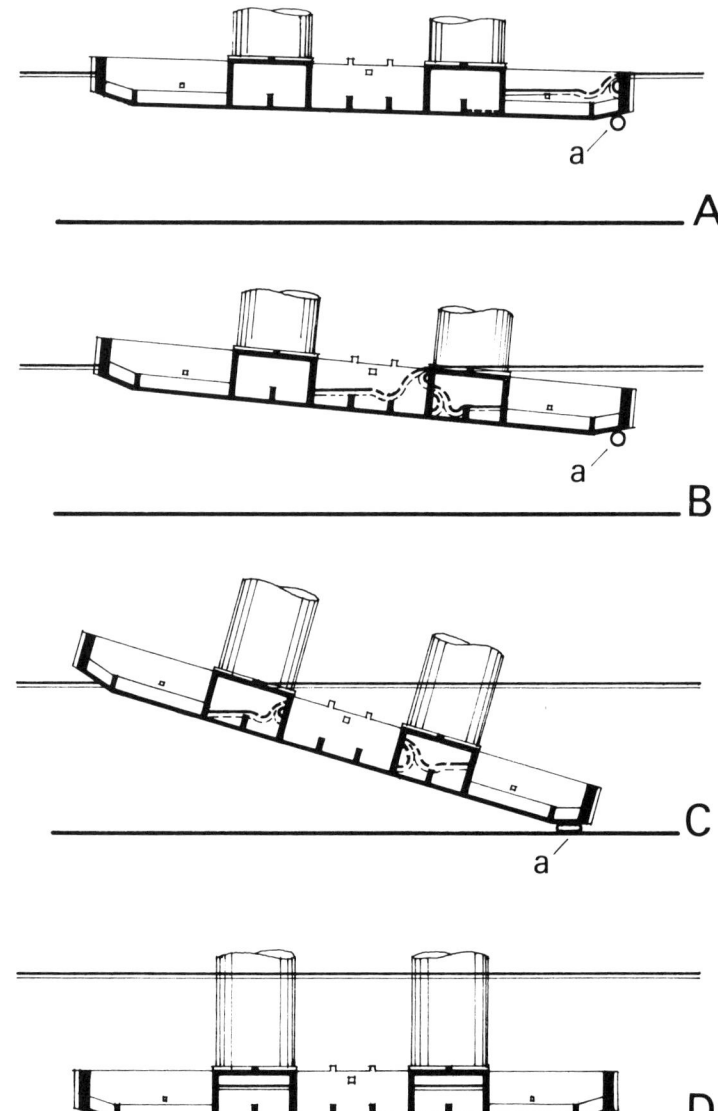

17. Naval fort being floated into position.
18, 19. Naval fort during the sinking operation:
   A. 15 minutes after opening of seacocks the bow dips below surface of sea and water pours into forward end of pontoon.
   B. Bow sinks rapidly and sea cascades into the centre compartment.
   C. Bow strikes sea-bed and the impact stresses on the structure are reduced by the crushing of the reinforced concrete buffer cylinder (a).
   D. Stern sinks and structure settles on sea-bed about one minute after stage A. Buoyancy chambers gradually filled with water.
20. Naval fort in position: buoyancy chambers still being filled.

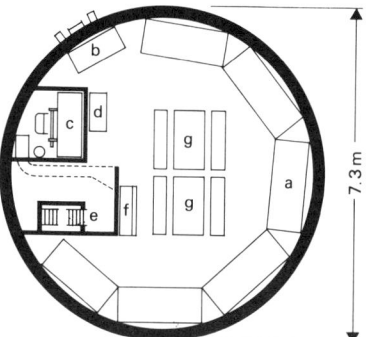

21, 22. Interior and plan of the crew's sleeping quarters and mess on a Naval fort. Some amenities such as showers were in the steel superstructure together with the officers' quarters.
    a. two-tier bunks
    b. suitcase rack
    c. ammunition hoist
    d. respirator locker
    e. access ladders
    f. mess rack
    g. mess tables and stools.

It can be seen that the forts were expected to meet a mixture of both army and naval functions. Furthermore, it is also interesting to note that though the outer artillery zone was abandoned in 1936, point one in fact fulfils its original objective. But Captain J. Hughes-Hallet points out:

"The true value of the Naval forts, however, was that they extended the radar cover over the Thames Estuary. That was what really mattered, and was much the most important part they played. They provided accurate radar cover on the surface, and at low heights, of a type not obtainable at that time with sets mounted on shore. The particular form of enemy action which was giving most trouble at that time was mine laying. If one had, under modern war conditions, radar cover over the area in which the aeroplanes which were laying the mines were operating, that was half the battle, because it was possible to plot where the mines were dropped and to know with comparative accuracy what parts of the sea had to be avoided. That was the duty of the forts to which the greatest importance was attached." (34).

Two types of forts were to be used: army forts, manned by AA Command and situated 6 to 9 km from the shore; and naval forts, manned by Marines and Blue Jackets, situated up to 15 km from shore. Both forts were built by the Admiralty and designed mainly by Maunsell who, after his initial proposals, had been entrusted with the project in 1941 by the Admiralty Civil Engineer-in-Chief due to his experience of similar construction methods on such projects as the Storstrom Bridge in Denmark (35).

Previously such a sea structure would have been constructed on site, using temporary caissons or dams, a programme which would have taken several months or even years. This method of construction was now impossible due to the danger of sea and air attack in the areas where the installations were needed. The forts were therefore to be built on shore in relative safety, then quickly brought into position by floating, and finally sunk into position on the sea bed. The naval forts built in the Thames estuary were the first to be built in this way and, although floating concrete structures had been built previously and concrete ships by this time were not a new idea, the design marked an important development in the use of concrete for sea structures. Mulberry Harbour and, more recently, floating airports and estuary barrages; all came a step nearer feasibility through their design.

Basically the naval forts had a simple structure. Two concrete towers, based on a concrete caisson, carried the steel superstructure with two gun decks and central control tower. The concrete caisson was shaped like an open barrage, and equipped with seacocks and a concrete buffer at one end. After opening the seacock, this end would sink until water could flow over the side. To reduce the speed of descent to a 'controllable' level, the rectangular box sections running across the barge under each tower were covered, allowing water to enter only through small holes in their side. Sinking one end first, the concrete buffer, crushing under impact with the sea bottom, would reduce the stress on the rest of the structure, after which the whole caisson would settle down on the sea bed. Brilliant in concept, the naval forts deserve detailed description.

Each of the two circular concrete towers had seven storeys, mostly under water, the lower levels of both towers containing stores—fuel in one and fresh water in the other. The next level contained food, other general stores and munitions. Above these decks came the crews' quarters, four levels in all. Each circular room, 6.7 m in diameter, minus room for access ladder and munition hoist, contained two-tier bunks, mess-tables and stools as well as storage for the personal belongings of 12 men, approximately 3 square meters per man. Except for the open decks and the washing and toilet facilities in the superstructure, these were the only quarters provided for the men. To humanize these somewhat cramped quarters, the floors were covered in linoleum and the timber lining which had been left on the walls after casting, as permanent shuttering, was painted. Care was taken in the choice of colours in order to provide a 'cheerful' environment. On the last level above the quarters two 30-kv generators were installed in each tower for power supply, complete with switchboards and workbench. Above the concrete towers the steel superstructure had three covered floors, the lower with one 3.7-inch AA gun in either end, and officer's quarters as well as an oil-fired galley, and showers for the men. This accommodation was enclosed by the steel walls which supported the deck above, carrying the two Bofors guns, and the ventilation equipment providing forced ventilation, temperature and condensation control throughout the lower decks. On top came the Control Room deck, similar to the bridge of a ship, with Lewis guns and radio equipment, and the radar on the roof. While the overall environment was

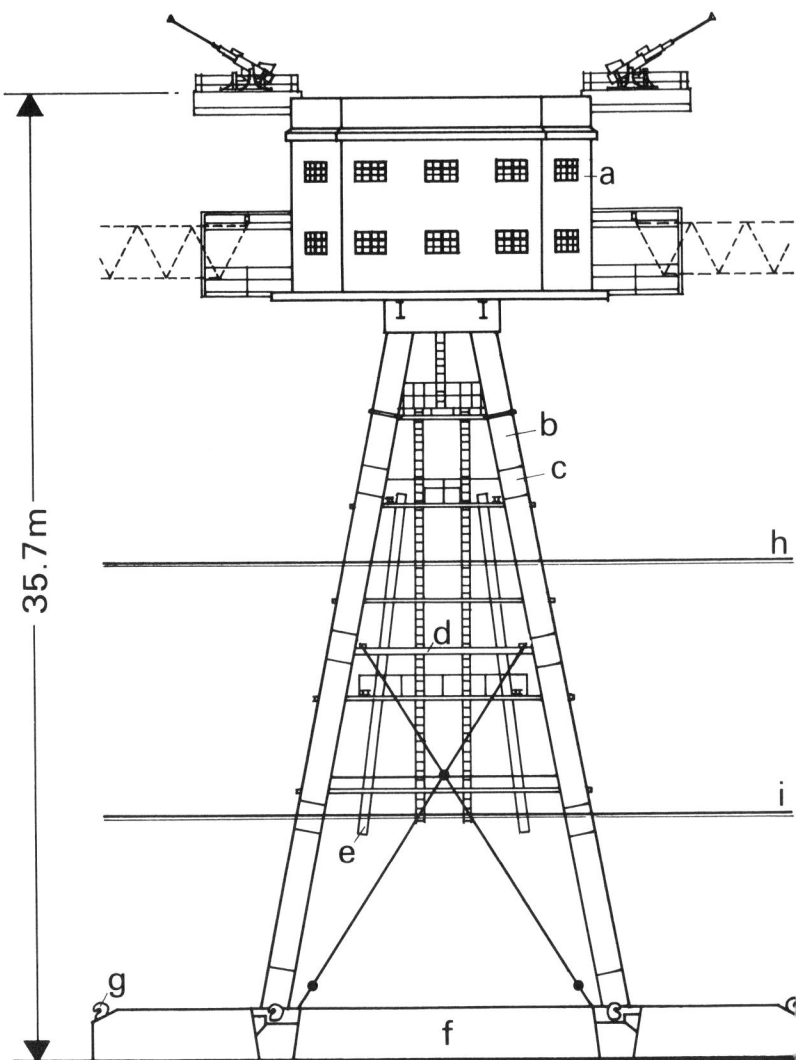

23. The army fort units during the final stage of construction.
24. Elevation of one of the towers (a Bofors tower):
    a. steel house
    b. pre-cast hollow reinforced concrete leg section
    c. in-situ reinforced concrete leg joint
    d. steel bracing frame
    e. fenders
    f. hollow reinforced concrete base
    g. lifting hooks
    h. high water
    i. low water

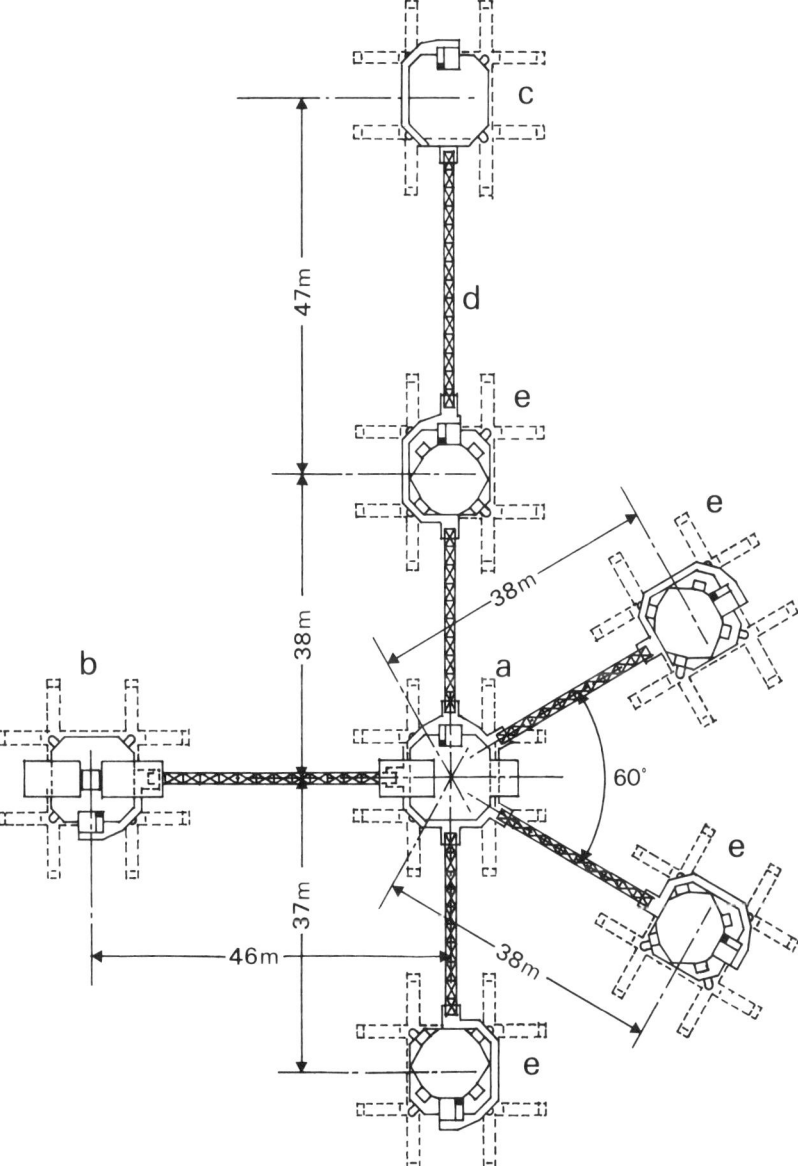

5. The towers during positioning
6. Close up of an army sea fort in position.
7. Plan of a complete army fort:
   a. control tower
   b. bofors-gun tower
   c. searchlight tower
   d. tubular steel bridges
   e. 3.7-inch gun towers.

therefore very similar to those found in the fleet, and almost completely dependent upon artificial lighting, the level of comfort for the men was probably slightly higher.

In order to cut down cost as much as possible, prefabrication and a 'total design' approach was used here as in many other wartime projects. The construction was thus looked upon as a series of fluent stages: the construction of pontoons, the construction of tower and superstructure, the fitting out, the floating out and finally the sinking.

Starting in a dry dock, the flat-bottomed pontoon was constructed. With a central section carrying the load of the towers, the overall shape was streamlined to prevent the scouring from tidal flow. From the dry dock the pontoon was floated to shallow water, where it was allowed to settle on the beach when the tide withdrew, and where the caisson was later flooded to prevent it floating with next tide. Here the two towers were built simultaneously, both 7.3 m in external diameter and 18.3 m high. Pre-assembled sections of reinforcement were lowered into position, between steel outer shuttering and timber inner shuttering, in 2.4 m sections. The timber was left as permanent shuttering acting as insulation and lining. For each section of casting a precast floor unit was lowered into position and cast in with the next concrete operation. These floor slabs were cast in stack assemblies, using building paper as separation. Finally the concrete was waterproofed with bitumen, reinforced with canvas and camouflaged with paint. The superstructure, prefabricated in sections, was lowered directly on to the towers and the main cross girders concreted into position. When the structure was finished, water was pumped out of the pontoons at low tide, and the 'fort' was floated to deep water where it was fully fitted out and the timber and lightweight steel Dolphin constructed for mooring re-victualling vessels. The fort, complete with crew, was then towed by tugs to a previously marked site, its 'commission', where the sinking took only one minute for even the deepest sites (12.8 m of water). Finally an underwater cable was laid, connecting the fort with land.

The detailed construction programme, with its assembly line, timed operations, re-usable shuttering, and prefabrication, is interesting as it shows some of the practical basis for some areas of post-war civil construction. When rehousing of bombed families became one of society's main aims in 1945, these same techniques were attempted as both cost and time were still the main objectives (36). The stubborn belief that this must still be the right approach today can thus be seen partly as a result of wartime success.

The army forts were also built by the Admiralty, but here changes in design criteria altered their shape. They were to be built initially in the Liverpool Bay where continuously shifting sand rendered the approach previously adopted useless. Instead of sitting on top of the sea bed the foundations had to bury themselves below its upper moving surface. The self supporting (i.e. buoyant) pontoon type of foundation was found to be useless in this new situation and instead a foundation based on four intersecting hollow concrete beams was used. This base could be constructed in dry dock, and floated to a position where further construction was to take place, but it would not keep the whole structure floating. Instead a pair of concrete barges were tied to hooks in either end of one pair of beams at low tide. When the tide rose, the barges lifted the whole structure off the construction site ready for floating to its destination. From this foundation four hollow concrete legs rose from each beam intersection to take the superstructure.

Whereas most of the accommodation in the naval forts was integrated into their main structure, the army forts had all their accommodation in a box-like superstructure, with a plan shape which was basically square with its corners truncated at an angle of 45 degrees. Three storeys high, the top storey was open. The army had always emphasized that their forts should conform as nearly as possible to the AA batteries on the coast; their men had already been trained in the use of this equipment. When the sea bed necessitated a different foundation approach, and thus a different solution altogether, it was therefore reasonable that the forts should be completely redesigned. Instead of using one self-sufficient unit it was subsequently decided to connect several units to form one self-sufficient assembly, each unit being specialized in one task, i.e. searchlight, Bofors 3.7-inch AA guns and control towers, forming a similar layout to the land emplacements. A complete fort thus divided accommodation for 120 men and officers between seven units.

The box-like shape and the greater area available gave more scope for a comfortable planning solution than in the naval forts and, as all the accommodation was above water, it could to a larger extent depend on natural ventilation and lighting. The main quarters were on the intermediate floor directly below the upper open equipment deck. Here, sleeping and messing were separated in different rooms in accommodation similar to a barracks. Below this level the officers' quarters, kitchen, toilets, bath, showers and stores were placed. The proportion of different accommodation at each level varied according to the function of the individual units. Common to all, however, was the 3 metre steel box running 6 metres vertically through the units transferring the floor load onto the substructure, the central heating plant on the lower floor (37), and the much larger circulation area than that found on the naval forts. Horizontal communication between the units took place at the lowest level by open steel bridges and increased the extent of circulation even further (38).

The construction procedure was similar to that of the naval forts. After having towed the base to the pedestal construction site, the legs, each precast in three sections, were hoisted into position and joined by insitu concrete with steel bracing. These legs were then topped with a 4.3 m square cap of (insitu) concrete, 1.2 m, with a central hole, 1.8 m in diameter, for access. Into this cap steel joists, 13.1 m long, were cast to form the base for the steel superstructure, which was lowered into position in prefabricated sections and bolted to this base. On completion, and after having been floated to position between the two barges, they were gently winched to the sea bed. To avoid stresses arising in the hollow base beams due to the difference in external and internal pressure, compressed air was pumped into them as they descended. This solution was used rather than allowing water to enter as this would have increased the load by 300 tons on the lowering tackle. With the light foundations wind resistance had to be countered and the 45 degree corners and light open bridges were a result of this.

It is interesting to note that although large numbers of private contractors such as Messrs. Holloway Brothers, the Cleveland Bridge and Engineering Company, Messrs. Crittall, Messrs. Dorman Long and Company, Redpath Brown and Company, Consett Iron Company, the Limmer and Trinidad Asphalt Company Ltd, Messrs. H. J. Cash and Company Ltd, Norris Warming Company Ltd, and Matthew Hall and Company (39) were all used on the Sea Forts project, confusion such as had occurred in other fortifications, was here avoided by the emphasis on research and central control. This could well have been a valuable lesson for the building of later fortifications but, although the army forts originally intended for the Mersey estuary were later also erected in the Thames estuary to supplement the existing naval forts, by this time the emphasis was no longer on defence. By 1943 the roles of Germany and Britain had been reversed. The Allies were now the potential invaders. The vast build-up for the final Allied offensive was under way and accommodation for the increasing flow of American manpower and machines became the primary objective for the construction industry. Britain was becoming an 'armed camp'.

Events

1938 Dec. 9 Fritz Todt is made General Bevollmächtigter für die Regelung der Bauwirtschaft (Plenipotentiary General for the Regulation of the Construction Industry) by Goering. Plans for making OT an army auxiliary body studied by Todt.

1939 Sept. 1 Invasion of Poland and OT put on war footing as a Wehrmachtsgefolge (Army Auxiliary Body). Bautrupps (Construction Detachments) and the conscription of firms.

1940 Feb. Todt made Generalinspektor für Sonderaufgaben des Vierjahresplanes (Inspector General for Special Tasks of the Four Year Plan) and later Reichsminister für Bewaffnung und Munition. Invasion of Scandinavia, Low Countries and France. Cap Gris Nez batteries started to back invasion of England.

Oct. 'Sea Lion' abandoned, fortifications for the U-boat bases begun.

1941 Todt becomes Generalinspektor für Wasser und Energie, and later Reichsleiter.

June 21 Offensive in Russia starts.

July 19 Directive No. 33, Channel Islands and Norway.

Oct. 20 OKW order for defence of Channel Islands.

Dec. 8 US enters war.

Dec. 14 Orders to fortify coast to West Wall standard.

1942 Feb. 8 Fritz Todt dies: Albert Speer takes over all Todt's responsibilities.

March 'Führerweisung 40'.

May Atlantic Wall work begins.

1944 Jan. 'Rommel Bar' begins.

April Hitler includes Normandy on list of likely spots for an Allied invasion.

JUNE 6 D-DAY

● The West Wall was not the only major defensive construction undertaken by Hitler but, while the West Wall and the Maginot Line had been built primarily in one operation and as a unified concept during peace time, the Atlantic Wall construction and design went through several modifications during the course of the war. Generally four such major phases can be identified: the offensive stage following Britain's rejection of Hitler's peace offer on July 22nd, 1940; the initial defensive stage concerned with the areas considered strategically and politically important following the start of the Russian campaign June 22nd, 1941; the attempt at defence of the whole coast following America's entry into the war on December 8th, 1941; and finally the desperate attempts of Rommel along the invasion coast following OKW's fear of an immediate invasion, when the Red Army was advancing rapidly and Italy had declared war on Germany, in late 1943. Throughout this time the army, navy, air force and OT never managed to integrate their efforts. Jealousy and inter-force conflict, followed by corruption and favouritism, became the backdrop for this amazing building project.

Following the British refusal of the German peace offer OKW (Oberkommando der Wehrmacht—High Command of the Armed Forces), with the aid of the army and navy, set about planning operation 'Sea Lion'—the invasion of England. Though planned as a gigantic river crossing, it was realized that the British sea superiority would be a great handicap. The supplies following the initial invasion force could be particularly affected. To overcome this, giant coastal batteries were planned at Cap Gris Nez. These were to deliver the initial bombardment of the landing beaches and keep the British navy at bay.

The initial plan, however, was short lived. In October 'Sea Lion' was postponed for an indefinite period, the Luftwaffe having failed to gain air superiority during the Battle of Britain, and Hitler now planned an eastward expansion. Instead of an invading army the navy, under Grand Admiral Raeder, was to bring Britain to her knees by unrestricted 'U'-boat warfare. Fortifications were erected at the submarine bases and the large batteries (their original objectives no longer existing) were kept in order to prevent convoy traffic through the Channel. Thus the first fortifications, the large batteries—Lindemann with three 16-inch guns, Todt with four guns, Grosser Kurfürst with four 11-inch guns and the Railway battery between Cap Gris Nez and Boulogne, as well as

---

◀ *1. One of the massive casemates of the Lindemann Battery.*

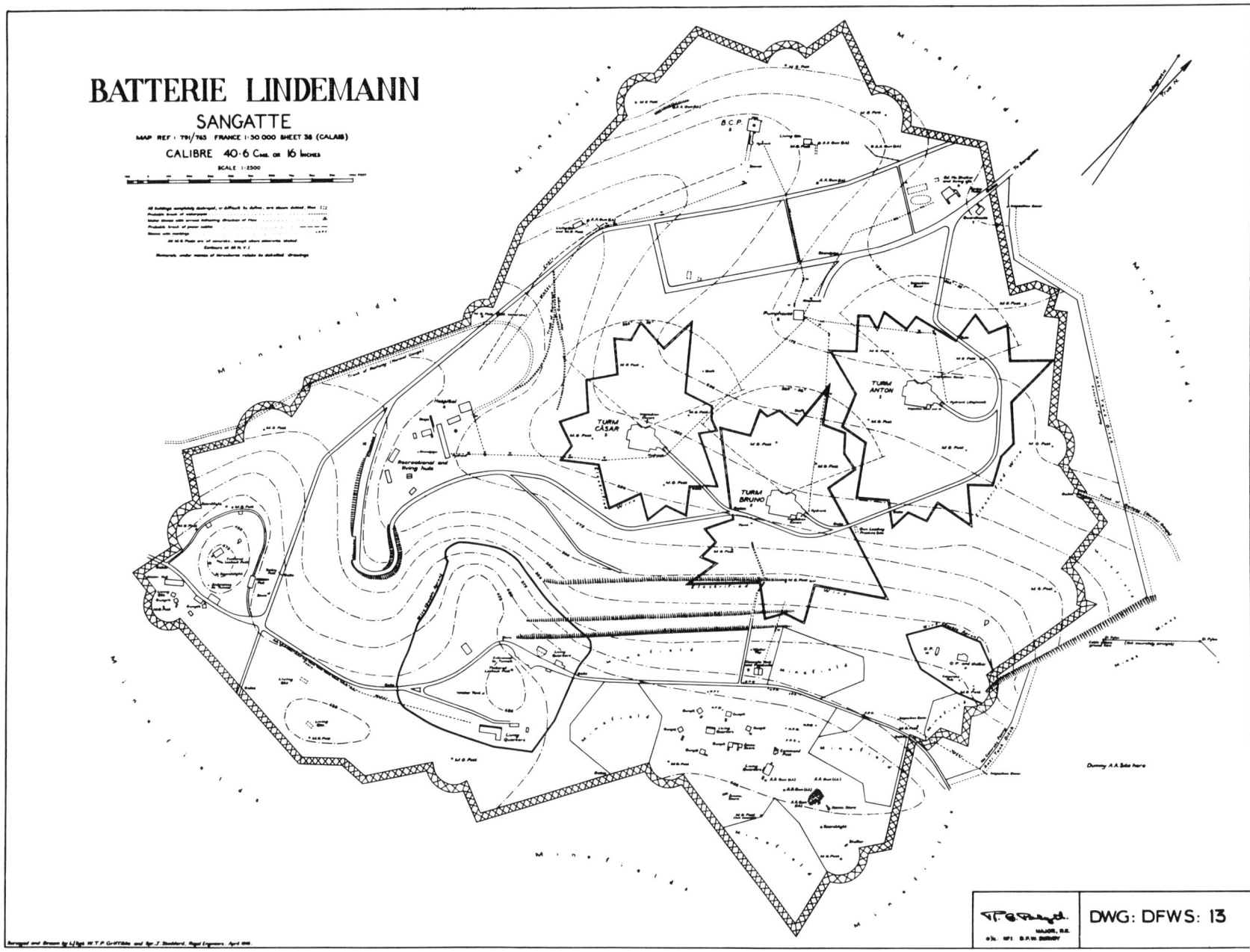

2. Layout of the Lindemann Battery. Later batteries and strongpoints on the Atlantic Wall were to be on a similar pattern: the closely grouped casemates, observation and fire control post, and varying numbers of support shelters, emplacements, minefields, wire and anti-tank ditches.

the numerous smaller flak and coastal artillery installations at the U-boat bases, were the product of offensive theory rather than defensive. Initially, therefore, the development of the Atlantic Wall started with Operation Sea Lion and the U-boat campaign (1).

Following the introduction of rifled artillery in the nineteenth century, coastal gun emplacements had been generally constructed on an open principle, 'en barbette', but by 1940 the threat of aerial attack had prompted a return to the construction of more complex and massive casemates. When the expansion of the Reich started, therefore, much of the responsibility for the construction of the large coastal installations was gradually shifted from the navy and air force to the Organization Todt, a government agency under ministerial control, which co-ordinated the technical abilities of the German building industry with government resources and requirements.

In September 1939 the OT had been put on a war footing and officially declared to be 'Wehrmachtsgefolge', an army auxiliary body, remaining ultimately under ministerial as opposed to military control. It was in theory to take over the construction tasks of the forces, and, while contractors had previously joined voluntarily they were now conscripted, forming the OT nucleus of mobile units. Forced manpower was attached to these units from Poland and later from other countries. By 1945 the OT, with a labour force of over one million, had become the sole agency responsible for army, air force, and navy construction—and the entire war production programme in the Reich, in as far as it was affected by Allied air raids. But in 1940–41 the scale of its work was more limited and was confined to the occupied territories. Its work on the coastal installations in the West suffered from several weaknesses, including poor central administration. Fritz Todt, though still in command of the OT, had outgrown participation in individual construction projects. Instead an autonomous group, 'Building Industry Economy Group', took over the day to day management. They issued their orders from headquarters, directly to each OT firm, and strict control over these firms soon became impossible. In addition, the OT's operational sphere in the West was limited to army projects. Although the navy and air force took advantage of the OT's resources when convenient, they bypassed the OT, subcontracting directly to local contractors or privately to individual OT firms. Amid this situation the OT firms in the West, probably not enthusiastic about conscription anyway, started to hire out their equipment to local contractors and carry imaginary staff on their payrolls. Instead of becoming an integral part of the forces they developed as a separate body—a tendency which was reinforced by Hitler's decision to fortify the Channel Islands. The OT were to execute this work under direct orders from Hitler himself.

In June 1941, the German attack on Russia began and on July 19th the first orders on defensive strategy were issued by OKW in Directive 33:

"... in the West and the North the possibility of British attacks on the Channel Islands and the Norwegian coast must be borne in mind." (2). This directive was followed by an order issued on October 20th, 1941, giving detailed instruction for the defence of the Channel Islands:

"... large scale English operations against the Western occupied areas remain unlikely ... (nevertheless) account must be taken of the possibility that the English may at any time carry out isolated attacks as the result of pressure from their Eastern allies and for political reasons; in particular they may attempt to recapture the Channel Islands which are of considerable importance for our escort traffic." (3).

In the last sentence 'political and military status' could well have been substituted for 'escort traffic'. Hitler, on whose inspiration the order had been issued, had all intentions of maintaining the islands, the only occupied part of the British Isles, under German control.

This order issued by OKW went into immense detail, specifying numbers and types of installations and thicknesses of concrete. It subsequently by-passed both OKH (Army High Command), OKM (Navy High Command) and OKL (Air Force High Command), who would normally have been responsible for such decisions. Through OKW Hitler placed himself in direct control of the building project and to keep himself up to date he kept large-scale maps, on which monthly reports were marked, in his own desk. Although Hitler was also to pay detailed interest later in the Atlantic Wall fortifications, this was the only time the army, navy and air force were completely by-passed. This unique situation was probably the reason for the unusual developments in observation post design which occurred.

While observation posts on the mainland were basically of rectangular shape, and fused into the landscape by low skylines and turf covering, those on the islands were circular in plan and four to five storeys high. The forerunner for their design was probably the Martello tower which was used on Corsica in 1793 (4). During the Napoleonic war several of

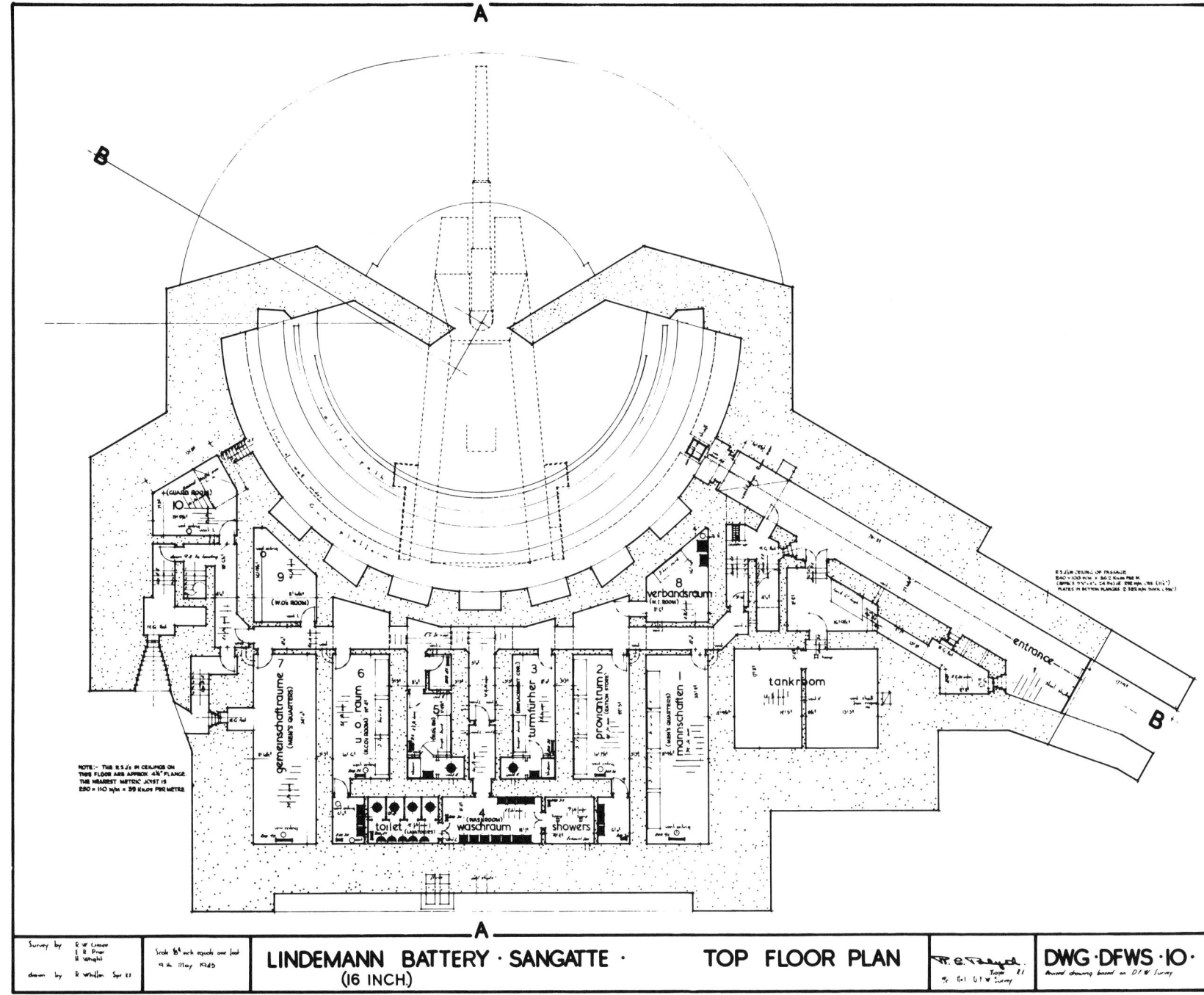

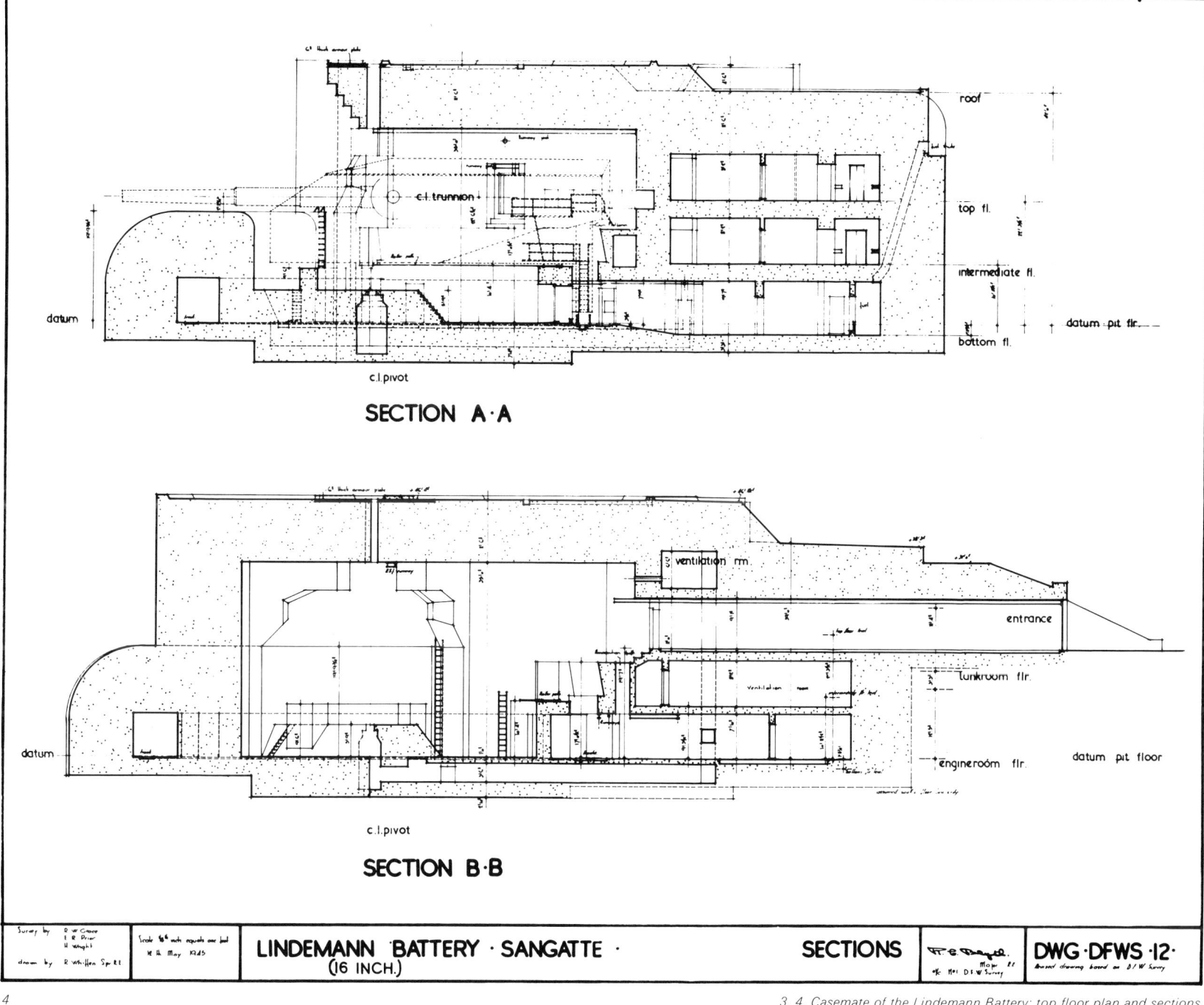

3, 4. Casemate of the Lindemann Battery: top floor plan and sections.

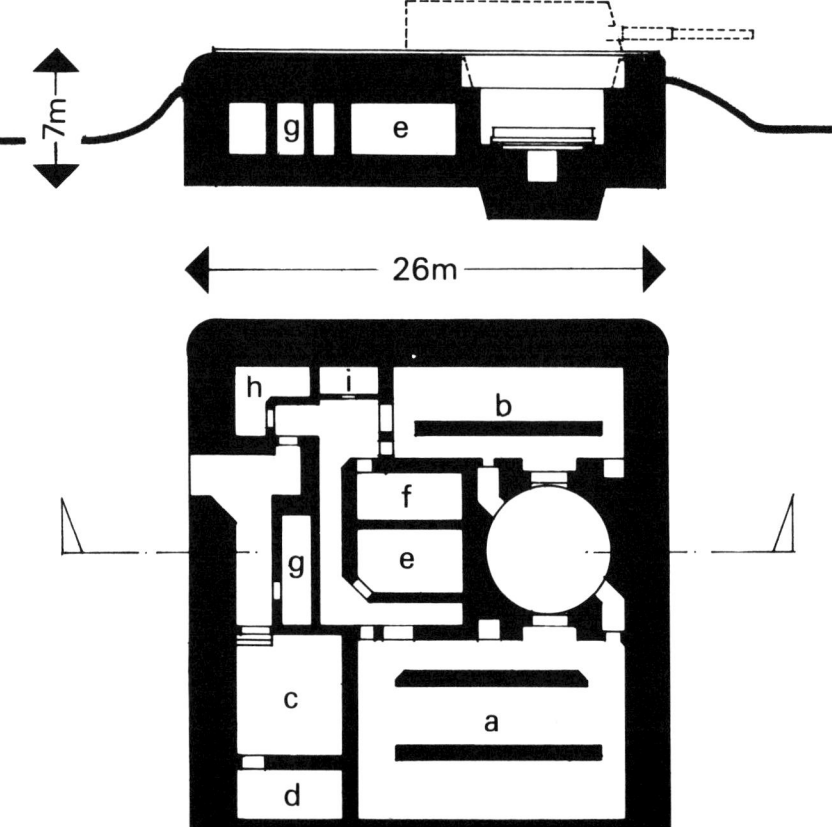

5. The Todt Battery.

6. Plan and section of the No. 2 Gun (11-inch) at Grosser Kurfürst Battery, Framzelle:
   a. cartridge stores
   b. shell store
   c. engine room
   d. water tanks
   e. ventilation room
   f. crew's quarters
   g. oil and paint store
   h. heater room
   i. toilets.

7, 8. Railway Battery: one of the shelters for railway mounted guns in the Cap Gris Nez area.

these were constructed on the Channel Islands (5). Up to 15 metres high and nearly 12 metres in diameter, in brick or masonry, they provided a strongpoint for troops as well as an open coastal artillery emplacement on the roof. The shape offered a unique chance for the Germans. By copying the shape in concrete and painting with a masonry-like pattern, high observation posts could be built and effectively camouflaged. The main protection, however, lay in their heavy monolithic structure, which was unbroken except for the narrow slits which gave an uninterrupted view of the sea and coast. Apart from the heavy concrete the inhabitants were undefended except for hand weapons and although the roof of a tower was sometimes fitted with a flak gun, a radar set was more usual. Quarters were directly attached to the towers and often covered by turf. The entrance was flanked by a rifle embrasure. Some of the gun casemates, like the observation posts, differed from those found on the mainland for similar reasons. The gun platforms at Jersey were typical examples. The gigantic underground 'Mirus' battery on Guernsey and the underground hospital on Jersey were also fairly unique to this part of the Atlantic defences. Dipl. Ing. Xaver Dorsch, later head of OT Zentrale in Berlin under Albert Speer, was to take great interest in such constructions (6).

With the Channel Islands the practice of obeying special 'Führer Directives' (7) within OT was begun and OT's independence of the armed forces developed, a feature Rommel was later to complain bitterly about. In addition the OT's strength and independence were reinforced through the promotion of Fritz Todt. In October, 1939, he was made Major General in the Luftwaffe, and in February 1940, he was made General Inspector for the Four-Year plan. Unfortunately both positions were under Goering who, being Commissioner of the Four-Year plan, tended to meddle in all building projects due to the rivalry within the party, and Todt, his subordinate, had to obey 'orders'.

To sum up, at the end of 1941 heavy German batteries stood opposite Dover; all U-boat bases were equipped by light artillery and flak; large fortification systems had been started on the Channel Islands and at Narvik in Norway; and some assault obstacles had been built along the coast due to British commando operations. To the two formerly independent forces responsible for coastal fortifications, the navy and air force, a third had been added, namely the OT. No detailed study of overall defence policy for the newly acquired territories had yet been undertaken. In 1942, to add to the confusion, a fourth partner was going to be involved in the building project—the army.

On December 8th, 1941, America entered the war. One week later the first general directive for defence of the Atlantic coast from North Cape to Spain was issued. This directive recommended fortifications of West Wall standard to be started immediately along the whole coast and gave the priority of locations. Due to the length of coast it was decided that the army, air force and navy should divide the responsibility between themselves. The order was however only general. 'West Wall standard' referred primarily to the thickness of concrete to be used which was 3.5 or 2 metres instead of the field standards, 1.5 metres of concrete and 60 cm of concrete or earth covered timber used so far. It also referred to the general strategy, such as the use of central reserves and a mobile defence, but to re-use individual installation designs from the West Wall would have been impossible. Not only were there too many such designs used in the West Wall, but the large number of new artillery calibres, both of German construction and captured ones, had made the designs of 1938 obsolete. The main military importance of the order was the priority given to different sections of coast. Norway and the ports between the Gironde and Brest were given top priority, the Channel Islands and the ports between the Seine and Schelde came second, while the open coast in Normandy and Bretagne—ultimately the invasion coast—came last.

A more detailed directive was not issued until March 23rd, 1942, 'Directive 40'. Fred Majdalany, author of 'Fortress Europe', describes it:

"Behind the formal military language—yet in terms comprehensible to any layman—it enshrined a breath-taking concept. It was a design for the defence of almost the entire continental landmass of German-held Europe." (8).

The most important sections of this remarkable document were as follows (Hitler's italics):

"Directive No. 40.

Ref. Competence of Commanders in Coastal areas.

I. General Considerations.

The coastline of Europe will, in the coming months, be exposed to the danger of an enemy landing force.

The time and place of the landing operations will not be dictated to the enemy by operational considerations alone. Failure in other theatres

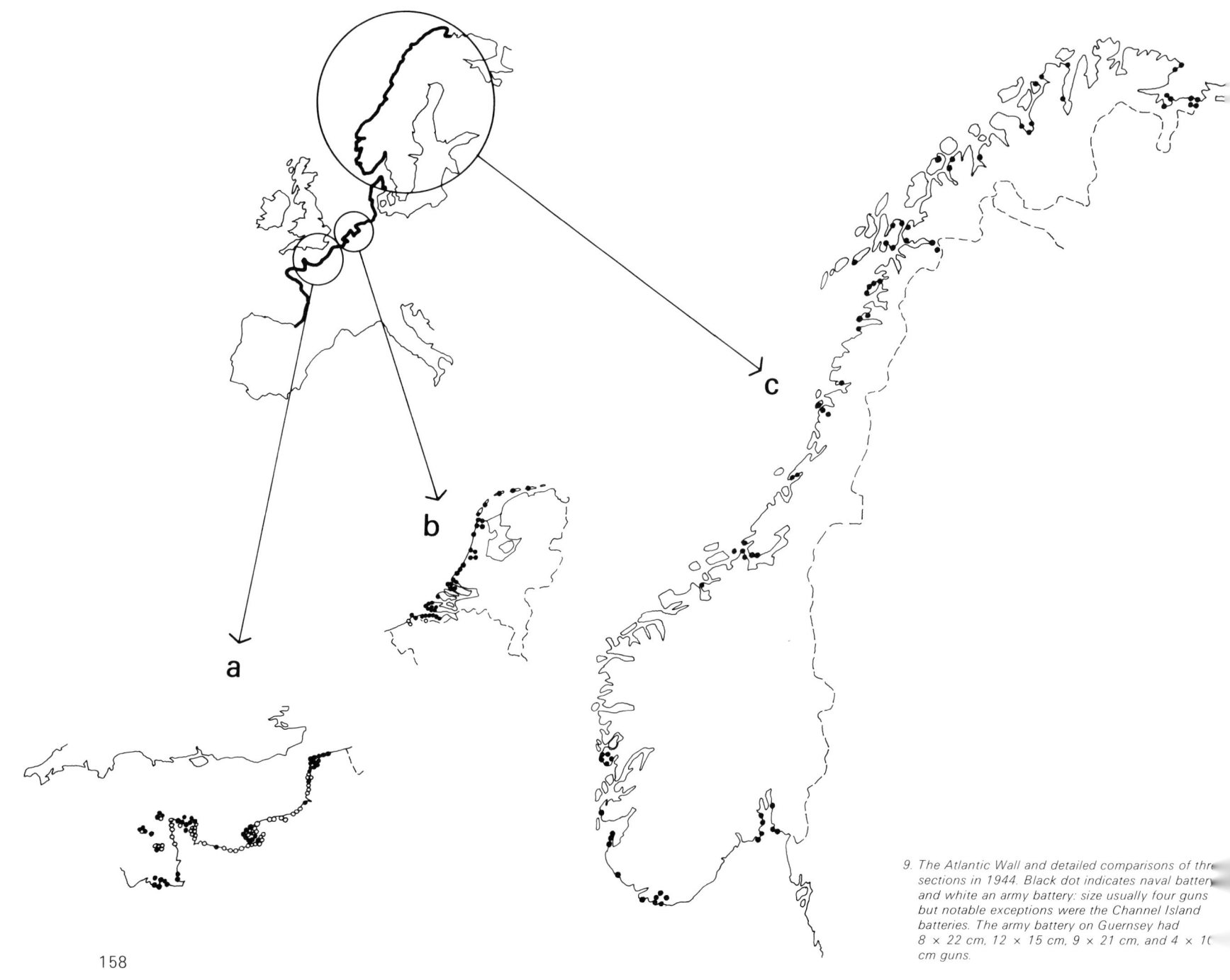

9. The Atlantic Wall and detailed comparisons of three sections in 1944. Black dot indicates naval battery and white an army battery: size usually four guns but notable exceptions were the Channel Island batteries. The army battery on Guernsey had 8 × 22 cm, 12 × 15 cm, 9 × 21 cm, and 4 × 10 cm guns.

10, 11. Recent photographs of a coastal observation and control tower at Les Landes, Jersey: remains of the radar equipment still visible.

12. Recent photograph of the strengthened Martello tower of Fort Saumarez strongpoint Guernsey: typical of the skilful German adaptation of existing fortifications.

13. Observation tower at La Corbière, Jersey: camouflaged with paint to resemble the masonry of the Martello towers.

14, 15. Observation and fire control tower at Noirmont Point, Jersey, as it remains today: this tower was equipped with a gun platform instead of the more usual radar.

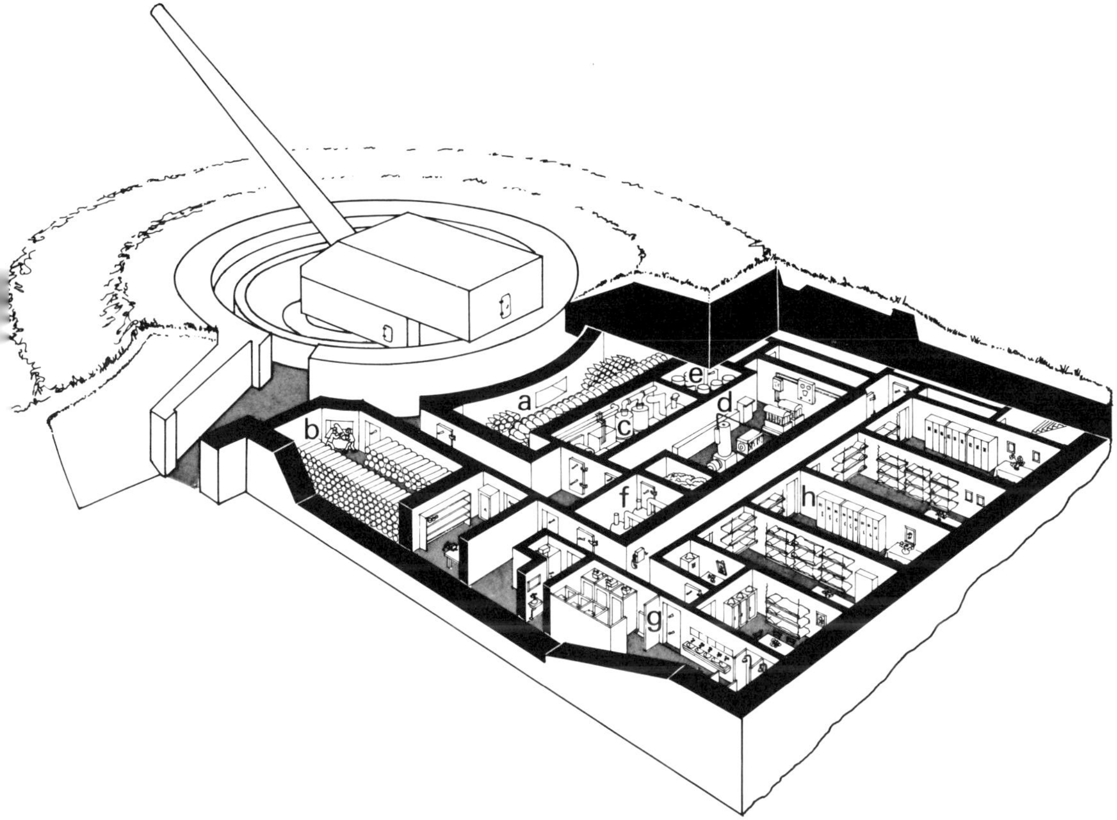

16.

17.

16. One of the giant Mirus guns on Guernsey: similar batteries were erected at Narvik in Norway at the same time. The only visible part of the Mirus, the gun itself, was camouflaged by the construction of a dummy house around it:
    a. shell store
    b. cartridge store
    c. ventilation plant
    d. plant room
    e. oil store
    f. heating plant
    g. toilets and showers
    h. quarters
    i. camouflaged entrance.

17. Fire control tower: Alderney.

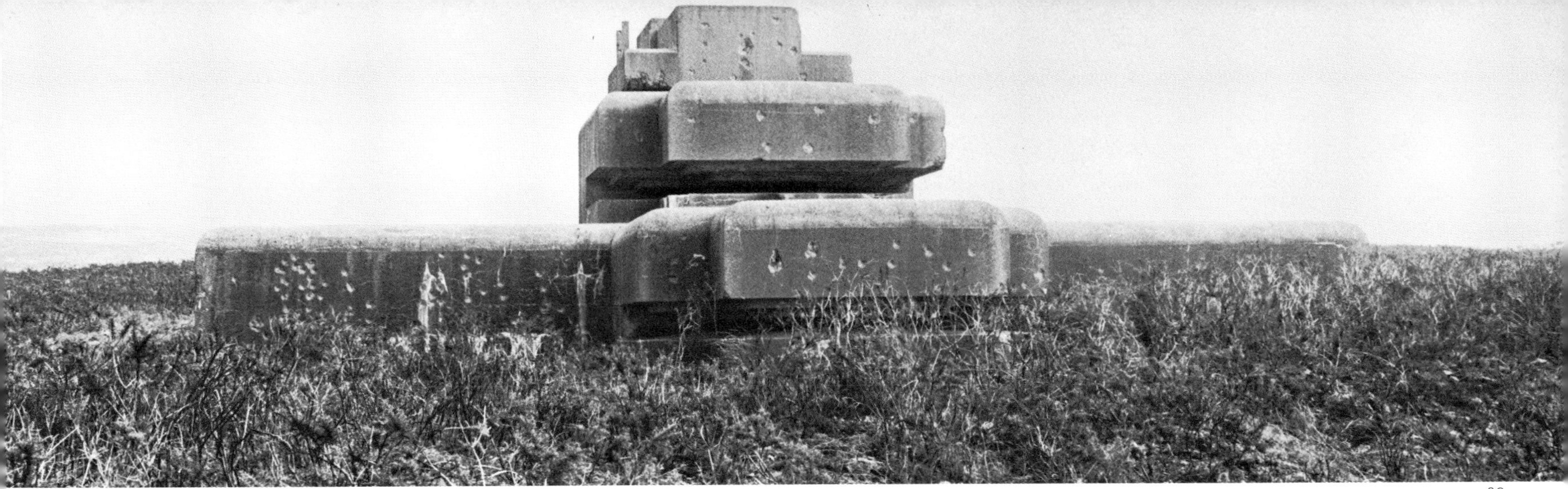

19

20

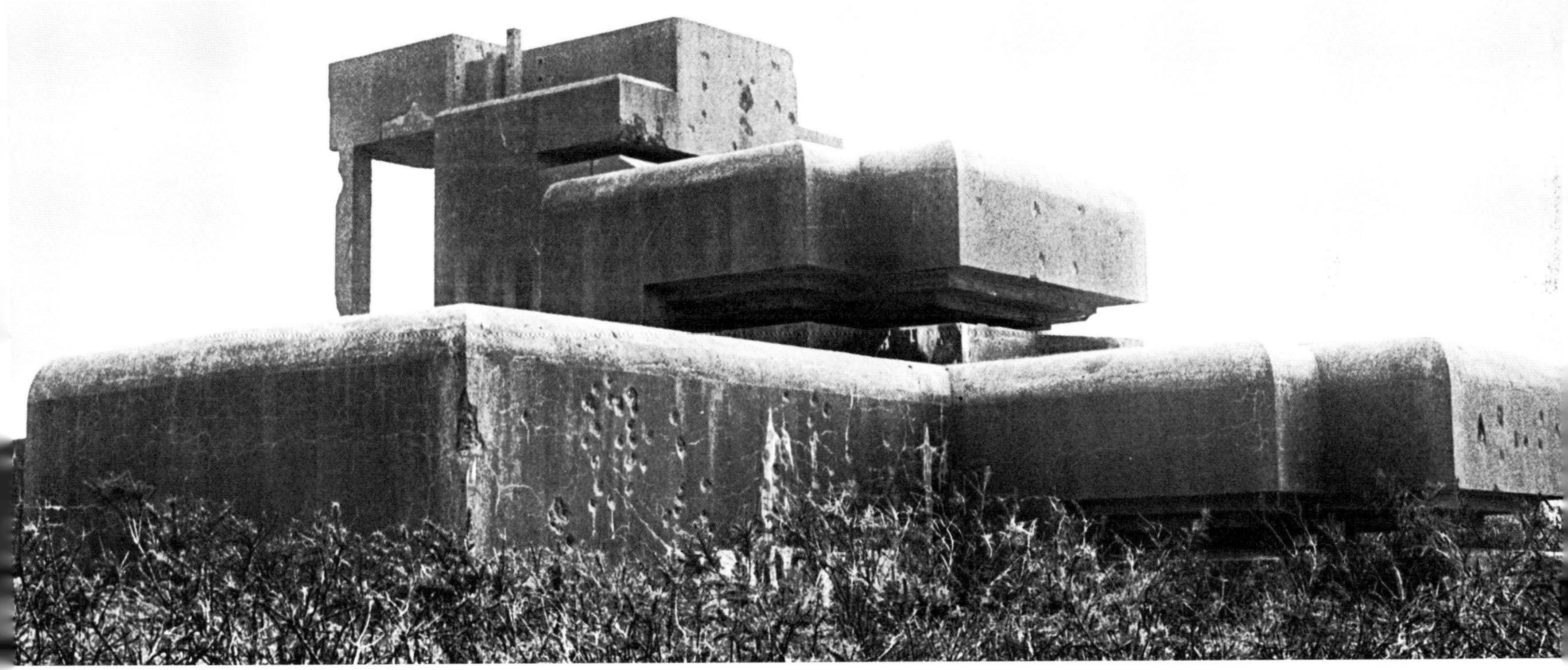

18, 19, 20. Naval observation post and range finder emplacement overlooking the coast near Cherbourg.

21

22

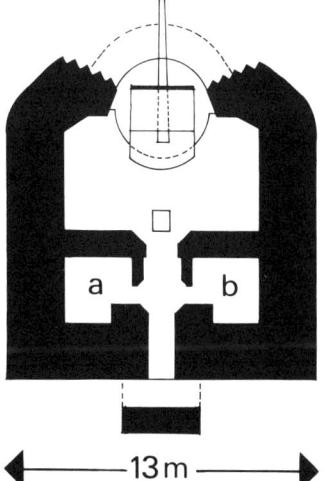

21, 22. Longues Naval Battery Normandy (Gold Sector): one of the four casemates for the 15.2 cm guns. Camouflage: low silhouette and pitted concrete texture formed by newspaper balls set in the concrete.

23. The Longues observation post and range finder emplacement set into the cliff top about two hundred metres from the line of casemates: detail design similar.

24. Plan of a Longues casemate:
  a. shells
  b. cartridges

25, 26. One of the four casemates for the 15.2 cm guns of the naval battery at Honfleur, Le Havre, with its adjacent range finder and observation post, as they remain today. Both naval and army casemates had roof reinforcement of closely spaced I beams (as much as 800 mm) supporting steel ceiling plates.

of war, obligations to the allies, and political considerations may persuade him to take decisions which appear unlikely from a purely military point of view.

Even enemy *landings with limited objectives* can interfere seriously with our own plans if they result in the enemy gaining any kind of foothold on the coast. They can interrupt our coastal sea traffic, and pin down strong forces of our army and air force, which will therefore have to be withdrawn from areas of crucial importance. It would be particularly dangerous should the enemy succeed in capturing our airfields or in establishing his own in areas which he has occupied. The many important military and industrial establishments on the coast or in its neighbourhood, some of them equipped with particularly valuable plant, may moreover tempt the *enemy to undertake suprise attacks of a local nature.*

Particular attention must be paid to English *preparations for landings* on the open coast, for which they have at their disposal many armoured landing craft, built to carry armoured fighting vehicles and heavy weapons. The possibility of *parachute and airborne attacks* on a large scale must also be envisaged.

II. General operational instructions for coastal defence:

1. Coastal defence is a task for all armed forces, calling for particularly close and complete co-operation by all units.

2. The intelligence service, as well as the day-to-day reconnaissance by the *navy* and *air force*, must strive to obtain early information of *enemy readiness and approach* preparations for a landing operation.

All suitable sea and air forces will then concentrate on enemy points of embarkation and convoys, with the aim of destroying the enemy as far from the coast as possible.

It is however possible that the enemy, by skilful camouflage and by taking advantage of unpredictable weather conditions, may achieve a complete surprise attack. All *troops* who may be exposed to such surprise attack must be in a state of *permanent readiness*. . . .

3. *In defending the coast*—and this includes coastal waters within range of medium coastal artillery—*responsibility for the planning and implementations of defensive measures must,* as recent battle experience dictates, lie unequivocally and unreservedly in the hands of a single commander . . . (9)."

In its entirety this directive, far from clarifying the situation, only developed the previous tendencies further. The command situation ordered was chaotic. The navy was to have command while the enemy was still at sea; the army was to take over command when the enemy had landed. When was this change to take place . . . 100 metres off the coast, when the first soldier put foot on land, or later? Similarly the air force was to co-operate with the force in command when the enemy was attacking by sea; while still in the air it was to act independently. If any decisions were to be made they would have to be by mutual agreement, but there was little likelihood of this within a system where personal prestige and jealousy were deep-rooted factors. Instead of co-ordination between the forces and their strategies, therefore, it separated them even further. Hitler, having completely taken over OKW, was quite content to preserve a situation where only he had complete and overall control.

The navy set about preparing for a sea battle. Large coastal artillery pieces were placed well forward on the coast and protected by concrete. The author of 'Fortress Europe' writes that:

"The navy tended to regard the headlands of the French coast as just another kind of ship. They sited their batteries in too far forward and exposed a position as if in a battleship gun turret, so that most of them could not fire directly on the shore but only out on to the sea." (10). As they were reassured by Hitler in person that 'the Allies could not make a successful invasion unless they were able to take a sizeable port' (11), harbours received their main attention. Only between Le Havre and Cherbourg were two naval batteries constructed on the open coast. This also followed the orders of December 14th, 1941.

As these forward batteries were extremely exposed to bombardment, both naval and aerial, much thought was given to concrete protection. Steel turrets were out of the question due to the steel shortage. Instead experiments were carried out using a rotating gun tower of reinforced concrete (12). But the process of perfecting this was slow, and meanwhile other means of protection had to be found. Huge casemates were built and Grand Admiral Raeder only managed to get enough concrete for these through the OT, by lobbying Hitler, at the expense of the army. These casemates, which also included ammunition stores, generally had rounded off corners and were camouflaged by slowly sloping earth banks. Below ground level a large concrete collar prevented shells from affecting the underneath of the casemate and added further stability in

27

28

27. Personnel shelter: Utah beach.

28. Command post with Tobruk emplacement and flanking embrasure: part of an army battery in Normandy. Some of the coastal posts were equipped with cupolas.

29. Recent photograph of a beach flanking strongpoint in the Pas de Calais area. The gentle mound on the roof indicates the position of a Tobruk type emplacement: a small concrete pit 1.4 m in diameter for close combat and sentry use.

30. Recent photograph of an open artillery (3.7-inch Vickers AA) emplacement on a beach east of Cherbourg with thin casemate superstructure added at a later date. This naval strongpoint had three such casemates, which were connected by underground concrete tunnels to Tobruk emplacements and underground quarters.

31. Open concrete searchlight platform with access ramp and shelter under: Cap de la Hague (Cherbourg).

32. Recent photograph of an open artillery emplacement on the cliffs 20 km north of Le Havre.

the case of near misses. All inward sloping surfaces around the embrasure were stepped, in order to stop shells being deflected into the interior of the casemates, a theory which had also been applied in the West Wall of 1939. The resulting casemates therefore provided extremely strong protection.

By adopting the casemate, however, the navy had been forced to make a compromise. The larger the embrasure, the more vulnerable the bunker became. The solution adopted was to use an embrasure giving a 120 degrees field of fire. By placing several such casemates together at an angle to each other, forming a curved axis, a 180 degrees field of fire for the complete battery could be achieved. Even so, the coast on either side could only be covered by one gun and the land behind not at all. This was to be a large drawback in the defence of the ports, after the Allied landing in June 1944, when Cherbourg and the other French ports were attacked by land (13). Meanwhile, the army was pursuing techniques which had been developed on the West Wall—but on a thoroughly unsatisfactory basis. In the harbours they constructed small installations and bunkers, together with the navy and air force, both along the water-front and to the rear. Along the coast, however, it was impossible to achieve any sort of density. Albert Speer, head of the OT from 1942, states that the construction of:

"a complete line of pill-boxes spaced close enough to offer mutual protection would have far exceeded the capacity of the German construction industry." (14).

But, in any case, the army did not have sufficient personnel to man such a defence.

The strategy adopted by Field Marshal von Rundstedt, now Commander in Chief West, was to provide only light resistance along the coast itself. During a major attack these troops would provide only initial opposition and then fall back in front of the enemy. Bringing along their equipment, artillery and machine guns, they would delay the enemy as much as possible until a major counter-attack could be organized by a large central reserve of mobile units. Von Rundstedt's strategy for a mobile battle conflicted, to a degree, with Directive 40, which read:

"II. 3. The Commander responsible must make use of all available forces and weapons of the branches of the armed forces, and of our civil headquarters in the area, for the destruction of enemy transport and landing forces. He will use them so that the attack collapses if possible before it can reach the coast, *or at the latest on the coast itself.*

II. 6. The fortified area and strong points must be able, by proper distribution of forces, by completion of all-round defence, and by their supply situation, to hold out for some time even against superior enemy forces.

Fortified areas and strong points will be defended to the last man. They must never be forced to surrender from lack of ammunition, rations or water." (15).

Despite this conflict the strategy of mobile defence, which was by now almost a traditional one in the German army, was maintained until the arrival of Rommel.

As a result of von Rundstedt's persistence with a mobile strategy a weak string of coastal observation bunkers and infantry shelters were the only defences which were constructed by the army. At locations of particular strategic importance clusters of bunkers, field defences and pill-boxes were built; sometimes artillery was included in open emplacements to the rear. Generally the soldier defending the coast was expected to fight in the open or from field fortifications. He was only provided with concrete bunkers for shelter during the initial bombardment. According to Albert Speer some 15 000 small bunkers were constructed along the coast for this purpose. He also comments:

"Hitler planned these defensive installations down to the smallest details. He even designed the various types of bunkers and pill-boxes, usually in the hours of the night. The designs were only sketches, but they were executed with precision. Never sparing in self-praise, he often remarked that his designs ideally met all the requirements of a front-line soldier. They were adopted almost without revision by the general of the Corps of Engineers." (16).

In addition to these defences, the army was also expected to construct coastal batteries protecting the foreshore and beaches along the open coast. Here some degree of overhead concrete protection was aimed at but it was found that this was usually impossible as the navy, through personal influence with Hitler at OKW, had forced the OT to issue most of the concrete to the naval batteries. When concrete was used for artillery installations, therefore, open emplacements were often all that could be provided. Needless to say, this increased inter-force rivalry even further.

It was into this situation of inter-force conflict: where army, navy and

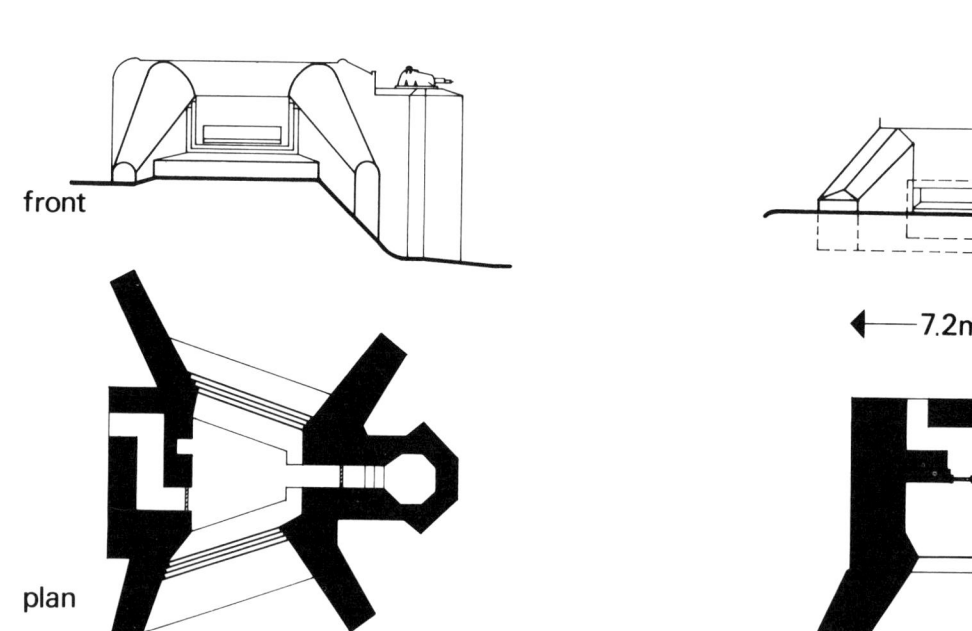

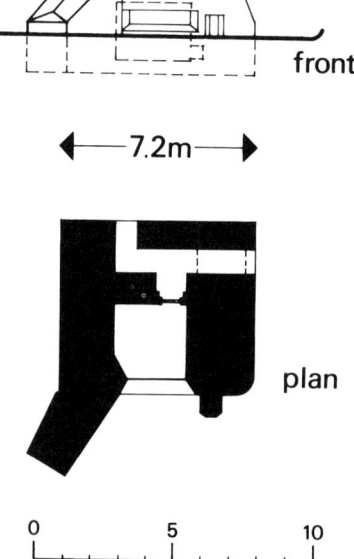

33. Casemate for 88 mm gun situated well behind the coastline following the original army strategy.

34. Beach flanking casemates for 50 mm guns. The larger casemate had a Tobruk emplacement fitted with a French Renault tank turret on its seaward side: a common adaptation on the Atlantic Wall.

air force were following different strategies; where completely different objectives were aimed at; and where Hitler, the would-be architect, was by-passing everybody by designing bunkers to satisfy his own ego (17); that the OT, now under new management, tried to introduce standardization.

When Fritz Todt had died, in the mysterious explosion which wrecked his plane on February 8th 1942, Albert Speer was asked to take over all of Todt's responsibilities. Together with Dipl. Ing. Xaver Dorsch, who had been one of Todt's close associates, he replanned the OT management (18). While Todt had probably been happy to think of the OT as a task force working for the armed forces, Speer tried in every way to increase productivity. To do this, with the corruption of the OT which had developed in the west, standardization and central control would be necessary. A central office in Berlin, OT Zentrale, was set up under Dorsch, and from this orders were to run through various levels of management down to the individual building sites.

In theory initial directives would be issued by Hitler's OKW to the High Commands of the army, navy and air force: OKH, OKM and OKL respectively. They would then plan their own strategy, based on the directives, which would be sent to the various groups, such as Army Group West with Commander in Chief von Rundstedt, or Western Naval Group with Admiral Krancke. These groups would then plan the detailed layouts together with the equivalent OT group or 'Einsatzgruppe' which, in the west, was led by Oberdirektor Weis. Using a mixture of army, navy and OT engineers, blue-prints would be drawn up and sent to OT Zentrale, for sanctioning. From OTZ the plans would then be sent through OT channels to the basic OT sectors, the 'Oberbauleitung', the lowest level of liaison between OT, Wehrmacht and Party agencies. Here ultimate management of the project would be carried out. As the importance of incorporating local knowledge into the design was recognized, this level was granted a fair amount of design responsibility. Modifications could therefore be carried out before the design was passed, through sub-sectors and local supervisory staff, to the actual construction site—the OT firm.

Theoretically this method of control could have helped to standardize both the design work and the tactics used for the defence. In reality, however, the OKW orders on which the management pyramid was reliant was by-passed by both Raeder and Goering. Similarly the army was often by-passed by Hitler, both at OKW and OKH level, when he gave orders directly to local commanders. The idea of keeping a bank of standardized designs at OTZ, for use in all theatres of work, did not work either. The army and navy, between them, possessed too many different pieces of artillery for this to be feasible. Their artillery, collected over five decades from 10 nations, included 28 different calibre groups. Instead OTZ became a distribution group for new designs, such as the 'Tobruk' emplacement which was first used in North Africa and introduced in the west through OTZ. A further handicap to any attempt at standardization were the modifications carried out at Oberbauleitung level. These modifications were not controlled by the OTZ and local control of design continued as did the practice of contracting to local firms by the army and navy.

If anything was achieved as far as the design of fortifications was concerned it was the strengthening of OT as a separate group of designers, additional to the army and navy, in those areas where they were mainly concerned: the Channel Islands, opposite Dover and northern Norway. It should be remembered, however, that the OT as part of the Speer Ministry, responsible for all war production in the Reich, was mainly concerned with the allocation of materials and manpower. Specifications were set out when the original blue-prints reached Berlin and to change them would have been difficult. It is probable, therefore, that as long as modifications were kept within these standards both Speer and Dorsch would be happy.

The two men who did have a chance of standardizing and co-ordinating design along the coast were Lieutenant-General of Engineers Schmetzer, who was Inspector of Fortifications in the West, and Inspector General of Engineers Alfred Jacob—but they were both more interested in individual detail (19). Even if they had been interested, there were far too many commands and too much red tape in the west: Commander in Chief West, Field Marshal von Rundstedt; Commander in Chief 3rd Air Fleet, Field Marshal Sperrle; Commander in Chief Western Naval Group, Admiral Krancke; The Military Commander, France, General von Stülpnagel; The Military Commander Belgium—Northern France, General von Falkenhausen; Flag Officer in Charge of Western Sea Defences, Admiral Breuning; Flag Officer Channel Coast, Admiral Rieve; Flag Officer Atlantic Coast, Admiral Schirlitz; the German Embassy in Paris; the Vichy French Government; and finally,

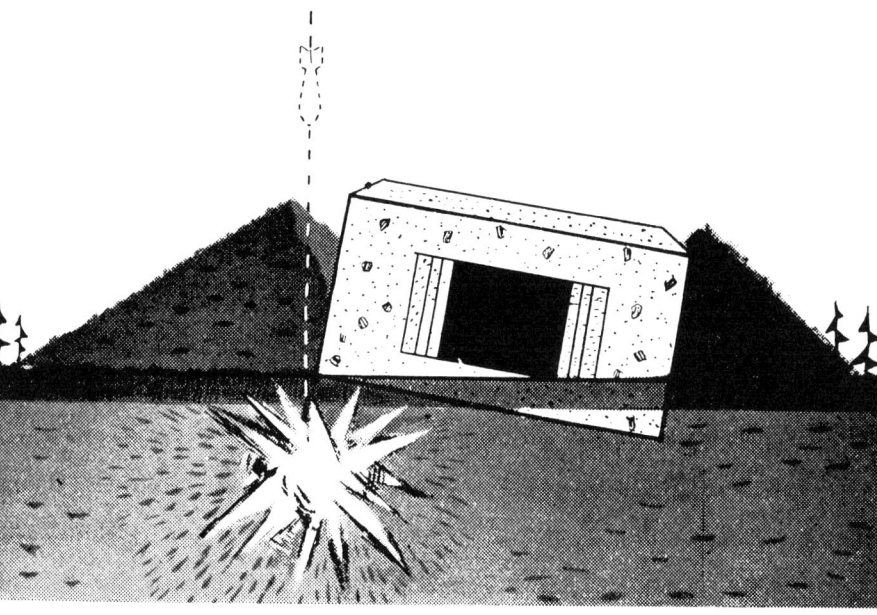

35. The difference between naval (top) and army casemates according to Bertil Stjernfeld, an officer in the Swedish Coastal Artillery. By 1944 some of the army casemates, such as those at Gatteville, were in fact more sophisticated with heavy aprons and revolving steel turrets.

36. Von Rundstedt's second line of defence in the Pas de Calais area: started but never completed.

now as an independent body, the OT, responsible through Speer to Hitler alone. Thus, F. Majdalany writes:

"Within this top-heavy command structure, additionally confused by the interference of the Nazi party organization, there was almost limitless scope for inter-service conflict and personal jealousies and frustrations. Directive 40 which should have resolved the conflict in fact perpetuated it." (20).

It is not surprising if work was soon slowing down as a result of this confusion. Von Rundstedt did start the construction of a second line of defence, in order to strengthen the weak coastal crust, but the project never got off the ground. Large naval batteries were being built but the army did not even have enough concrete to build the few installations, which they did construct in concrete, to West Wall standard. Meanwhile the military situation steadily deteriorated. Russia was advancing and Italy declared war on Germany. By the end of 1943 the chances of a major invasion in the west had increased.

To speed up construction, and generally strengthen the position in the west, OKW issued a new directive. The opening sentences of Directive No. 51, dated November 3rd, 1943, read:

"The danger in the East remains but a greater danger appears in the West: an Anglo-Saxon landing. In the East the vast extent of the territory makes it possible for us to lose ground, even on a large scale, without a fatal blow being dealt to the nervous system of Germany. It is very different in the West. Should the enemy succeed in breaching our defences on a wide front here the immediate consequence would be unpredictable. Everything indicates that the enemy will launch an offensive against the Western Front of Europe, at the latest in the spring, perhaps even earlier." (21).

At the same time Rommel was appointed Inspector of Fortifications. Together with Vice-Admiral Ruge, as his advisor on coastal artillery, he toured the complete coast from the Skagerrack to the Spanish frontier. After some early delays the tour was started in December. With his experience from North Africa Rommel was appalled by what he saw, as Desmond Young writes in his biography:

"The German navy, had indeed, erected batteries for the protection of the principal ports. These had been linked up to some extent, by batteries of the Army Coastal Artillery. But whereas the naval guns were in steel cupolas, the army artillery was merely dug in and had no overhead protection against shells or bombs." (22).

France had become a holiday camp for tired generals. The high quality troops had been sent to the east. In the navy, the U-boat fleet had been given priority. The OT, previously engaged at least in the construction of larger installations, had become absorbed in the repair of bomb damage. Even elementary precautions such as obstacles, mines and wire had been neglected.

After submitting his report Rommel was made Commander in Chief of an area extending from the Netherlands to the Loire, in January 1944, and an OKW order was issued at about the same time demanding that all artillery positions should be given overhead protection. While the army had maintained that the only effective defence was to keep a large central reserve, to be released for the crucial counter-attack once the enemy objective had been discovered, Rommel believed that as long as the Allies had air superiority such movement of troops would be impossible (23). Instead he wanted all forces to be moved forward to the coast line. In order to achieve this he was ready to sacrifice depth and von Rundstedt's second line was abandoned. According to Rommel the enemy must be beaten while he was still in the water: afterwards it would be too late. Better to have one tank on D-Day than three tanks on D-Day plus three. In this he was opposed by von Rundstedt, Guderian and Geyr von Schweppenburg, the Commander of Panzer Group West. Hitler and OKW solved the dispute by taking the edge off both plans—a compromise. Half of the present forces were to be kept as a central reserve while the other half was weakly spread along the coast.

In Rommel's resulting construction projects his two main concerns were beach defence and overhead protection for the artillery but the resulting battery construction compared unfavourably with the naval batteries. Speed and quantity became the primary factors at the expense of quality. In the Seine bay just over half of the army artillery installations, which had so far been open ones, were given overhead protection by D-Day in this way (24). These casemates were usually box-like structures, with shallow foundations, lacking the underground concrete collar of the naval casemates. Corners were cut off at 45 degrees instead of rounded and, together with the steep earth bank side protection, the casemates not only proved easier to spot from the sea and air but also more vulnerable to near misses.

37. Recent photograph of the control tower at the army battery of Riva Bella on the flat mouth of the R. Orne in Normandy: blockwork was used as permanent formwork for many of the army projects.

38, 39. Type observation posts in the Pas de Calais: recent photographs from the book 'Architecture Cryptique' by Paul Virilio.

While ammunition storage was provided inside the naval casemates, such storage in the army casemates was only sufficient for the first few rounds. After this had been fired the ammunition had to be collected from special underground ammunition stores which were often as much as 50 metres away from the bunker. This practice was probably a tradition left from the time, prior to 1914, when open concrete emplacements had been normal. Now, when the landscape was converted to a series of bombcraters, it caused severe inconvenience. But the real Achilles heel of the casemates was the embrasure. This problem was common to both navy and army installations. These openings had been envisaged as being protected by heavy armour, but with the shortage of steel, only thin armour could be found. This did not give any protection against direct hits, but even so it gave splinter protection and prevented the delicate instruments of the artillery from being covered by soil during bombardment.

The construction of personnel shelters had, as previously mentioned, begun already in 1942. Within the batteries several of these were constructed underground, and this process was speeded up by casting 40 cm concrete directly onto prefabricated arched shuttering plates which had been placed in trenches. These were later covered by earth. This type of construction was considered bomb-proof at the time but, during the preliminary air attack, battery personnel preferred the safety of the ammunition stores within the casemates. Observation and fire control installations also had to be built along the coast, and had been started well before 1944. Meanwhile, however, the size of the rangefinder equipment had increased, and now often had to be placed outside bunkers—unprotected.

Following the policy of overhead protection, the Germans attempted to connect as many of the installations as possible with underground communication tunnels. According to Bertil Stjernfeld, a Captain in the Swedish coastal artillery, these underground communications and the many small pill-box installations covering the position 'gave the Allies some unpleasant surprises during the battle.' (25). If protected communication routes were constructed at all they were, however, more likely to be open trenches. In addition to the pill-boxes small Tobruk emplacements and other emplacements, usually for one man only, were used. All bunkers and casemates had also some sort of protection in the form of flanking rifle embrasures by their entrances or machine gun installations, sometimes in steel turrets, on their roofs.

Rommel's main interest lay in beach defence. He often put forward detailed proposals supported by his own sketches. Rows of obstacles, mined or otherwise, were erected below high water mark. Wire, more obstacles and mines were laid along the beach and around all strong points. The positions which had previously been equipped with only concrete shelters and field fortifications were now strengthened with additional concrete bunkers for anti-tank and machine guns, sometimes in cluster pattern, but usually in a weak discontinuous line of single installations immediately behind the beach.

Obviously standardization of design, not enforced earlier, did not take place now, although similar designs were used within each of the short sectors of coast which were under one commander. Bertil Stjernfeld comments:

"Three batteries similarly equipped by French 155 mm field guns between Barfleur and Orne all have different bunkers (casemates) . . . (and) . . . on the complete Normandy coast not two observation posts are exactly equal." (26).

Moreover, the large-scale building project started by Rommel, soon to be known as the 'Rommel Bar', was not generally supported by the forces. His request that the air force's 3rd Flak Regiment be put under his command was rejected. When he complained to Hitler that the Channel Islands had received 38 guns for 11 batteries while the coast from Dieppe to St. Nazaire, 950 km, had only 11 batteries with 37 guns altogether, he was forbidden to ever mention the subject again.

This was not Rommel's only problem. The army largely resented his strategy, and the OT was occupied elsewhere as Lieutenant-General Hans Speidel, Chief of Staff to Rommel, recalls:

"The Todt building organization was also independent and obeyed so-called 'Führer directives' issued by the Reich Minister of Arms and Munitions (Albert Speer) and the Army High Command. The Commander in Chief in the West could indicate his requirement to the Todt Organization, but he could not give it orders." (27).

As a result, despite Rommel, interest was concentrated on the large fortifications. There was an average of 63 OT workers per coastal kilometre in the Pas de Calais area whereas the Normandy and Brittany area where Rommel was now carrying out his large-scale operation, was only allocated 16 OT workers per coastal kilometre. To speed up operations he made a compromise and used troops, including artillery crews, for construction purposes.

In addition to Rommel's other problems a major difficulty, as D-Day approached, was caused by Allied bombing. Timber shuttering, always used previously, had to be abandoned. It was too vulnerable under bombardment and a slower method using concrete blockwork as permanent shuttering was adopted. Similarly during three days in May only 47 of the 240 railway trucks of cement required for the Normandy coast arrived. Soon afterwards all supplies stopped. Time was running out. Even so an amazing amount had been done, despite corruption, jealousy and favouritism. In one year, 1942–43 17 million cubic metres of earth had been excavated by a labour force of half a million men. Four of the largest casemates at the batteries 'Lindemann' and 'Todt', each needing 14 400 cubic metres of concrete, had been completed with living quarters in ten weeks (28). On the 1400 kilometres of coast, taken over by Rommel in January 1944, 9,300 fortified works were finished by June 1944.

To make 3000 kilometres of coast, as the crow flies, impregnable, would, however, have been impossible. Milton Shulman, who as a Canadian intelligence officer interviewed von Rundstedt in a prisoner-of-war camp, aptly called the Atlantic Wall 'so many knots in a piece of string' (29). Hitler's decision to hold the coast at any price, putting his prestige foremost as he had done at Stalingrad, the Don Battles, the Crimea, Sicily and in Italy, was not only to cause his defeat in these battles, but also the loss of the Reich. Hans Speidel, Chief of Staff to Rommel, quotes Frederick the Great:

". . . he who defends all, defends nothing." (31).

But despite the eventual military failure of the Atlantic Wall, its casemates and observation posts stand out, together with the National Socialists' flak towers, submarine pens and civilian bunkers, as some of the most formally 'architectural' military construction of the 1939–45 war.

9 THE ARMED CAMP

Build up of invasion forces in England and the resultant vast hutting programme; aspects of prefabrication.

Events

1936 First steps towards rearmament taken by British.

1937 Britain has standing army of only 164 000.

1939 March 29 Britain decides to double her territorial army.
April 27 Britain reintroduces conscription.
Sept. 1 Britain and France mobilize: Invasion of Poland.
Sept. 3 Britain and France declare war.

1940 April BEF in France total 394 165 men.
May 338 226 men from the BEF evacuated from Dunkirk.
July Churchill establishes Combined Operations Command to prepare for return of British forces to the Continent.

1941 Dec. British army to have reached the maximum size planned in 1939: 2 300 000.

1942 Feb. Decision to launch the first Anglo-American effort against Germany, not Japan.
Spring Cross Channel invasion proposed for 1943: 'Operation Roundup'.

1943 Jan. Cross Channel invasion postponed until 1944: main objective is no longer one of establishing a second front in Europe, as desired by Moscow, but one of dealing a fatal blow to Germany.
Spring Large-scale build up of American forces in Britain starts.

1944 June Total of 3 500 000 Allied troops assembled in Britain; 33 per cent of these are American.
June 6 D-Day.

● In the inter-war period, Britain's military barracks, like her civil buildings, depended on slow heavy forms of traditional construction. The Second World War brought a drastic change in this situation, particularly in the case of military hutting. Rapid increases in the strength of the armed forces and shortages of both labour and materials

1. Erection of one of the large camps necessitated by the build up for 'Overlord' in 1943: group of Romney huts for an American depot.

forced correspondingly rapid changes in approach.

These changes consisted of a series of phases evolving primarily out of the military situation. From 1939 until 1942 Britain's army underwent a drastic programme of growth which made the existing numbers of barracks completely inadequate. The problem was increased as sources of raw materials were cut by German military action: for example the steel which did get through the U-boat packs in the Atlantic was needed for munitions. Apart from the labour needs of the mushrooming war industries, the growing requirements of defence construction made manpower a further problem. The second major phase began in 1943. In that year it was finally decided to postpone the invasion until 1944. The objective was now no longer one of establishing a second front in Europe, as desired by Moscow, but one of dealing a fatal blow to Germany. For this an even greater number of troops were needed in Britain. From summer 1943 until spring 1944 the preparation of the invasion army was supplemented by $1\frac{1}{2}$ million American troops. In Britain by June 1944 some 3 500 000 Allied troops were poised for the assault and breach of Hitler's Atlantic Wall.

These main phases of the accommodation problem were affected technically by other military actions. In steel and timber-starved Britain, during 1939–43, concrete was almost the only available material. When the Battle of the Atlantic was won in May 1943 America was able to ship steel across to Britain more freely. The predominance of concrete hutting in the first years of war, followed by steel hutting in post 1943 such as the Iris and Romney huts, was the type of pattern which resulted. With such rapid changes in the availability of both labour and materials the military hutting programme, with its fixed military objectives, achieved almost as an incidental by-product a level of constructional innovation unknown in peacetime.

The first phase of this innovation began in 1939. In 1937 Britain's standing army had numbered only 164 000 men, of whom 38 000 were abroad. By 1939, when the threat of Nazi aggression emerged conclusively it became clear that the existing number of troops would not be sufficient for the defence of Britain or the Dominions, let alone European intervention through a British Expeditionary Force. In March that year it was consequently decided to double the territorial army and on April 27th, four months before the invasion of Poland, general conscription was reintroduced. It was estimated that through these

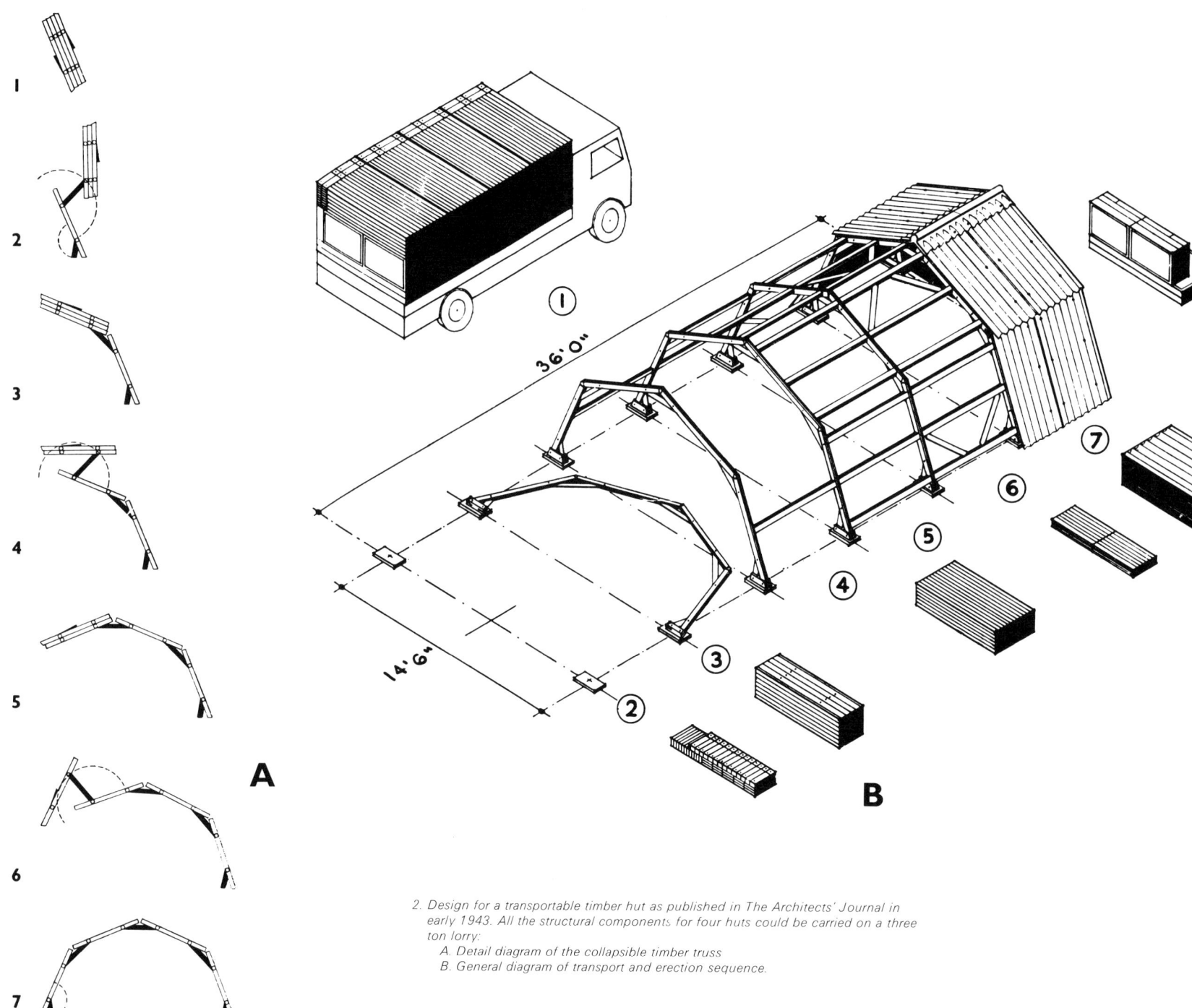

2. Design for a transportable timber hut as published in The Architects' Journal in early 1943. All the structural components for four huts could be carried on a three ton lorry:
    A. Detail diagram of the collapsible timber truss
    B. General diagram of transport and erection sequence.

measures Britain would have an army of 2 300 000 by, but not before, December 1941 (1). But now existing accommodation and the traditional methods available for increasing this accommodation would not match the rapid growth of the military establishment. Instead there was an urgent need for quickly erectable hutted camps. The situation which had existed on the Western Front three decades previously was recreated. Once the war had begun the combination of the shortage of skilled labour together with the rapid inflation in the size of the British army emphasized these requirements: ease and speed of erection, low cost and opportunity for mass production. The Nissen hut would have become a favourite once more had not two new factors made previous designs obsolete: the shortage of steel and timber and the increased number of uses to which hutting was put in the new war. It was the former which was to cause the main changes in design: earlier designs which had relied on these materials were rendered obsolete.

Nearly all the hutting used by the British army in France during 1914–18 had relied on steel and timber. They were also the two main, if not the only, tensile materials of that time. Though concrete frame buildings had been built during the inter-war period which needed much less steel than a steel frame, and a paper on the pre-stressing of concrete had been read at the Institute of Civil Engineers in 1936, such 'new' ideas had found little response. The building industry as a whole had, after an initial approach in this direction following the 1914–18 war, gradually swung back to using steel frames for large buildings and timber beams in smaller structures. Now the available steel was being swallowed up by the munition industry and timber, which had previously been imported, became scarce when Scandinavian supplies were cut following the German occupation of Denmark and Norway. Later the unrestricted U-boat campaign intensified this already existing situation. New materials, such as plastics, were introduced as substitutes but they were not far enough developed for large-scale production. There was no time to wait and old materials were used in new ways instead. Concrete, clay, glass, plaster, asbestos-cement, building boards and various patented modifications such as sawdust concrete were to fill the gap. The need for tensile materials was either reduced through such methods as pre-stressing and short-span design, or ultimately avoided altogether through a new structural approach such as the reinstatement of the arch in the C'tesiphon hut (2).

The increases in the uses to which hutting was put during the Second World War, other than new barracks, complicated the design problem. This increase was largely caused by the extension of demand into the civilian sector. Mr. R. Fitzmaurice from the Building Research Station summarized the variety of requirements:

"New factories and extensions to factories producing commodities necessary to the military effort. Barracks and camps for the armed forces. Storage space for munitions of war, equipment, etc. Structures required for defence. Housing for workers engaged in the production of munitions of war." (3).

To these could well have been added: fire stations, emergency hospitals and hospital extensions, temporary office buildings (TOB's), refugee housing and post-war rebuilding. In 1941 Alvar Aalto wrote in the RIBA Journal:

"The fact is that the technique of the present war destroys more buildings in non-military areas than it does human beings. The population feels the full weight of the present war first of all in this indirect way—through the destruction of its homes." (4).

Though this increase in the functions that a hutting design might be expected to fulfill was in many ways less important than the scarcity of materials, the former did lead to a certain increase in the flexibility of designs. This was perhaps particularly so since the British army largely relied on civilian contractors and designers whereas in Germany many civilian contractors were absorbed into para-military organizations, notably the Organization Todt (5). In Britain the civil building sector now tried to meet both military and civilian demands; in some cases post-war use emerged as an additional consideration on top of the immediate war requirements.

In addition to the problems caused by different types of war-time and post-war use there were the problems created by the differing requirements within the army itself. A study of the influence of war on army hutting, started by the Directorate of Fortifications and Works in 1939, concluded that any steel or timber which was available should be used for portable hutting, as these materials provided the only suitable solutions (6). Other materials such as asbestos-cement or concrete were considered to be either too fragile or too heavy, inevitably limiting the portability of the hut. They were therefore only suitable for static hutting where steel and timber could be reduced to a minimum. In his

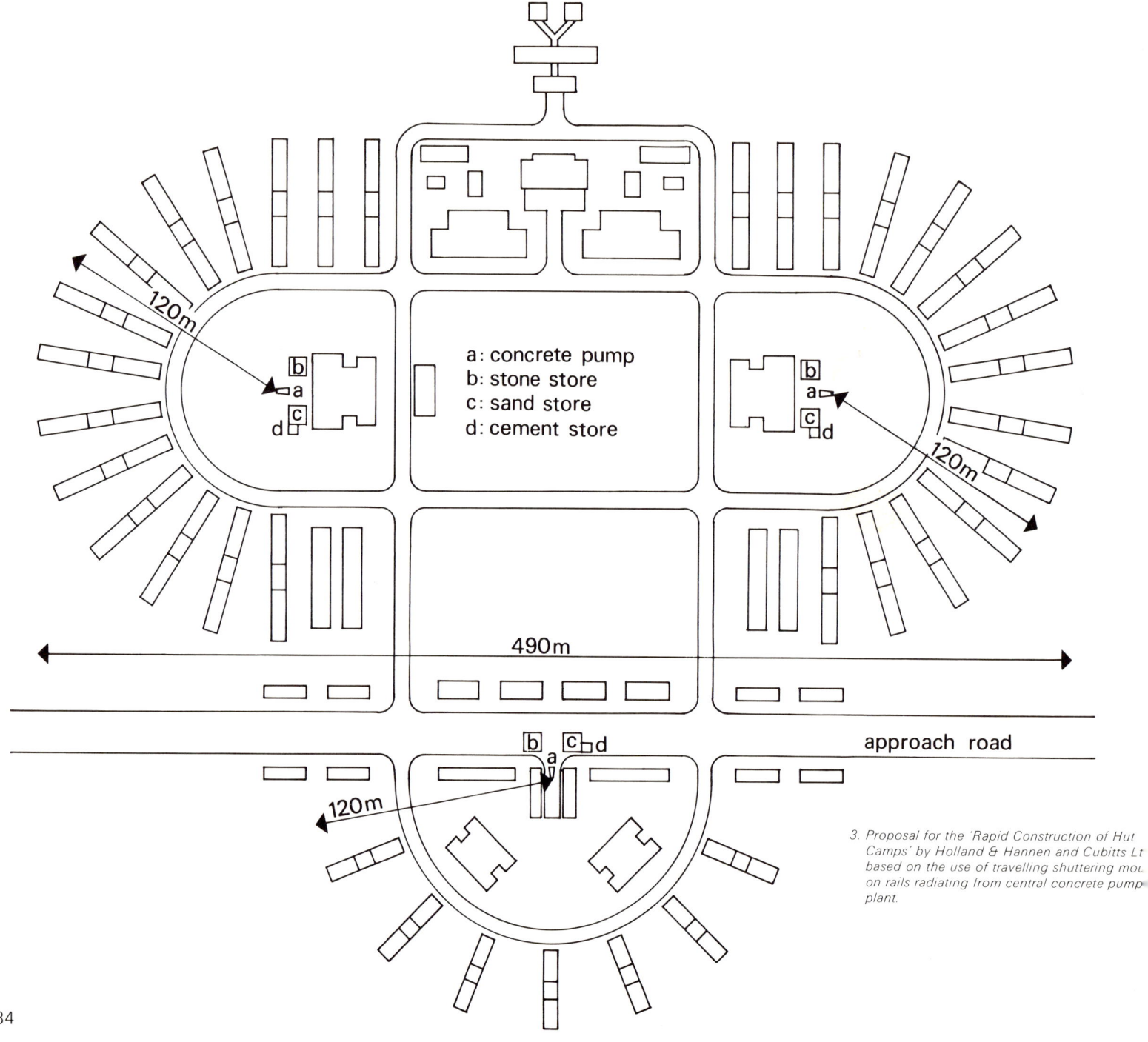

3. Proposal for the 'Rapid Construction of Hut Camps' by Holland & Hannen and Cubitts Lt based on the use of travelling shuttering mou on rails radiating from central concrete pump plant.

4, 5. Two of the prefabricated concrete huts which were erected in 1939 by the Cement and Concrete Association in order to show possible wartime military and civil use: application of the Mopin system (4) and the Plycrete system (5). The Mopin system had already been used in 1937 for the Quarry Hill flats at Leeds.

6. Diagram of the Stancon system:
   a. 400 mm precast concrete beam
   b. precast concrete secondary beam
   c. precast concrete cap bolted to column
   d. 200 × 200 precast concrete centre column
   e. 100 mm concrete slab
   f. precast concrete cill
   g. metal window
   h. precast concrete mullion
   i. precast concrete wall slabs 1830 × 230 × 25
   j. 19 mm plasterboard
   k. 38 mm precast concrete roof slabs
   l. waterproof roof membrane.

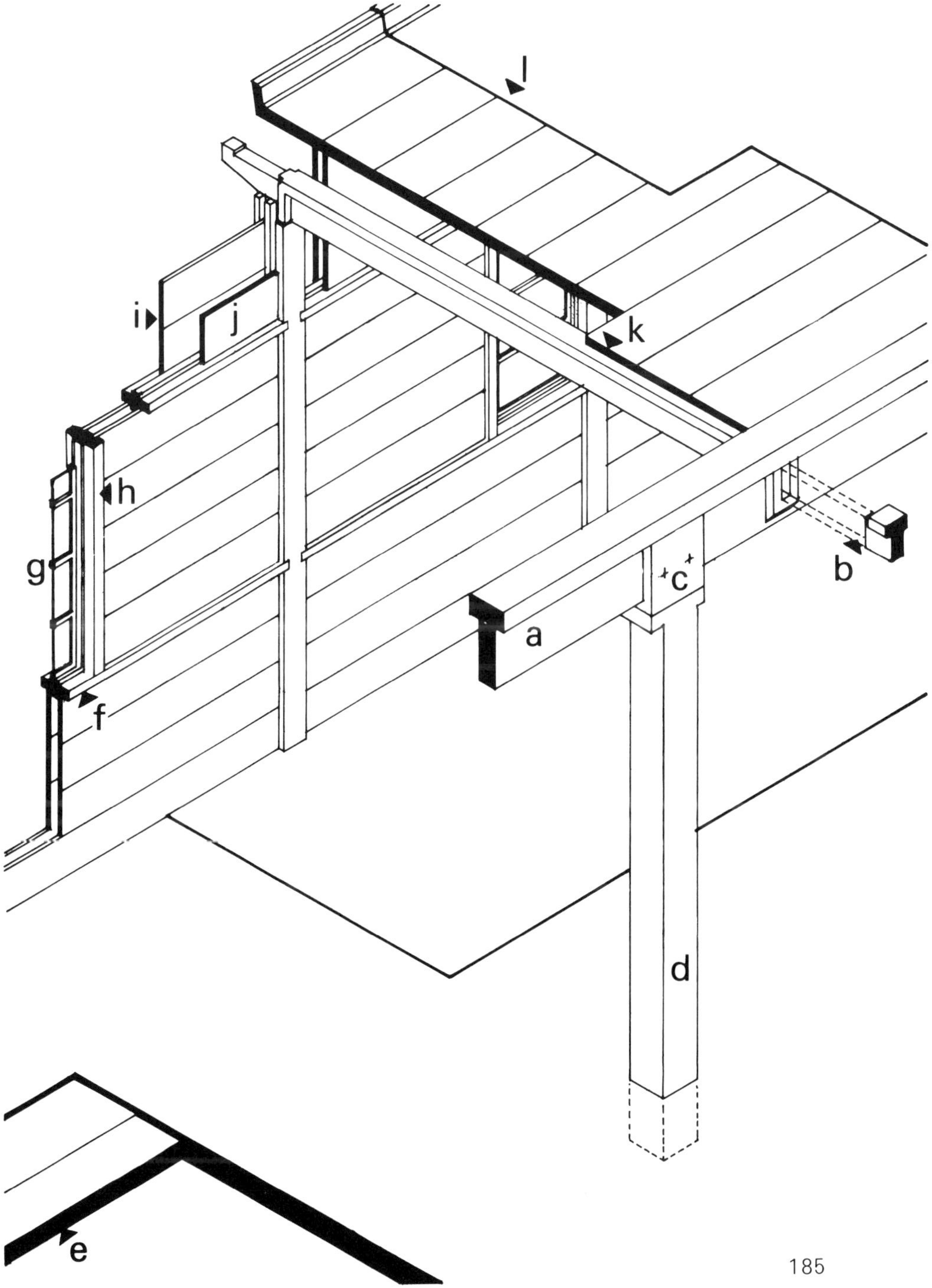

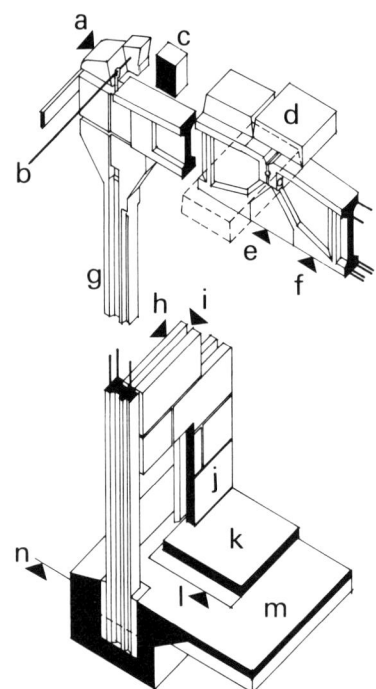
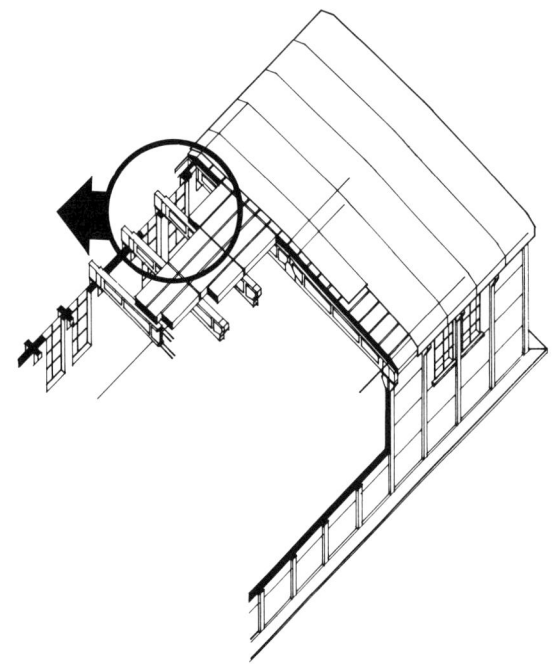

7. Basic construction of the BCF hut:
   a. eaves slab
   b. longitudinal tie bar with in situ concrete
   c. wall closer
   d. roof slab
   e. centre joint
   f. roof beam
   g. outer wall post
   h. 38 mm impervious outer wall slab
   i. 38 mm air space
   j. 50 mm breeze inner wall
   k. 50 mm screed
   l. bitumen
   m. 38 mm in situ concrete on hardcore
   n. ground level

8. BCF hut being erected at an American camp in 1943.

report on this study Major C. M. Singer of the Royal Engineers classified the two types of hutting required as: 'Portable hutting for mobile searchlight positions, and for use in the field generally' and 'Static hutting, for the housing of troops at home in general' (7). This effectively meant that the small amount of steel and timber available for hutting would be tied up in overseas theatres of war and that the hutted camps for the new conscript army in Britain would have to be designed to the same design criteria as similar civilian structures. A large and guaranteed market for development in this field was effectively opened and the civilian building industry, including the architectural profession, was not slow in answering this demand (8).

During 1939–42 three trends emerged. Firstly the modification and rationalization of traditional methods occurred to meet the new situation: insitu brick and concrete construction, but with modifications reducing the need for steel or timber as shuttering or tensile members. Secondly the use of prefabricated structures was developed using cement, asbestos or other materials. Thirdly new structures were introduced to suit new building requirements. Though all three trends emerged simultaneously they can be analysed in this sequence.

Some designers suggested a return to the use of the traditional blockwork arch and buttress construction, such as a design published in the 'Builder' in 1940 (9), but more outstanding developments were made in the field of concrete shuttering and handling. Re-usable steel and timber shuttering were made possible through design standardization. 'Concrete Publications' published a design for concrete shutter plates avoiding the need for scarce materials for this purpose altogether (10). Similarly the problem of speedy construction was looked at critically. An example was a proposal for the 'Rapid Construction of Hutted Camps' (11). The designers of this all-concrete system, Messrs. Holland & Hannen and Cubitts Ltd., proposed to construct on site a number of army camps in a semi-circular layout. This layout was based on the use of concrete pumping plant at the focus of each semi-circle. Re-usable shuttering was to be mounted on rails running radially from this plant and extending to a maximum radius of 120 metres. The shuttering would be moved along the track in stages as each section of the 'hut' was cast. When one radius was completed, the track and shuttering would be moved to the next radial position.

This system can be seen, in principle, as an attempt towards the industrialization of on site construction work. The price was calculated on 1940 costs to be approximately $8\frac{1}{2}$d. per cubic foot and it was suggested that this could be cut further by the use of a 'central heating house instead of the separate units installed in most camps'. The designers even proposed to compensate the short-term demand for army barracks by converting the buildings to holiday camps after the war. An article in 'The Builder' in 1940 described this proposal which, in the light of post-war criticism of holiday camps, appears an unfortunate one:

"For use as a holiday camp the huts would be partitioned off . . . swimming pools, lawns, tennis courts, gardens and other amenities would be provided on the areas reserved for parade grounds and other open spaces; and some of the buildings would be used as a cinema, writing and reading rooms, etc." (12).

While the requirements of the first phase of the hutting programme forced such changes in on site construction, they also forced similar developments in component production. Prefabrication, neglected before the war, became a common aspect of hutting design but here also, as in the 'Stancon' system, attention was paid to post-war use. In 1940 the 'Architect & Building News' wrote:

"Cost; speed of erection; economy of scarce materials—these are primary factors which must be considered in designing war-time buildings. Added to this there are considerations of use after the war for peace-time buildings, (and) high salvage value. . . . It was with these aims in mind that Mr. Hamp set out to design the system (the 'Stancon' system). . . ." (13).

Though this design was estimated to cut down on the cost of traditional methods by $12\frac{1}{2}$ per cent, 'the appearance of the finished buildings had not been neglected' (14). It was therefore assumed that by dismantling the structures no longer used by the army, these could be reassembled as post-war hospitals, schools, recreation buildings and housing. Indeed designs for such structures had already been drawn up and to ease such dismantling no mortar was used to fix the wall slabs. Instead pre-cast wedges pressed the panels against bituminous tape, and sealed them with the concrete frame.

The concrete frame was standardized in design with a flat roof which permitted a large amount of flexibility. It allowed floor dimensions of any proportion as long as they fitted into the grid lines. This was an important design development but an equally significant feature of the Stancon

9. The basic structure of the MOWP hut 1942.

system was its use of pre-stressing. This feature alone must have made the system advanced for its time, 1940, for pre-stressing is not generally recognized as being used in this country before it was investigated for use as railway sleepers in 1942 (15). Similarly most of the other features incorporated in this early design seem to have been forgotten by the designers who later tried to compete for army contracts. When the British Concrete Federation exhibited their 'BCF' hut design, the Architects' Journal wrote in April 1942:

"Prefabrication offers the possibility of escape from the limitations imposed by obsolete buildings.... With very light systems of construction like the plywood hut, the plan may be difficult to adapt (partitions take the place of roof beams which are eliminated altogether) but there is no limit to the number of times the hut can be dismantled and moved—a relatively easy process. With heavy materials, concrete for instance, which demand a strong and rigid framework, the difficulty of dismantling the framework, at any rate, is great.... A permanent framework can be designed to leave the floor space free so that though the building cannot be moved the inside can be replanned again and again to suit different requirements. Intermediate systems which are neither very heavy nor very light should be able to combine both advantages (the Stancon system two years previously). The BCF hut unfortunately possess neither...." (16).

By 1942 dozens of huts and systems had been designed to meet the needs of the growing army, such as the 'Maycrete', 'Nashcrete', 'Orlit', 'Plywood' huts and others (17). Though some of them did incorporate the advantages of lightness or flexibility suggested in this criticism, the BCF hut had one big advantage over these: its system of wall panelling. So far concrete designs had suffered from a common drawback: the difficulty of transporting the complete system from a central production plant. The BCF hut went a long way in easing this situation. By using paving slabs for wall panels a large part of the production would be done by local contractors as most contractors had the necessary formwork for this, which was at present laying idle. Site manufacture thus became possible on a large scale.

But still the flexibility of existing designs left much to be desired. In 1942 the RIBA Journal wrote:

"Hitherto hut needs have been met, chiefly, by designs exploiting one or another material to the exclusion of others." (18).

A design might not be suited in an area if the supply of appropriate raw materials was lacking locally. Or if a design was used under such circumstances the subsequent transport could delay construction and increase the cost. The design developed by the Ministry of Works and Planning, the 'MOWP Standard Hut', was the first attempt at providing a basic frame structure in which several types of cladding could be used. When the MOWP hut was introduced in late 1942, towards the end of the growth of the new army, it provided a flexibility not previously experienced. It not only saved on scarce materials but, except for its three pin portal frame, the materials could be freely chosen according to their current local availability as they were independent of the structure.

The previous examples were not the only examples, nor the first, using prefabricated concrete construction. The Cement and Concrete Association had sponsored such huts as the 'Mopin' and 'Plycrete' systems already in 1939 (19). The developments made in this type of construction during the expansion of the British army during 1939–42, prompted by shortages of labour and materials, had significance beyond their immediate military context. In a book on the resistance and response to change in the British building industry, Professor Marian Bowley writes:

"During the war ... the pressure of necessity in the form of changed conditions of supplies of resources and in demand for products, and government control of the economy, worked on the motives of patriotism, professional pride and the desire for profit. The result was the introduction and development of innovations at a rate quite abnormal in the British building and construction industry." (20).

But while development in the basic principles of prefabrication occurred during the first major phase of the army hutting programme other secondary problems connected with this development were of similar post-war significance. Marian Bowley writes:

"Pre-casting also involved problems of jointing between the members; this problem had received little attention in this country (Britain) since the time of the First World War. By the end of the Second World War much knowledge of the newer foreign techniques, as well as knowledge from experimental developments in this country, had been acquired." ... and ... "Reference must ... be made to the fact that wartime construction under conditions of great shortage of manpower necessarily led to the accumulation of experience in the use of mechanical

10, 11. One of the original parabolic Tarran huts in 1940: constructed with cement sawdust panels.

12. The 'improved' Tarran system with vertical walls: 1942.

13. Blister hangars.

equipment and in the organization of site work." (21).

Surprisingly these developments find no mention in the standard histories of civil architecture. Professor Bowley is, incidentally, a political economist.

These examples, in retrospect, undoubtedly show the birth of prefabrication as a generally acceptable form of construction. But the prefabrication was traditional in that the frames were generally based on post and lintel principles. Innovation 'at a rate quite abnormal in the British building and construction industry' is best seen in the more advanced *structural* developments which resulted from the new material restrictions during 1939–42: the third aspect or trend in this period.

The investigation conducted by the Directorate of Fortifications and Works in 1939–40 had found that:

"Theoretically, the only form of roof in which *all* tensile members are eliminated is the parabolic arch, springing from the ground. Several designs based on this principle have been examined, the most successful being one built up of pre-cast sections of a cement-sawdust composition, which are readily assembled into 19-feet span living huts. The Nissen hut is an example of a design which approaches this ideal, though its semi-circular section throws some bending stress on the ribs and the covering of corrugated iron or asbestos sheeting involves the use of purlins." (22).

The idea of the parabolic arch was not new. The reinforced concrete airship hangar built by Freyssinet at Orly in Frnce in 1916, for example, was based on this principle. Now the same principle was incorporated in hutting. The first designs, however, did not actually abolish the use of tensile materials altogether. The design found by the Directorate of Fortifications and Works to be 'most successful', the 'Tarran' system, did in fact use both timber and reinforcing wire.

The Tarran system did however manage to reduce the amount of steel and timber required while maintaining the mobility and lightness of timber hutting. According to the Architect and Building News:

"Taking a standard prefabricated W.D. timber hut, 60 feet by 19 feet as a basis of comparison, the Tarran hut employs only half a standard of timber and 54 lb. of wire. . . . This system, it is claimed, effects a saving of 30 per cent in structure weight as compared with standard prefabricated timber construction, and 60 per cent as against brick construction." (23).

The hut was made out of pre-cast structural panels with windows set in as necessary. Each panel was made up of 'Lignocrete', a mixture of cement and chemically treated sawdust, cast on wires running between two arched timber ribs. For erection the panels were screwed together through these ribs and covered externally by roofing felt. The main advantages of the hut were claimed to be: the economy of materials, economy of transport and erection, low cost, fire resistance, low wind resistance and protection against incendiary bombs due to the shape and simplicity of detail. Five lorries could carry a hut, and semi-skilled labour could erect it in nine hours.

The shape needed hardly any tensile materials but this fact was curiously not stressed by the manufacturers, Tarran Industries of Hull (24), nor by the journals of that time. The limited importance paid to this feature is further emphasized by the modifications made to the system two years later. In an exhibition at the Conway Hall in London, during 1942, the original huts of 4.9 and 5.8 metre span were kept without any major modifications or further saving in timber and steel. In addition to these original designs two 'improved' (25) designs were launched which, if anything, were a departure from the ideas of an all compressive structure. One version used straight walls with a barrel vault and timber ties at ceiling level; the other, which was intended to fulfill the need for large span structures, used a shallow circular arch construction. For large span structures it was argued that a parabolic arch would give excessive headroom. The use of an all compressive thin concrete shell had to wait until Major J. H. de W. Waller developed the principle for camp hutting (26). Instead the shallow arch philosophy was allowed to dominate the buildings based on arch principles.

The shortage of steel in 1939–43 made steel arch construction a luxury, but the curved asbestos hut was a direct continuation of the Nissen hut principle. These principles were, however, soon also to be implemented on larger structures: the 'Blisters'. The name arrived from the typical shape of the first military hangars built by this method (30). While the needs for army accommodation had grown in 1939–43, so had the need for long-span structures. After being successfully used for hangars, the same type of construction was soon used for other large span structures such as factories, assembly rooms, halls and even refugee housing. Here, as in the original Nissen hut, the reduction in the number of tensional members required was not considered the major advantage

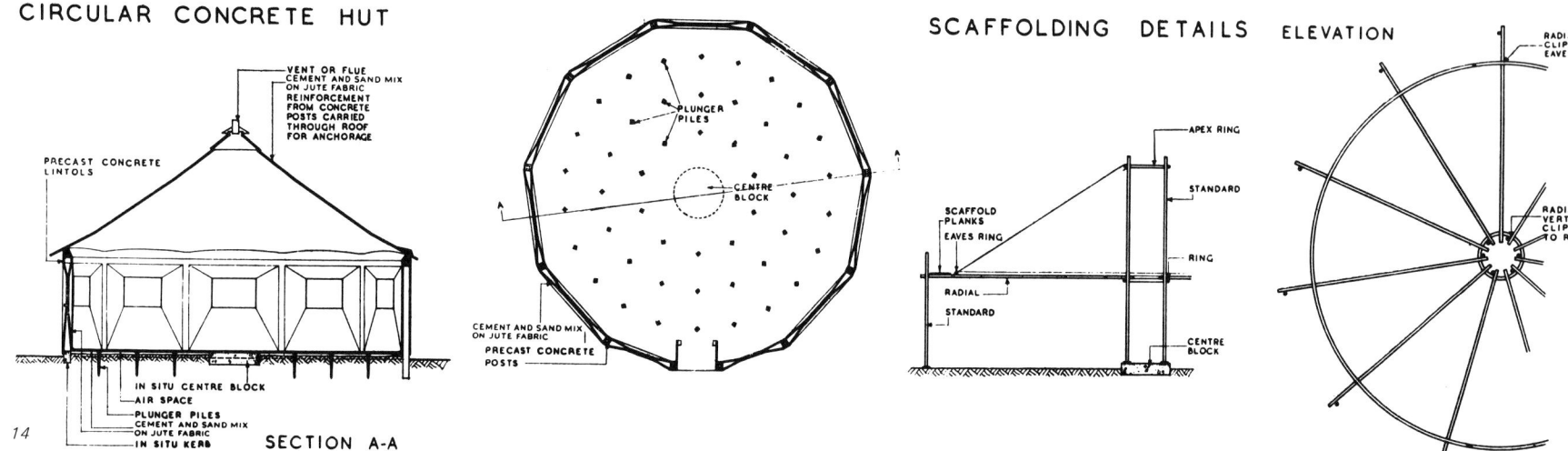

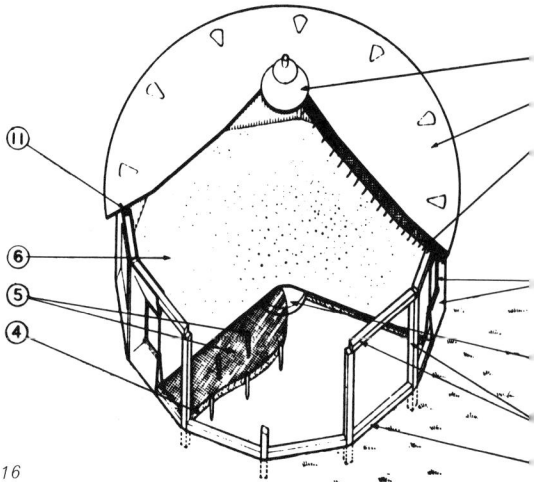

14, 15, 16. Duodecagonal hut (8.23 m diam.) by Major J. H. de W. Waller 1941. Key to isometric:

(1, 3, 4, 5, 6) unreinforced floor construction
(2) precast concrete frame
(7, 8) jute fabric stretched on both sides of wall frames and rendered with cement–sand mix, outer and inner layer sewed together at window openings and left unrendered
(9, 10) conical roof formwork constructed by stretching wires between apex ring and radial ring, jute fabric stretched over, rendered, reinforcement laid after initial coat
(11) gap between roof and lintels built up, shuttering wires and rings removed
(12) central vent or flue

17

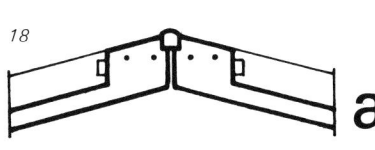  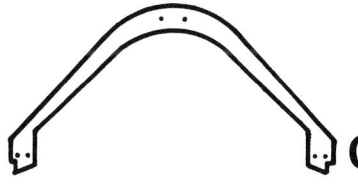

18

17, 18. De W. Waller's method of construction used for a prefabricated hut—'The Patrick Portable Building':
  a. bolted ridge junction: sealed with mastic
  b. section through unit at the heel and ridge
  c. section through unit at knee.

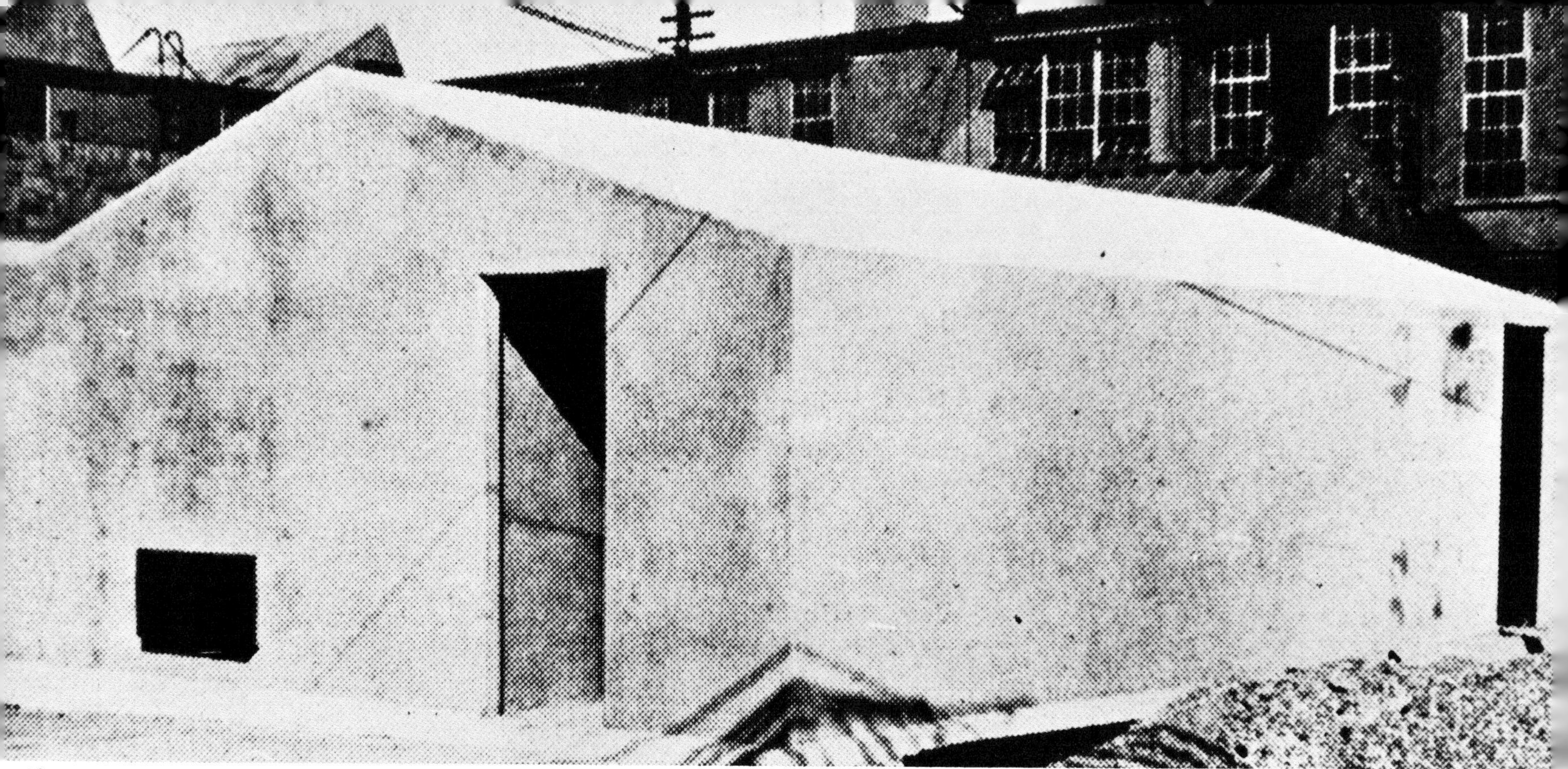

19

20

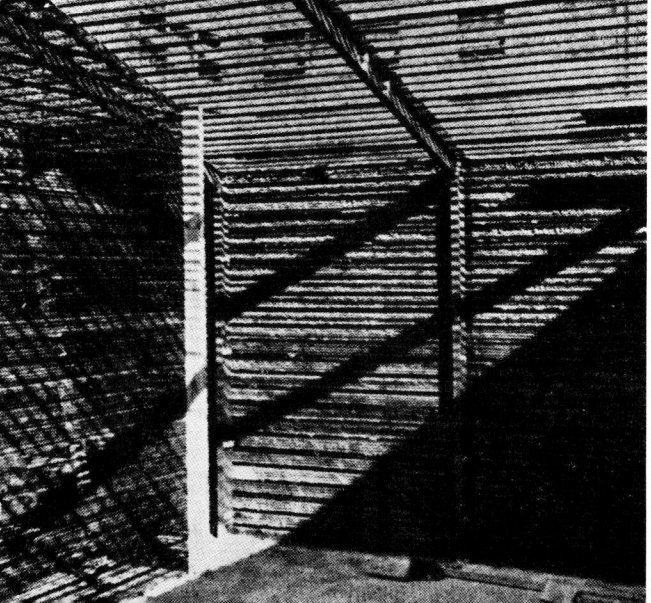

19, 20. The 'Padmos' hut, produced in 1943 by the Ministry of Supply PAD using rendered steel mesh on steel frames: cleansing station during and after construction.

of the shape. The blisters were in any case much shallower for reasons of camouflage (28). In 1941 when Graham Dawbarn had published an article in the RIBA Journal on Blister construction, he considered only two types of materials to be suitable: timber and steel (29). Speed had become increasingly important and the large-span structures were not considered able to satisfy this demand if built in concrete. Instead economies in the use of scarce materials were attempted by reductions in the overall weight, and subsequent reductions in the size of structural members. Only when the question of protection was considered essential was concrete used. Designs were shown incorporating a bomb bursting layer and heavy overhead protection but, according to Graham Dawbarn, the use of concrete generally implied 'a degree of permanence which is somewhat alien to the original intentions of blister design' (30).

The idea leading to the all-compressive thin concrete shell developed by Major J. H. de W. Waller did in fact not originate from the development of arched structures at all. Waller was primarily interested in the problems of shuttering. In 1941 his design for a duodecagonal hut with a conical roof, which had been used for army hutting, was published (31). Here, using cement grout applied directly to jute fabric, shuttering was avoided altogether. With the jute stretched on a simple concrete frame, construction was both quick and easy. The need for steel was not overcome, however, and reinforcement bars were used for the conical roof.

A few weeks later a second design was published where this system, cement grout directly applied on fabric, was again used, but this time for prefabricated structures. Sacking was stretched between side frames to form the structural panels of three-pin-portal constructions. Each panel was shaped with the deepest part of its 'U' section at the eaves. After casting, the units were bolted together and sealed with mastic. The particular advantages of this system were described in the 'Architect and Building News'. The system was claimed to obtain:

". . . the maximum advantage from the use of reinforced concrete by forming the covering or infilling structures in the framework. . . . Using conventional methods of construction, such a shape would require great skill in the placing of reinforcement and concrete." (32).

Later the Ministry of Supply used a similar technique in order to avoid using shuttering: 'Hy-rib' construction (33). Having noted how well the rendered metal mesh used as a substitute for glass in skylights stood up to blasts, the 'Padmos' hut was constructed by stretching metal mesh on a steel subframe and by rendering in cement grout (34). Both systems were still reliant on steel as reinforcement, but in 1943–44 with the vast build up of forces for a cross channel invasion more camps were needed. During the massive construction programme which resulted Waller was able to implement his discovery of a construction method, which allowed a greater plasticity in design, on the theory of the all compressive parabolic arch using no reinforcement (35).

The invasion decision in 1943 had marked the start of the second major hutting phase. The rise in manpower due to conscription had been expected to reach its maximum in December 1941. But no sooner were the new hutted camps needed for this nearing completion, than the decision was taken to build up a vast force in England for a cross Channel invasion. Initially this meant a build up of British and other Allied forces on a reasonable scale but, from 1943 onwards, the increase would be drastic and would call for a new approach to camp design. The British army had been expected to number 2 340 000 men by December 1941, but of these a large section would be abroad. Even in 1944 only 1 750 000 British soldiers were on British soil. These could largely be integrated into a local environment, and catered for by local entertainment facilities. This was not the case with the increasing number of 'foreigners' on British soil. Between summer 1943 and the first month of 1944, 750 000 men arrived in Britain from the USA alone; between January 1944 and D-Day the total was brought up to 1 500 000 men. Added to these were other forces which had built up over a longer period: 175 000 Dominion forces and 44 000 from other countries, Czechs, Poles, Norwegians, Dutch, Belgians and French. This gave a total of 3 500 000 men being housed, fed and entertained on British soil, of which less than half were British.

The main problem was the large American section. It was somehow found possible to house 60 per cent of them in existing accommodation but the remaining 600 000 men had to be provided with new camps. These were partly built by British (27 per cent) and partly by American (73 per cent) labour (36). The fact that British contractors seemed to take a minor part in this project is amply explained by the many other construction projects taking place at the same time—such as the massive Mulberry Harbours.

The new camps had to cater not only for the soldiers but for the

21. The prefabricated timber Harvard Field Hos[pital] shipped to England from the USA.

22. Interior of a Nissen hut hospital at a US Ar[my] Engineer camp in Britain: 1943.

environment they brought along. Concentrated mainly west and north of Southampton, the one and a half million Americans had their own newspaper, filmshows, radio and colour bar which they immediately grafted on to the local British culture. Ice cream parlours, pubs and canteens were opened in towns, some for black and others for white G.I.'s. The British military historian Liddell Hart writes that:

"British uniforms were so rare in the south-west, apart from soldiers on home leave, that this part was ironically described by its inhabitants as 'occupied England'...." (37).

Though these foreign forces swamped the local countryside, not all of their needs could be catered for by existing towns and such buildings as hospitals and clubs had to be built within the camps. Similarly the decision to concentrate manpower along the south coast led to the growth of clusters of camps which, in many ways, raised their own problems of urban structure. Usually each camp site held approximately one thousand men. In November 1942 a cluster of five such camps, with a 750-bed hospital and a quartermaster's store, was described as the largest single project of its kind yet in Britain (38). But by 1944 the south-west coast could be described as one continuous cluster holding not thousands but millions of men. Within this massive build up it is not surprising that important changes were taking place in hutting design.

Two major factors led to this change of attitude: the demand for increases in the speed of construction and changes in the availability of materials. The first had been seen as a gradual change and probably found its main turning point when Hitler turned east. Then the emphasis was no longer on defence but on a speedy recovery and build up for the attack. The second probably centred around two main events: the entry of America into the war as a supplier of materials, the winning of the battle of the Atlantic in May 1943 and the consequent reductions in the dangers to shipping. The result was that more and more steel was available for construction purposes. When the New York editor of the 'Engineering News Record' visited American camps in Britain he made a list of 'the most plentiful materials in Britain', and subsequently found them to be: 'brick, concrete, asbestos-cement and *steel*' (39) (italics by the authors). The critical material now was timber (40).

In the USA the opposite had always been the case. When scarcity of steel became apparent during the early stages of the war, the large amount of home-grown timber was generally used as a substitute as well as concrete. Considerable research went into laminated timber beams and plywood girders (41). Large-span timber structures were developed, such as the 46 metre trusses developed for an aircraft plant (42). The design of military hutting also relied extensively, if not completely, on timber (43) and when the American Red Cross sent over a hutted Harvard field hospital it was therefore in timber. The RIBA Journal had commented in 1941:

"... the sight of so much timber so prodigally used makes it seem rather a dreamland affair to a timber-starved British architect today...." (44).

By 1943 this difference in available materials had manifested itself even further and it was decided to use British designs rather than American for the new camps in order to utilize local materials. The editor of the 'Engineering News Record' wrote:

"The structures are all of British design, because they are for the British, and we are merely going to use them. They appear to be permanent constructions only because *they are built of the most plentiful materials in Britain*." (italics by the authors) (45).

As previously seen steel was now one of those materials and most of the earlier hutting designs had become obsolete. Instead the old Nissen hut and modifications of this design, the 'Iris' and 'Romney' huts, formed the main part of the hutted accommodation within the camps. When it was felt that savings in steel ought to be made the 'Asbestos Arch Hut', using semi-circular asbestos sheets without framing (similar in appearance to the Nissen hut), and the BCF and MOWP huts were used. The latter two were chosen mainly where the span of the Nissen hut was insufficient. A variety of other systems were used, but mostly on a small or local scale. One such system is worth further description, however, the C'tesiphon hut by Waller: the first use of an all compressive thin concrete shell structure.

During the feverish construction of the American camps, in preparation for the offensive, Waller had been given the go-ahead to construct hutting using the method of cement grout on fabric, which completely avoided the use of steel reinforcement. A temporary set of steel or wooden arches was erected over which fabric was stretched. When rendering was applied the fabric would sag between each arch, thus providing corrugation of the shell and further stability. The cement rendering would be applied until it reached a final thickness of 50 mm, after which the temporary arches were removed. Spans of 4.9, 8.5

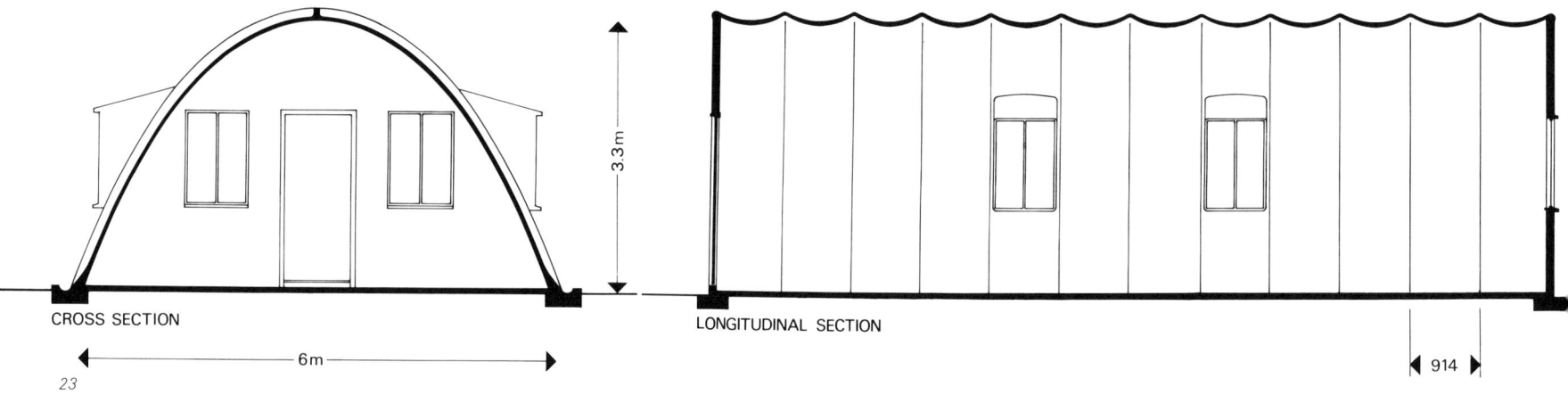

23, 24. A C'tesiphon hut during construction and sections through a hut with vertical brick end walls.

25. Exterior view of one of the C'tesiphon huts at an American camp in Britain during 1943: end wall also of C'tesiphon construction.

and 11.0 m were built. One hut, 4.9 m wide and 11.0 m long, was completed by 16 men in 12 hours. Later this type of construction was also successfully used in forming the ends of blister hangars. A cylindrical segment was built between a semi-circular base and the arch of the hangar.

This development opened up a whole new field of structural approach. But, in addition to this and other structural developments taking place, the experience gained in the management of these huge building operations was also significant. The administrative success of the hutting programme certainly led Churchill to attack the housing shortage after the war as a military operation (46). Even after he lost office in the postwar election, the chief reliance for new accommodation during the 1945-47 period was placed on the temporary prefabricated bungalows scheme (47). The influence of the hutting build-up and its structural solutions was great.

By spring 1944, using the remarkable innovation shown in the hutting programme, an invasion force of 3 500 000 men was prepared in Britain for the assault on the Atlantic Wall. The Official History states:

"Britain had become a huge store house, workshop, arsenal, armed camp, and aircraft carrier. It was claimed facetiously at the time that only the great number of barrage balloons floating constantly in the skies kept the island from sinking under the seas." (48).

The ingenuity and scale of this programme was now to be matched by the constructional innovations used in the preparation of the floating prefabricated harbours for the breach of this 'Wall': the Mulberry Harbours.

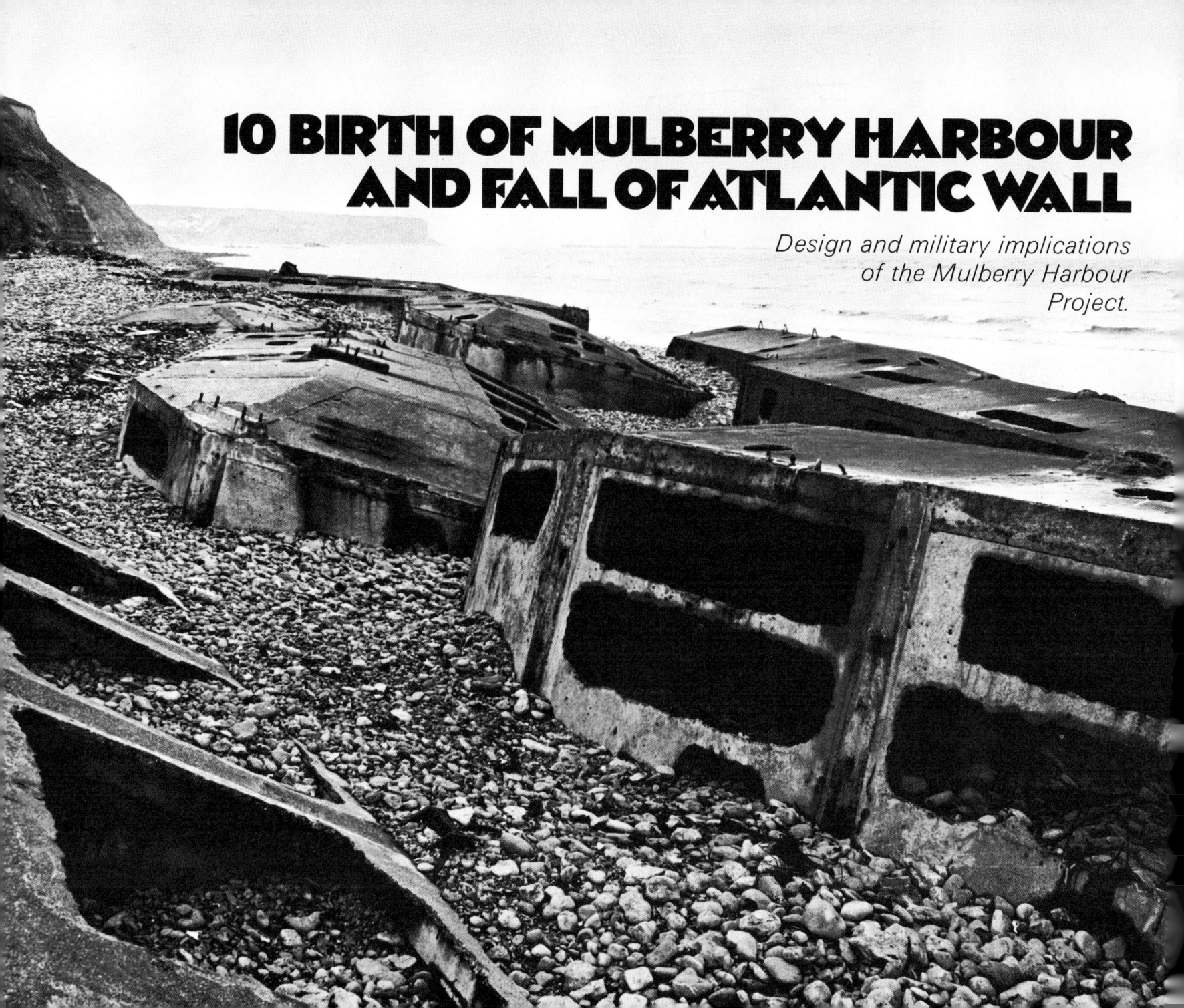

10 BIRTH OF MULBERRY HARBOUR AND FALL OF ATLANTIC WALL

Design and military implications of the Mulberry Harbour Project.

Events

1917 Churchill proposes artificial harbour as naval base.

1940 July Churchill establishes Combined Operation Command to prepare for the return of British forces to the Continent.

1941 Summer First of a series of telegrams from Stalin reach London, urging the establishment of a British second front to ease German pressure on Russia.

Dec. 17 Japanese attack on Pearl Harbour and Americans enter war.

1942 Feb. Decision to launch Anglo-American effort against Germany first; Japan is to be attacked later.

Spring Marshall proposes 'Operation Roundup' across Channel in 1943.

Summer Mountbatten suggests use of artificial harbour in Cross Channel offensive.

1943 Jan. Cross-Channel offensive postponed to 1944; main objective is no longer one of establishing a second front in Europe, as desired by Moscow, but one of dealing a fatal blow to Germany.

Summer Decision made to use artificial harbours.

1944 June 1 Block ships leave Britain for French coast.

June 6 D-Day.

June 7 D-Day plus one: Foothold established and first harbour units positioned.

June 13 D-Day plus seven: Blockships and floating breakwaters in position. Phoenix units and piers proceeding according to schedule.

June 18 D-Day plus twelve: Gale destroys Mulberry A and causes heavy damage to Mulberry B. Salvaged units from Mulberry A later used to reinforce Mulberry B.

July 1 Cherbourg captured: other ports fall in sequence.

Sept. France liberated.

---

1. *Remnants of Mulberry 'B' at Arromanches today. In the foreground are some of the wrecked Beetle pontoons.*

● Plans for the establishment of a second front in Europe were first prepared with the aim of relieving the German pressure on the Russian front. The intention was to make an invasion to achieve this in spring 1943, but when Churchill and Roosevelt met at Casablanca, in January 1943, the objectives for such an invasion were changed, instead of merely establishing a second front, Allied forces were to be landed in such force as to liberate Western Europe, striking into the heart of Germany through the French coast. To prepare such a large-scale scheme it was necessary to delay the invasion until 1944. For both tactical and strategical reasons it was felt that a landing in Normandy offered the best chance of success. But the invasion planners at COSSAC (Chief of Staff to the Supreme Allied Commander), pointed out one major drawback which had to be considered:

"... some time must elapse before the capture of a major port." (1).

The ports were the major fortified points in the Atlantic Wall—the 'knots' in the 'piece of string'. Their strength had been proved by Churchill's tragic 'reconnaissance in force' on Dieppe in August 1942 (2). Plans had therefore to be made under the assumption that even if Cherbourg could be seized within the first two weeks, the clearing of mines and repair of demolition would take at least two months. This was to cause quite a headache for those responsible for supply and movement. Without at least one port in the first few days, they concluded that the invasion project was not feasible. It was when discussing this in the early summer of 1943 that the then senior Royal Navy representative at COSSAC headquarters, Commodore John Hughes-Hallett, pronounced jokingly:

"Well all I can say is if we can't capture a port we must take one with us." (3).

Little did he know then that this was exactly what was to happen. In his book 'Overture to Overlord' Lieutenant-General Sir Frederick Morgan sees this as the birth of Mulberry Harbour, as the project was later to be called, and he believes that the credit for the entire concept goes to Hughes-Hallett (4). The idea of an artificial harbour had a certain amount of precedent. Certainly there is some comparison to be made with the sea forts project with which Hughes-Hallett had also been involved but, as mentioned previously, even Maunsell's ideas were not completely new.

In July 1917, Churchill had written a proposal for the capture of

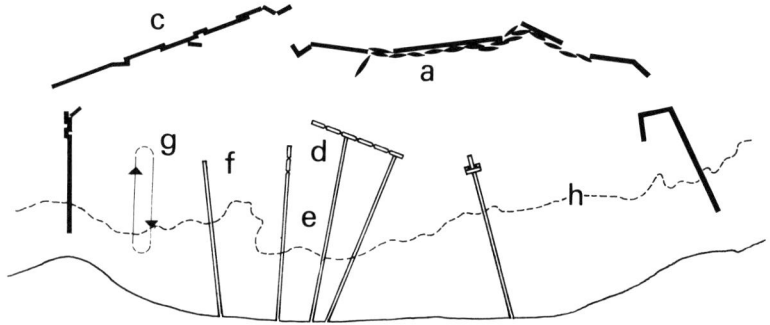

2, 3. Aerial photograph and plan of the Mulberry 'B' Harbour at Arromanches:
    a. blockships forming the initial breakwater
    b. position of the row of Bombardons—floating outer breakwater
    c. Phoenix breakwater units sunk in position
    d. Spud pierhead units on telescopic legs
    e. Whale bridges on floating pontoons or pontoons on telescopic legs
    f. Swiss Roll—the Royal Navy's floating roadway
    g. ferry
    h. line of low tide

Borkum and Sylt, in the Friesian Islands, for use as naval bases. Churchill, without any technical assistance, had included the following additional proposal:

"*Para. 30.* One of the methods (for erecting a naval base) suggested for investigation is as follows:

A number of flat bottomed barges or caissons, made not of steel but of concrete, should be prepared in the Humber, at Harwich, and in the Wash, the Medway and the Thames. These structures would be adapted to the depths in which they were to be sunk, according to a general plan. They would float when empty of water, and thus could be towed across to the site of the artificial island. On arrival at the buoys marking the island sea-cocks would be opened and they would settle down on the bottom. They could subsequently be gradually filled with sand, as opportunity served, by suction dredgers. These structures would range in size from 50 feet by 40 feet by 20 feet to 120 feet by 80 feet by 40 feet. By this means a torpedo and weather-proof harbour, like an atoll, would be created in the open sea, with regular pens for the destroyers and submarines, and alighting-platforms for aeroplanes. This project, if feasible is capable of great elaboration, and it might be applied in various places. Concrete vessels can perhaps be made to carry a complete heavy gun turret, and these, on the admission of water to their outer chambers would sit on the sea floor, like the Solent forts, at the desired points. Other sinkable structures could be made to contain store-rooms, oil-tanks, or living-chambers. It is not possible, without an expert inquiry to do more here than indicate the possibilities, which embrace nothing less than the creation, transportation in pieces, assemblement and posing of an artificial island and destroyer base (harbour)." (5).

Twenty-five years later Churchill still had this idea in mind when, after the loss of Singapore, he ordered Admiral Lord Louis Mountbatten, the Chief of Combined Operations, to investigate the possibility of building an artificial harbour on an island in the Indian Ocean. In a minute to Mountbatten on May 30th, 1942, the Prime Minister wrote:

"Piers for use on beaches: They must float up and down with the tide. The anchor problem must be mastered. . . . Let me have the best solution worked out. . . . Don't argue the matter. The difficulties will argue for themselves." (6).

The Indian Ocean project never came beyond the research stage but Mountbatten mentioned the idea, as a possible way of entering Europe, at a meeting in the early summer of 1942. At that time plans for the invasion of German-occupied France were still only embryonic:

"If ports are not available, we may have to construct (them) in pieces and tow them in." (7).

This was exactly what was to happen two years later but Eisenhower, who was present, remarks that the suggestion was met by 'hoots and jeers'.

When the decision was made to go ahead with the Mulberry Harbour project in 1943 the performance specification was formidable. A harbour had to be erected in Normandy in 15 days which would be capable of a lifespan of at least three months. The harbour had to have a handling capacity of 12 000 tons of stores and 2 500 vehicles per day as well as offering protection for Liberty ships and landing craft. The solution was to build two complete harbours, Mulberry A and Mulberry B, and several small harbours known as 'Gooseberry Harbours'. The latter were to be simply sunken blockships, offering breakwater protection for small craft during the initial landing, whereas the Mulberries were to have a complete system of floating piers, protected by breakwaters, from which larger craft could be directly unloaded. Initially the War Office had planned the Mulberry harbours to take all the transport but the Admiralty pointed out that it would take at least two weeks to finish their positioning. If a wind from the north should blow up during that time, the whole expedition would find itself unprotected. It was with this in mind that Rear Admiral William Tennant, in charge of the naval co-ordination of the Mulberry harbours, had suggested the Gooseberry harbours as an additional measure.

Breakwaters were one of the main problems in the design of the Mulberry harbours. The initial reports to the Admiralty stated quite bluntly that:

"Nothing favoured the choice of this coast (Normandy) for the construction of artificial harbours." (8).

An important factor was the tide for, with up to 7.3 m difference between high and low tide, a breakwater in 9.0 m of water would have to be over 16.5 m high. It was decided to build an inner breakwater on similar principles to those suggested by Churchill in 1917. Huge concrete caissons, called 'Phoenix' units, were to be towed across the Channel and sunk on location. Due to the extreme conditions, however,

4. Lilo: the flexible sided breakwater unit.
5. The Hard Lilos or Bombardons during mass production.
6. The Bombardons moored in position at Mulberry 'A': pairs used between each buoy.

and the need for some shelter facilities in the deep water outside the main harbours, a second more flexible breakwater was planned outside the Phoenix units. The research on this was given to DMWD, Department of Miscellaneous Weapons Development, at the Admiralty.

The department was led by Commander Sir Charles Goodeve, but the specific research was led by Dr. White, Lieutenant-Commander Coulson and Lieutenant-Commander Lochner. Initially they tested existing ideas on bubble breakwaters. The ability of bubbles to calm waves had been first observed by an American swimming-bath superintendent. In 1916 bubble breakwater equipment was tried out on the California coast but, following difficulties and only limited success, the project was abandoned. When DMWD took it up again they came to much the same conclusions. The larger the waves, the more air had to be pumped out from the pipes on the sea bed. To build a complete harbour in Channel conditions was therefore nearly impossible—if only because of the vast amount of air compression plant needed.

Fortunately Lochner had a quite new idea which was to result in pneumatic, floating breakwaters, later known by the code-name 'Lilo'. The concept of a pneumatic breakwater was apparently first tested by Lochner and his wife, in the middle of the night, using a buoyant rubber mattress and two metal bars in their garden pond. Further tests showed this new breakwater principle to work remarkably well. Wave action only takes place near the surface of the sea and the pneumatic superstructure was to absorb this action while a concrete keel was to keep the superstructure upright and prevent it from moving other than vertically. Construction of full-scale prototypes was then carried out in conjunction with the Balloon Development Establishment, part of Beaverbrook's Ministry of Aircraft Production (MAP), who gave advice on the production of the vast pneumatic envelopes required. Bateman, a balloon designer at Cardington, helped in the design of these envelopes and Dunlop were to build them. Design of the concrete keel was to be carried out by the consulting engineer, Dr. Oscar Faber (9).

When full-scale tests of these prototypes were later carried out they were successful. The final Lilo was to consist of a pneumatic superstructure 61.0 m long and 2.4 m wide with three separate compartments. This was attached to a hollow concrete tube, 2.4 m in diameter, which could be easily pumped dry. When full of air it could float, carrying the inflated pneumatic, but when flooded this keel weighed 750 tons. By running the compartments lengthwise, the air pressure at three levels of depth could be matched with the external pressure. Thus an increase in water pressure on one side would cause that side to yield and in this way the wave would be prevented from transferring its energy to the other side. Lochner was not satisfied, however, as the thin membrane of the pneumatic was highly vulnerable to damage during construction and enemy attack. He therefore started work on several rigid-sided models of different cross-sections. Prototypes of this type of breakwater, initially known as the 'Hard Lilo', proved to be successful and on September 13th DMWD decided to go ahead with the construction of these in steel. The soft membrane Lilo was subsequently dropped.

The rigid breakwater, which was now given the name 'Bombardon', looked like a Maltese cross in cross-section. Each unit was 61.0 m long and just over 7.6 m wide with a draught of 5.8 m. The bottom and side arms were flooded to increase their weight while the structure was kept afloat by a series of separate watertight compartments along the top arm. Due to the shortage of time and labour, Bombardons were mass-produced as a series of prefabricated components which were bolted together during the final assembly. A bolted connection would not have been considered by a naval architect in peacetime but using this method 1700 men were able to build the required units at Tilbury and King George V Docks at Southampton in the six months now left before the harbours had to be ready (10).

Meanwhile development of both the Phoenix units (concrete caissons) and floating piers had been going ahead. The initial research on such units had been undertaken in 1942, on orders from Mountbatten. These orders did not relate to the Mulberry project but were based on the minute from Churchill concerning a harbour in the Indian Ocean. Despite this extra period of development the design situation which arose was not ideal; responsibility for the construction of the caissons, roadways and floating pier-heads being in the hands of the War Office who naturally had little experience of seamanship (11). The work on the Phoenix units called for a total of 15 000 workers and 630 000 tons of concrete. Although it was intended to use a battery-casting construction method and prefabrication, which had been developed in the USA, the construction methods varied from site to site and from firm to firm in order to retain the labour force available.

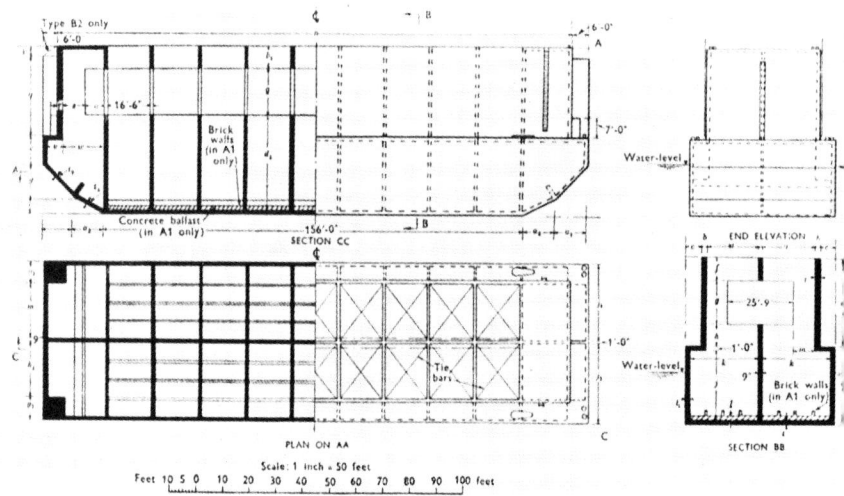

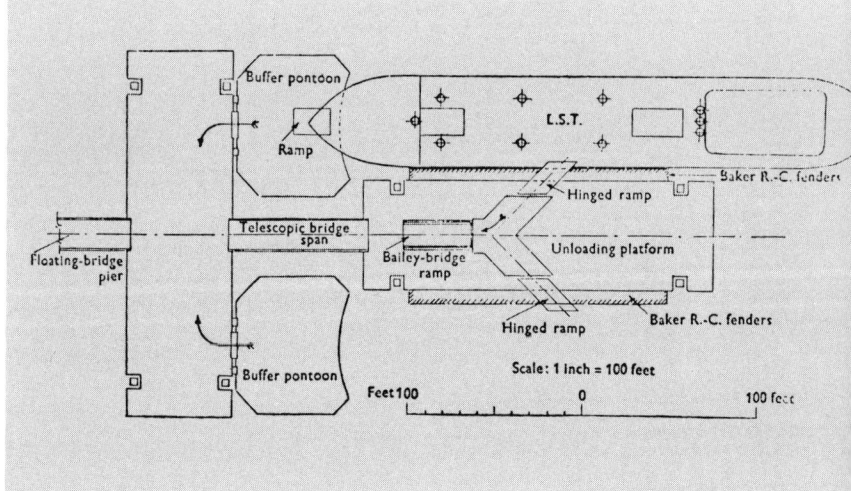

7. The Phoenix units, the massive concrete caissons forming the inner breakwater, in position with their gun platforms.

8. Diagram showing the concrete construction of the Phoenix units: the height varied with the depth at which they were to be deposited.

9. The Spud pierhead with telescopic legs and Baker R.C. fenders.

10. Plan of pierhead fully assembled.

One method developed, for example, was the use of plastered 'Hy-rib' reinforcement as permanent shuttering. Each unit was equipped with a small flak tower with quarters and stores below the gun platform. Their size varied according to the position they were to take in the harbour (12).

The pier-heads, known as 'Spuds', were self-contained units, 61.0 m long, 18.3 m wide and 3.0 m deep, which floated up and down with the tide. At a symposium on 'War-Time Engineering Problems' at the Institute of Civil Engineers, Richard Pavry explained the idea behind their design:

"The conception of the type of pier-head which was finally adopted arose from the knowledge of the behaviour of a particular type of spud-operated dredger. This dredger presses down its 'spuds' or legs on to the sea-bed and lifts itself partially out of flotation during dredging. The operation of 'spudding' affords a rigid anchorage and transfers the reactions of the dredging operation to the sea-bed by means of the spuds, which can be moved up or down, but which are restrained laterally by vertical guides. It was on record that a dredger of that type, built in 1923 by Lobnitz and Co., Ltd., shipbuilders of Renfrew, by using its spuds for anchorage, successfully withstood a severe storm in the West Indies when all other craft in the vicinity were wrecked or driven ashore." (13).

Subsequently a pier-head unit was equipped with telescopic legs in each corner which took part of its weight and served as an anchorage, as well as keeping the unit steady while loads were being moved about. The power plant necessary to manipulate the legs was contained within the units together with crew accommodation for 22 men. Although self-contained in this way, the units were linked by ramps with a system of buffer barges and floating berths to which the supply ships could unload.

The Spuds were connected to the beach by the 'Whale': a system of flexible bridges, 3.0 m wide, which was supported by pontoons. The system was ingenious but its design had a certain amount of precedent in the field of military bridging. Each Whale was 1.5 km long and consisted of 80 units. Every sixth span was telescopic to allow for the twist and sag of the pier in heavy weather. Due to the steel shortage the pontoons were of two types: steel spud-type pontoons for a firm sea bed, and fully floating concrete pontoons, known as 'Beetles', similar in design to the concrete barges used on the Thames. The latter were of pre-cast construction with 30 to 50 mm concrete panels mounted on a concrete frame. Their construction was designed, like the Bombardon, for mass-production and each unit could be finished in 10 days (14).

A second type of pier was also employed in the final design of Mulberry B. This was known as the 'Swiss Roll'. The principle was discovered in 1941 by Ronald M. Hamilton and a year later was perfected by him together with the DMWD. In accordance with Hamilton's theory of 'Rolling Dynamic Buoyancy' the roadway, with its flexible side extensions, floated horizontally when unloaded. When loaded at any point the flexible sides would fold up, thus in effect forming a 'boat' immediately around the load. The important feature was, however, that when a load such as a lorry drove along the road it would drive out of this depression, overtaking its own wave in the same way as a speed boat. The road was successfully tested in a heavy swell and included in the Mulberry project (15).

The Bombardons had meanwhile been tested for mooring. While ships would usually moor the same distance apart as their own length, this could not be done with the Bombardons as it would allow waves to pass through. The final design called for pairs of Bombardons mounted between buoys leaving gaps of 15.2 m. To reduce the wave action further, two separate rows were to be erected 240 km apart. Tank tests showed that this would reduce waves to a tenth of their original force and a full-scale test at Weymouth, under a force 7–8 wind, further proved the capabilities of the units. Waves 2.5 m high and 60 m long provided twice the force for which the units had been designed but still the double lines held. A US minesweeper managed to lower a small boat, which, sheltered by the Bombardons, rowed about quite safely.

By June 1944 the various components of the Mulberry Harbour project were ready to go into action. The initial impetus was a single bold concept but the final project represented the total of a considerable number of diverse inventions and developments. Nor, incidentally, was the Mulberry Harbour project the only development designed with the object of breaching Hitler's Atlantic Wall. Such devices as rocket-propelled grapnels for scaling the cliffs on the invasion coast, special landing craft with extending fire-escape ladders equipped with machine guns and prepared for the same purpose, and a vast number of special

11

12

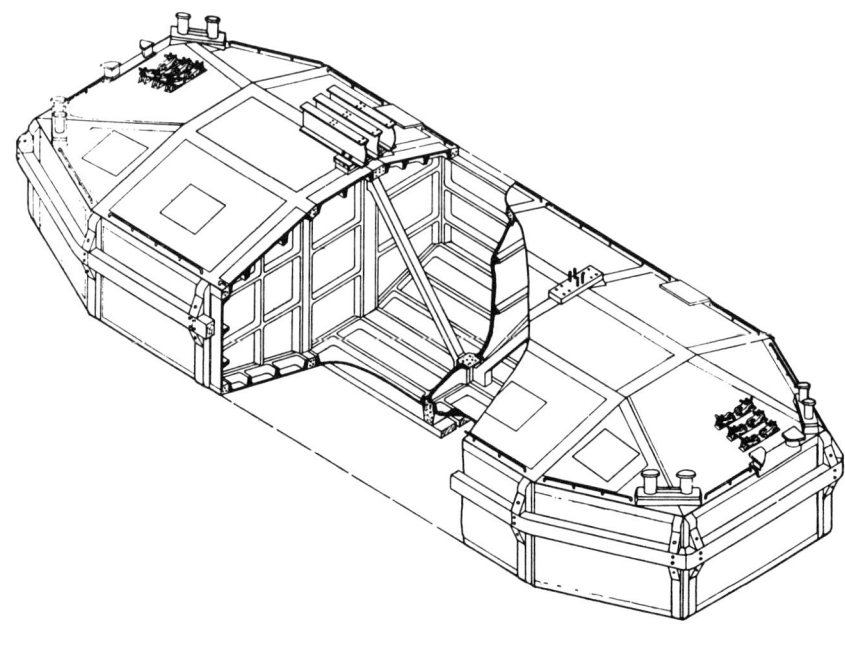

13

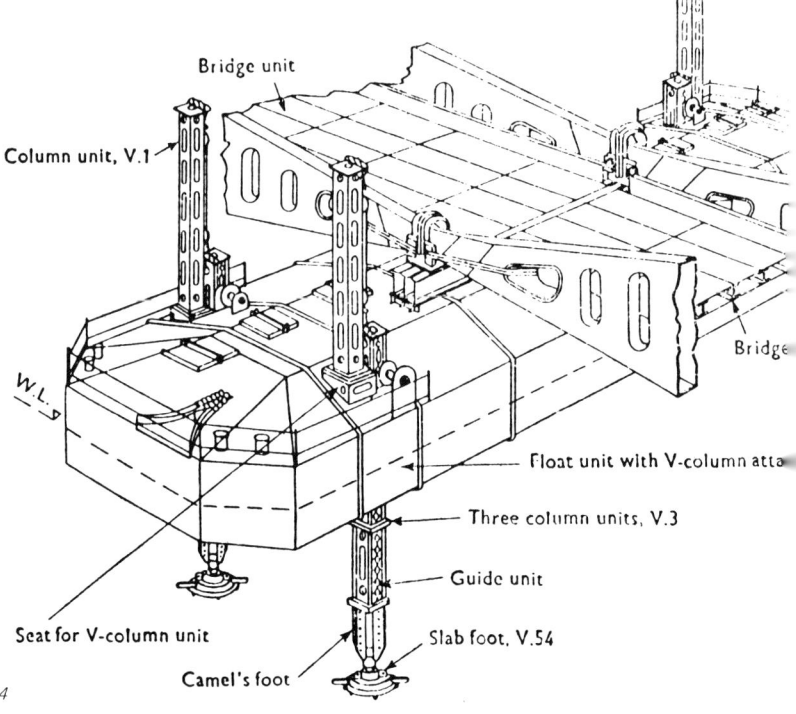

14

11. One of the floating bridges connecting the pierheads with land.

12. Telescopic bridge span being tested.

13. The fully floating Beetle bridge pontoon: diagram showing the concrete construction.

14. The steel pontoon with telescopic legs. The use of this pontoon was limited to firm seabeds due mainly to the shortage of steel. On soft sand the Beetle pontoon was used.

assault vehicles and tanks were designed with the sole object of smashing through the coastal fortifications of the 'Wall'.

On D-Day, June 6th, 1944, the invasion of Normandy began. Initial air bombardment of German defence positions had already started. Many of the field defences were flattened but Rommel's insistence on casemating all defensive positions had proved part'y successful. The naval batteries with their foundation collars proved much more resistant than the army batteries, however, and the naval battery 'Longues' stood unscratched even after a 2000 lb. bomb had swept away the whole of the embankment from one of its casemates. The army batteries at Houlgate and Pointe du Hoc were less fortunate. Guns in some of the batteries which did not prove safe were moved back and camouflaged while the Allies bombed the empty casemates. Generally the most successful effect of the bombing of the defences was the destruction of the six radar stations in the invasion area. Due to this and decoy 'echoes' created by the Royal Navy the Germans were given the impression that the assault would come in the Pas de Calais. Even after D-Day Hitler expected the major assault to arrive in the Pas de Calais area where his V-weapon sites formed the bait.

Batteries, strongpoints and bridges of special importance were captured by airborne troops on D-Day before the actual landing. Training for this purpose had been undertaken on full-scale mock-ups in England but the inaccuracy of these mock-ups, plus last minute preparations by the defenders, gave the airborne troops some unwelcome surprises (16). Ordinary batteries and strongpoints were left to the landing forces, but the navy were also involved in this. On D-Day the navy began bombarding the defences for 40 minutes before sunrise. The only battery that answered was Longues with its four 155 mm guns. Chester Wilmot describes the duel which followed:

"With its (Longues) first salvoes . . . this battery was tactless enough to straddle the ship which carried the Sector HQ and the Corps Commander. At once, the six-inch guns of HMS Ajax were brought to bear and in twenty minutes the duel was won. Three of the four casemates received direct hits and had their guns silenced, two of them by shells which entered the narrow embrasures (2.5 m by 3.85 m)." (17).

The vulnerable nature of embrasures, like the poor quality of construction of the army batteries, became only too obvious under the test of D-Day. The Atlantic Wall thus proved to have serious detail design faults even before the landing.

The initial bombardment by the air force and navy had not been as effective against the smaller fortifications which, unlike the casemates, were difficult to identify. These installations had been given overhead protection and concrete walls or earth embankments had been constructed, providing conditions of enfilade fire not unlike the flanking galleries of the nineteenth-century fortresses. Like the casemates, a layout designed for flanking fire in this way restricted the field of fire and the tactical advantages of flanking fire were offset by the fact that the guns in such installations were unable to fire directly towards the sea. When making this decision the Germans had assumed that an attack would take place at high tide, in which case the main fire should be brought to bear between high tide mark and the sea wall. They also assumed that, if protected against naval bombardment, the assault battalions would have no means of silencing the machine guns and anti-tank guns in their heavy emplacements. Thus they would wipe out the infantry before armour could be brought to their support.

The main assault was divided into five groups: Sword (British), Juno (Canadian), Gold (British), Omaha (USA) and Utah (USA). Within this structure certain differences of approach to the problems created by this system of defence became apparent.

On Omaha beach the system of defence worked well, but in the British and Canadian sectors Montgomery did not comply with the German assumptions. Anticipating Rommel's plan of beating the attack while still on the beaches, he sent a variety of armour in before the infantry. This armour had been designed with the sole object of penetrating the Atlantic Wall. A weird assortment of 'Crabs', 'Bobbins' and other specifically equipped armour cleared a path through the beach obstacles while amphibious armour such as the DD (Duplex Drive) Sherman conversion swam ashore to take care of the fortifications (18). As the larger German guns in their concrete emplacements were restricted in field of fire, this armour could drive in close and fire at point blank range through the embrasures. Even strongpoints such as Le Hamel and La Rivière, in the heavily defended Gold sector, were silenced by the special armour without too much difficulty. The Americans, under Bradley, found the penetration of the Atlantic Wall more difficult. According to Chester Wilmot, Bradley had a contempt for British 'over-insurance' (19) and excused himself on the pretext that training

15

16

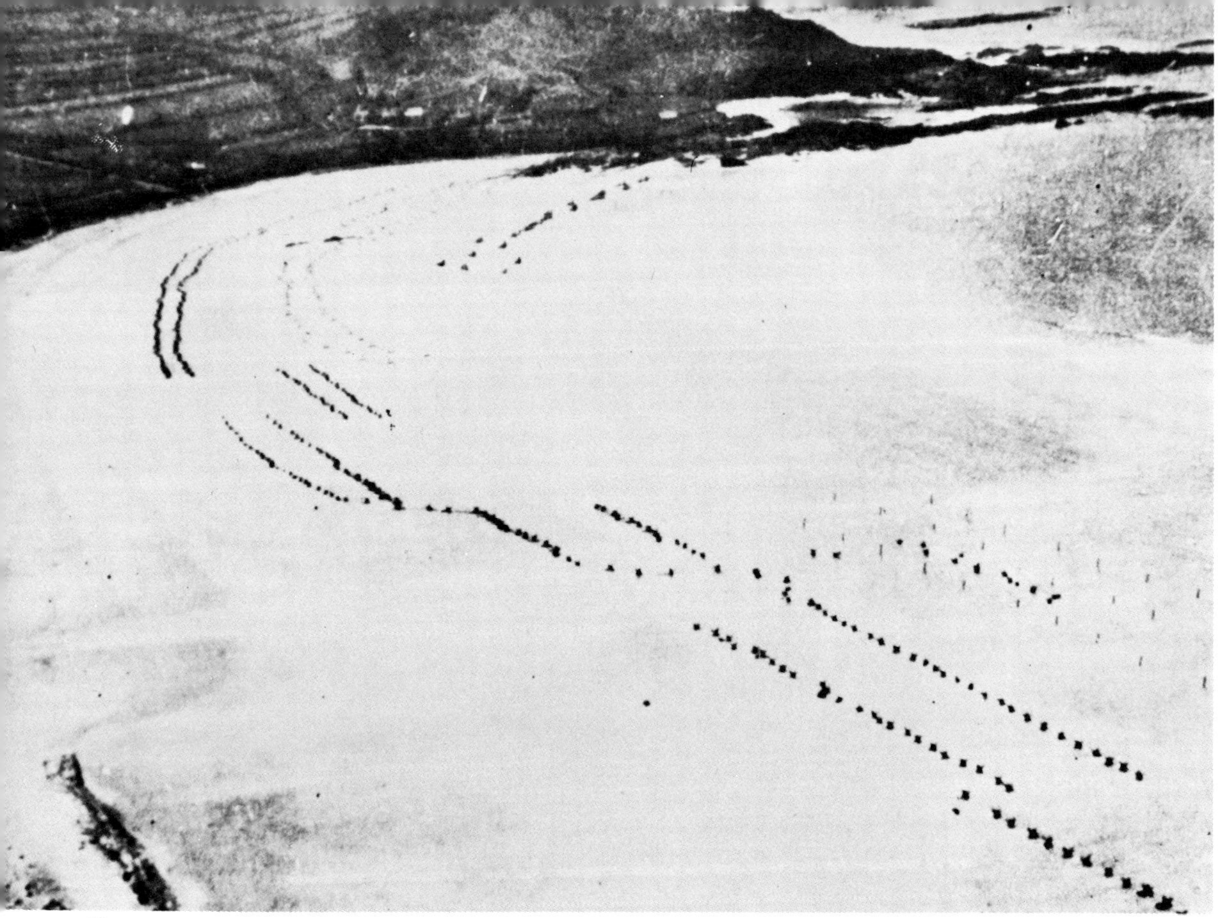

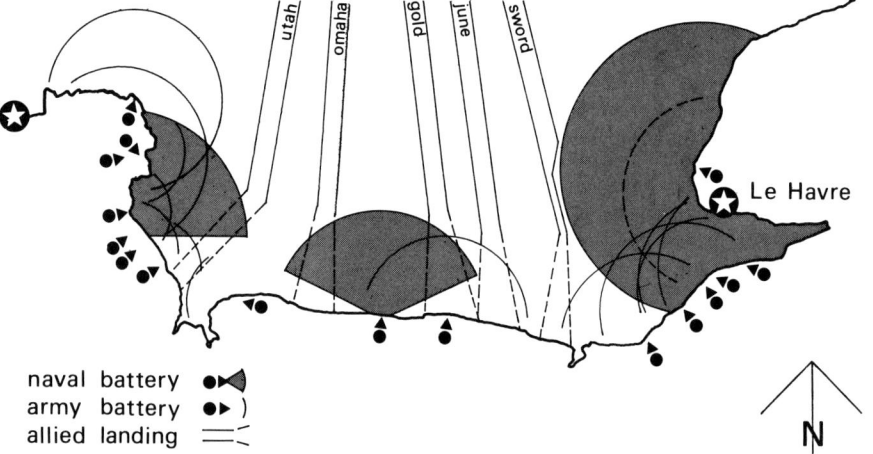

15. Aerial view of the beach defences built by Rommel on the Cherbourg Peninsula.

16. Atlantic Wall between Cherbourg and Le Havre: relationship of invasion plan to battery disposition.

crews for the special armour presented too many difficulties. Out of the various conversions he chose to incorporate only the amphibious DD tank in the assault. Even this tank was only used in small numbers.

In the Omaha sector, where only two such tanks were to reach the beach together with the first assault, the shortsightedness of this policy soon became apparent. This stretch of beach naturally lent itself to defence, and with steep cliffs on either side, the Germans had been able to concentrate their efforts, confronting and flanking the escape valleys with trenches and concrete emplacements for machine gun and anti-tank guns. The intention was that the defences at Omaha would be cleared by engineers using pole-charges and man-pack flame-throwers against pill-boxes, hand placed explosives against barbed wire and concrete walls, and clearing mines by hand. While the British had learnt at the Dieppe raid that this could not be done effectively without armour protection, the Americans had not. Delay and heavy casualties were the result. When the tanks did arrive there was not enough room clear to manoeuvre but in the Utah sector the Americans were more fortunate. Here 28 DD tanks had reached the shore and by a stroke of luck the current had moved the assault party east away from the heavily defended batteries by Marcouf. This sector was only weakly defended as the Germans believed that the double rows of swamp area would prevent the Allies from choosing it as an assault beach. The breakthrough, here helped by armour as in the British sectors, was thus made relatively easily.

Having won the tactical battle by knocking out the reinforced concrete fortifications and closing the gap between Gold and Omaha the Allies were now to exploit the overall strategic weakness of the 'Wall' by fetching in the Mulberry Harbour units for erection. In this way the main 'Fortresses' of the Atlantic Wall, the Channel ports, were to be strategically outmanoeuvred.

The first units had started across the Channel quite early in the invasion programme. These were the 60 blockships for the Gooseberry harbours. Some of them, such as the 'Centurion', 'Coubert' and 'Sumatra', had proved so slow that many had to start six days before the main fleet. By the time the last blockship had been sunk the Phoenix units started to arrive and the other units were soon to follow, but the siting and sinking of Phoenix units was a slow process. The blockships and floating breakwaters therefore provided most of the protection during the first few days. These were finished by D-Day plus six at St. Laurent, the American Mulberry A, and by D-Day plus seven at Arromanches, the British Mulberry B. All seemed to follow the schedule without a hitch (20).

June 18th, D-Day plus 12, was to change this. A calm day had prompted the navy to start 23 tows of various units in one batch but in the evening a gale blew up and only one and a half of the 23 tows survived. The gale soon increased in strength to a storm not equalled in the Channel for 40 years. Some of the landing craft managed to reach shelter behind the blockships, while those less fortunate were blown on shore. Both harbours suffered colossal damage. The Bombardons stood the pressure of waves 4.5 m high and 90 m long but after this they all broke loose. Those at Mulberry B bore down on the Phoenix units at Mulberry A like battering rams (21). At Mulberry B, however, all but four of the Phoenix units withstood the storm, probably because they were sited at a different angle to their American counterpart and received the storm head on. The vessels sheltering within the harbours, now out of control, smashed the Spud units and piers. The result was utter chaos.

After the storm the units which could be salvaged from Mulberry A were used to strengthen the British harbour, which was still not completely demolished. It is here interesting to note the point raised by Chester Wilmot on the result of the destruction of Mulberry A:

"Undismayed by the destruction of their artificial harbour, the Americans applied to the development of the Omaha and Utah anchorages their talent of invention and organization. In defiance of orthodox opinion, they beached coasters and unloaded them directly into Army lorries at low tide. The boom of Liberty ships were regularly overloaded and sometimes they broke. . . . On the Omaha foreshore they scooped up the shingle to make a road-bed only to find that, with the removal of a shingle bank, the waves came up and washed away the road. For the most part, however, their improvisations paid big dividends. The open roadsteads and beaches of Omaha were so organized that during July the Americans here handled more than twice the tonnage which passed through the British Mulberry. . . . This achievement had led to the suggestion that the vast expenditure of effort and materials on the artificial harbours was in fact unnecessary and that the same build-up could have been achieved far more economically with a few hundred

17. Concrete anti-tank wall with flanking casemate on the Atlantic Wall of 1944: Sentry in a Tobruk emplacement.

more landing craft and ferries." (22).

However, as Wilmot points out later, the value of the Mulberries was also psychological. Without any experience to show that such build-up over open beaches was possible and with no shipping to spare, it enabled the planners to go ahead without relying on the immediate capture of one of the heavily fortified harbours.

Even Mulberry Harbour and the American beaching method could not, on their own, provide enough supplies for a campaign of this magnitude; it was necessary to allocate forces for the capture of each individual port as the Allies advanced towards Germany. In trying to defend them the Germans found another weakness in the Atlantic Wall; for, having casemated the large coastal guns, they could no longer defend the rear of the harbours. Only the fortifications which had been built after the harbours were declared 'fortresses', in 1943, could be brought to bear. But even Cherbourg, where rear fortifications had been built in great strength, was captured on July 1st. The other ports, not so strongly defended in their land approaches, took less time. General Crerar's First Canadian Army took Dieppe and Ostend without a fight, Le Havre in 48 hours, Calais and Boulogne in six days. The main Allied problem proved to be not the conquest of the harbours, but the repair of their demolition. The slowness of this, together with a storm which further reduced the material intake of Mulberry B and Cherbourg, made it necessary for Montgomery to push all his resources into the clearing of Antwerp instead of following his original idea of a break-through into the Ruhr. In assessing the effect of the Atlantic Wall on the course of the war, therefore, it must be conceded that the strategy of defending the Channel ports, although originally successfully countered by the Mulberry project, proved in the end to have a considerable effect on the speed of the Allied advance.

Whether the Atlantic Wall justified the energy put into it is, however, debatable. Albert Speer, the then Reich Minister of Armament and War Production, has recently expressed his opinion:

"For this task (the construction of the Atlantic Wall) we consumed, in barely two years of intensive building, seventeen million three hundred thousand cubic yards of concrete (as recorded June 5, 1944) worth 3.7 billion DM. In addition the armaments factories were deprived of 1.2 million metric tons of iron. All this expenditure was sheer waste. By means of a single brilliant technical idea the enemy by-passed these defences within two weeks after the first landing. . . . Our whole plan of defence had proved irrelevant." (23).

II THE SHELTER CONTROVERSY

British air raid shelter provision

Events

1917 June 26 Fourteen twin-engined Gotha aeroplanes bomb London, causing 162 deaths.

1924 Committee of Imperial Defence form Air Raid Precautions sub-committee.

1937 The ARP Act makes provisions for ARP obligatory for British local authorities.

1938 Sept. The Munich Crisis marks the start of practical shelter programme.

1939 Sept. War declared. 1 500 000 Andersons now delivered but not much else. Government asks local authorities to provide 'brick and concrete surface shelters' for public use.

1940 April Anderson production suspended due to steel shortage, some 2 300 000 having been produced.
March Government starts production of concrete and brick shelters for domestic use.
July–Oct. 'Battle of Britain' as Luftwaffe attempts to destroy RAF prior to an invasion.
Sept. 'Big Blitz' begins.
Oct. Government allows limited tunnelling.

1941 Jan. Government announces 'Morrison' shelter.
Feb. Bunks and amenities being provided in shelters.
March–April Government strengthens existing shelters.
May Blitz ends. 40 000 been killed due to bombing: approx. 65 per cent of total over 1939–45. Luftwaffe switch attention to Russia.

1942 Spring Shelter force cut almost completely.

1944 Jan.–March 'Little Blitz'.
June Flying bombs start to drop.

◀ *1. One of the eccentric ideas developed in the late thirties, before the utilitarian nature of the British shelter programme established itself. The pyramid of concrete balls was to take the impact of a direct hit, the energy of the bomb being being absorbed in displacing the balls.*

● The bombing of London in 1915 had established that war no longer had boundaries. Post-1918 Europe was consequently air-raid conscious. In Britain the Committee on Imperial Defence asked the principal service experts to report on the problem of future air attacks on the UK assuming France as the hypothetical enemy. When they reported in 1923 the results were frightening. The French Air Force would be able to drop the same tonnage of bombs on London in three days as was dropped on England during three years of war. Such a powerful threat prompted the formation of a special Air Raid Precautions sub-committee under Sir John Anderson in 1924. Its first annual report concluded:

"... that in the next war it may well be that nation, whose people can endure aerial bombardment the longer and with the greater stoicism, that will ultimately prove victorious." (1).

Despite this warning the development of passive shelter provision in Britain was slow and hindered by secrecy. The only area which managed to progress successfully was gas protection (2).

In the rest of Europe the situation was quite the reverse. The subject of ARP attracted extensive research and publicity. In the Pas de Calais and other parts of France Marshal Pétain, as Inspector-General of Air Defence, had carried out full-scale experiments. Similar tests had been carried out in Königsberg and other continental cities. In 1932 two documents were published abroad which proved quite definitely that British preparations were far behind those of France and Germany in detail design. 'Reichsverband der Deutschen Industrie' illustrated a scheme for the protection of industry against air attack and in France 'Practical Instruction on Passive Defence against Air Attack' was put on sale to the public for four francs. Both were representative of the public awareness which existed on the Continent. This awareness was well illustrated in the proposals by the architect Le Corbusier for a 'Radiant City' ('La Ville Radieuse') in 1933. Using the 'infamy of war as a pretext' Corbusier justified these proposals:

"Given the present stage of city planning, only those cities which are conceived along the lines of the Radiant City are capable of emerging victoriously from an air war. ... Because in the Radiant City 100 per cent of the ground is free and only 5 per cent to 12 per cent of the surface has buildings on it; now, those buildings are 'on pilotis', which means that air circulates, that there is continuous space, that the wind will be able to dissipate the gas. ... Densely populated buildings (like those he

215

2. Steel shelter for key personnel unable to leave their posts.

proposes) covering only 5 per cent to 12 per cent of the city's surface can, with a normal expenditure, be equipped with bomb-resistant armoured platforms; and on the last floor there can be shock-absorbing devices to take care of falling projectiles. . . ." (3).

It took the scare of Abyssinia in 1936 and then the Munich crisis in September 1938, as Britain faltered on the brink of war, to prompt comparable interest in the subject of ARP in Britain.

Munich started not only a surge forward in rearmament, but prompted a realistic attempt to come to terms with the public shelter problem in Britain. As far as the actual provision of shelters in Britain was concerned, the position in 1938—despite 20 years of breathing space—was no different from that of 1917, when the Cabinet Committee had recommended a policy of dispersal. The ARP Act of 1937 had only allowed public shelters for people caught in the street. During the Second Reading of the Act, the Home Secretary justified this policy:

". . . neither this Government nor, so far as I know, any government in Europe, can protect a building, short of an overwhelming expense, from a direct hit by a high explosive bomb." (4).

Munich did not alter this policy, which was by then a long-standing one. The population was to be 'dispersed' as evenly as possible on the premise that a bomb dropped on lower densities would cause fewer casualties than one dropped on a high density area. Artificially created high densities—large shelters—were frowned upon for this reason, though the logic of this aspect of the shelter programme was later proved to be mathematically doubtful (5).

During the Munich crisis, in September 1938, trenches had been constructed for the use of people caught in the streets but the main protection under the dispersal policy was the home. After September 1938, however, the public were demanding some form of Government protection, regardless of cost, and it was in an attempt to satisfy this demand within the dispersal policy that the Lord Privy Seal, Sir John Anderson, sought technical advice on implementing some form of protection. On December 21st, 1938, he was able to announce the invention of a new steel shelter (6), which was to be given free to the poorer inhabitants of vulnerable areas. Based initially on an income limit of £250 p.a. this meant the production of sufficient shelter for an estimated 10 000 000 people, or about 2.5 million shelters.

The Anderson shelter, as it was known, was a well thought-out piece of mass-production. Like the Nissen hut of 1916, the shelter was designed from corrugated steel components and with a covering of earth it was intended to give protection for four to six persons from splinter and blast. The design overcame the cost problem, costing little over £5 per shelter to produce. Even unlined trenches had been costing £2 per person (7) and cost was, from the Government point of view, a not unimportant design criteria. Its original specification was that it should be suitable for erection inside the average small working-class house, but this idea was eventually rejected because of the danger of escaping coal-gas or fire to a trapped householder. During Munich householders had been asked to build trenches in their gardens, and the final plan was, therefore, for a shelter which would be an extension of this plan.

The first Anderson to be produced was 1.83 m high, by 1.37 m wide and 1.83m long, but during the steel shortage the length was reduced considerably. This was perhaps indicative of the shelter's failings; designed for the expected short daylight bombing, it did not meet the requirements for a shelter which would be suitable during the prolonged night raids which developed. The shorter versions were later supplied with extensions to make it possible to sleep in them, but even then the Anderson shelter was a cold and damp place to spend the night. The two exit holes were open and condensation dripped from the cold steel of its roof. Sunk 600 to 900 mm into the ground there was a tendency in some areas for water to flood the floor and the steps which had to be taken to avoid this often pushed up the true cost of the shelter considerably. When comparing the high level of discomfort with the low level of protection offered by the Anderson shelter, the conclusions were not always favourable.

The Anderson programme also contained other provisions for shelter in addition to the Anderson shelter. Local authorities were asked to supply public shelter for 10 per cent of the population. This was to be provided mainly by strengthening the trenches which had already been excavated during Munich (8). It was hardly comparable with the huge concrete public shelters being built in Germany at this time. Not surprisingly these limited measures undertaken by the Chamberlain Government did not satisfy public opinion. They led to considerable criticism being levelled at the whole principle of dispersal and in particular the resultant programme of partial protection.

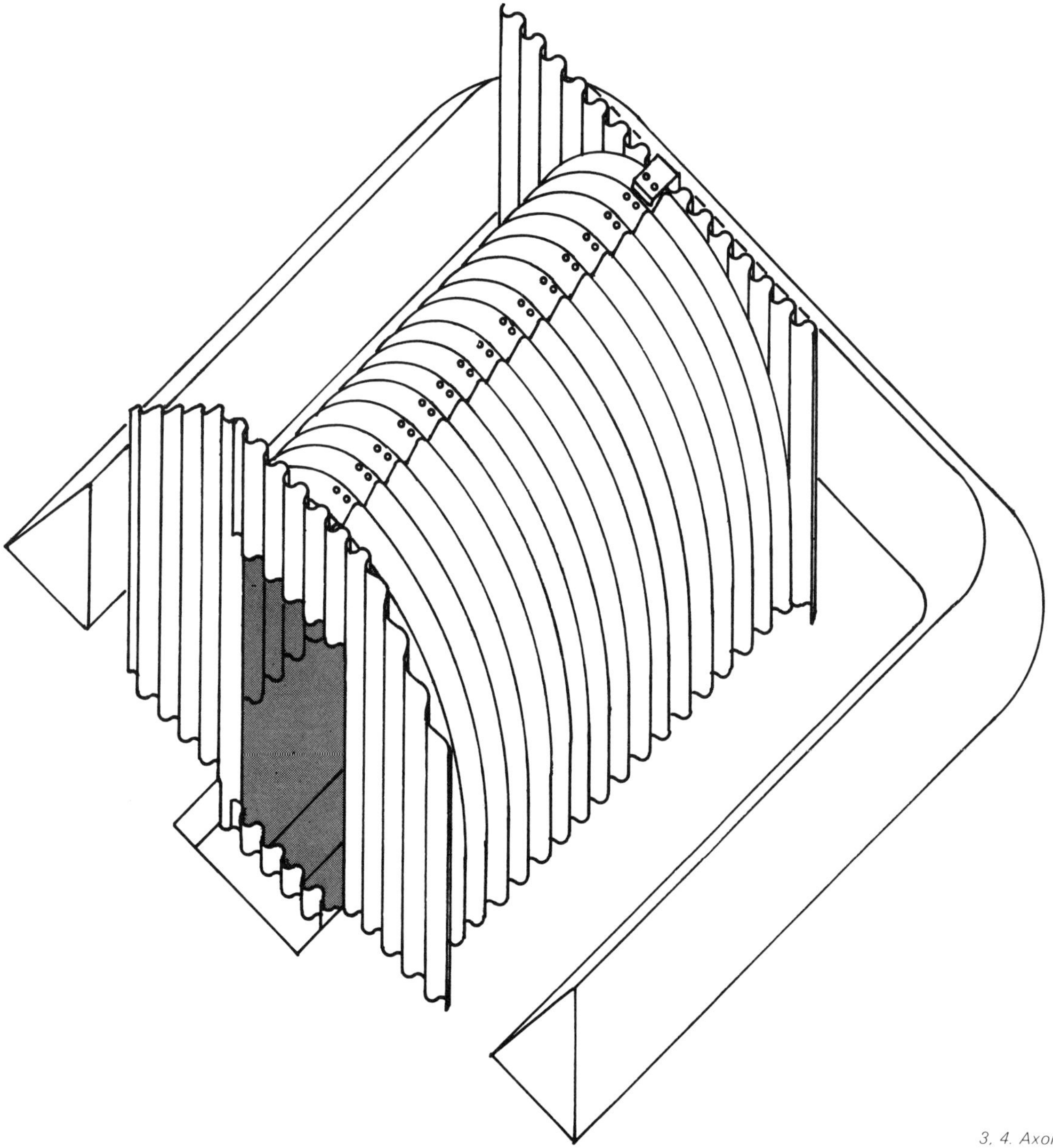

3, 4. Axonometric diagram of the basic Anderson shelter (3) and the shelter as incorporated into the family back garden (4).

Two such critics were the ARP Co-ordinating Committee and Prof. J. B. S. Haldane. In August 1938 the ARP Co-ordinating Committee, an unofficial body of architects, surveyors and engineers, had submitted a detailed scheme for a system of deep tunnel shelters in St. Pancras. Shortly afterwards Prof. Haldane had published a book which strongly criticized Government policy, and an article he wrote in 'Nature' in October 1938 showed how the dispersal policy was based on doubtful logic, that there was no reason for believing that bombs dropped at random would cause fewer casualties if people were dispersed rather than concentrated. Both Haldane and the ARP Co-ordinating Committee were part of the growing challenge to Government policy; they saw the provision of strong protection in the form of 'deep shelters' as being essential in the vulnerable areas. Politically the issue was first taken up by the Communists, but the 'deep shelter campaign' soon became a part of the long term policy of the more moderate Labour and Liberal parties as well. Thus in March 1939 the Finsbury Borough Council, a Labour stronghold, exhibited and publicized a scheme for heavily protected underground shelters, further increasing the deep shelter agitation.

After the Munich crisis Finsbury had commissioned the architects Tecton to design these shelters and they in turn had asked the engineer Ove Arup to assist them. The resultant proposals were that 15 large multi-storey shelters be constructed in the Borough, each based on a continuous ramp turning about a central column to overcome problems of access (9). The ramp idea must inevitably be compared, at least superficially, with Tecton's earlier design for the famous interlocking ramps of the penguin pool at London Zoo. Brilliant as the application of ramps may appear in theory, however, other contemporary authorities on the shelter problem were critical of its value. Donovan Lee wrote in his book 'Design and Construction of Air-Raid Shelters' published in 1940, that:

"Ramps leading to sunk shelters are unsatisfactory; they are wasteful of material and space, and dangerous if the floor is wet or frozen." (10).

Similarly the Design Panel of the Engineering Precautions (Air Raids) Committee, which was asked by the Government in 1939 to make recommendations on the design of bomb-proof shelters, proposed a circular plan which was comparable to the Finsbury scheme but which preferred the use of stairs and horizontal surfaces to ramps.

Whatever the detailed criticism of the Finsbury scheme, the principles behind its design are worth examination. In a report published in March 1939 Arup commenced by criticizing the dispersal policy:

"If we consider the distribution of shelters inside an area where the chance of a bomb falling is the same in every part.... So far as safety of each shelter is concerned, it does not matter whether many shelters are placed close together or whether they are evenly distributed over the whole area, unless the vicinity of other shelters changes the vulnerability of the particular shelter considered." (11).

The proposals in this report considered four different sizes of shelter, for 50, 250, 1,000 and 5,000 plus people respectively, the largest size considered being of 12 300 capacity. All had a circular floor plan, the report stating that 'a circular plan gives the minimum area of top protection (burster course) if this extends beyond the area of the shelter proper.... A circular plan gives the minimum area of outer wall. ... A circular outer wall is a better protection than a straight wall of the same thickness against bombs exploding outside the shelter.' (12). All of which seems so convincing that one wonders if the extra cost factor of curved construction justified the rarity with which circular plans were used for shelters in both Germany and Britain.

In the course of the study into Finsbury's requirements the team came to the conclusion that tunnels, i.e. 'deep shelters', were not justified on cost unless some post-war use such as underground motorways could be integrated into the scheme. Arup calculated that these large spiral shelters, which could be used as car parks in peace time, could provide equivalent protection for little more than half the cost per person of tunnels. This was because the multi-storey spiral shelters were not really underground. Arup described this point in 1940, when he had the advantage of retrospection:

"I should like to correct a popular misunderstanding about these 'deep bombproof shelters'. If we, as I suggest we should, define a deep shelter as a shelter which is situated deep in the ground and derives its protection mainly from the natural soil above, then the Finsbury shelters were not deep shelters at all. They were bombproof—that is against G.P. bombs (General Purpose) of a certain weight, and in this case $\frac{1}{2}$ ton to possibly 1 ton bombs. I should prefer to call them heavily protected shelters—the protection at the top consisting of a 10 feet thick reinforced concrete slab. They were multi-storey shelters and in

B7d

SHELTER FOR 7,600 PEOPLE.

B7a WITH STANDARD PROTECTION AGAINST BOMB 1
B7b 2
B7c 5
B7d 10
B7e 20

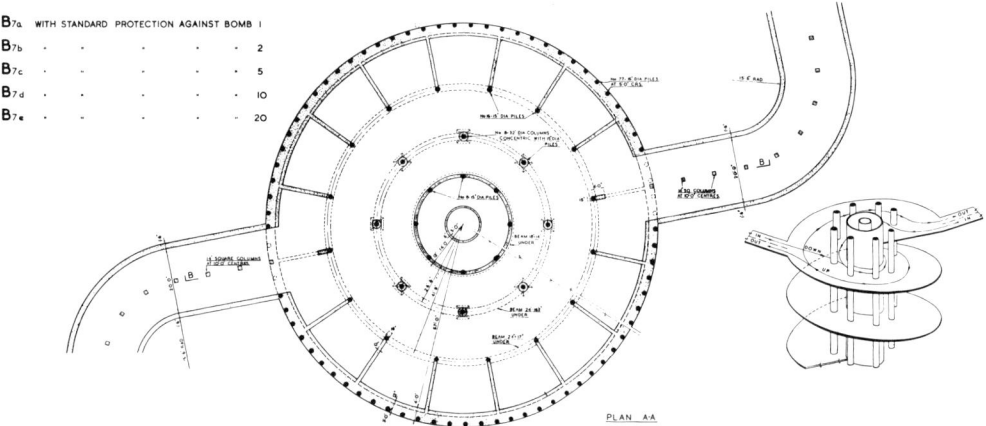

T8

EACH CIRCULAR UNIT HOLDS 50 PEOPLE.

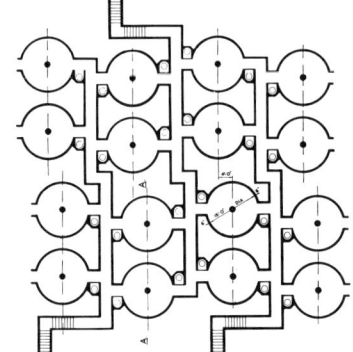

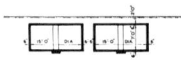

SECTION A.A

b : COMPOSITE
c : 6' WALLS
d : 9' WALLS

T9

EACH CIRCULAR UNIT HOLDS 50 PEOPLE.

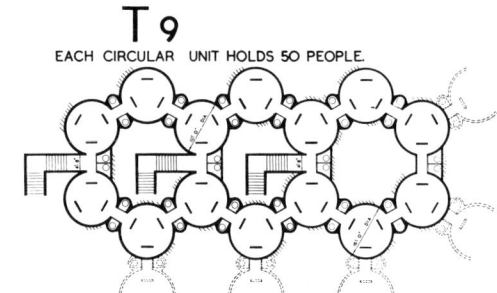

ALTERNATIVE LAYOUTS.

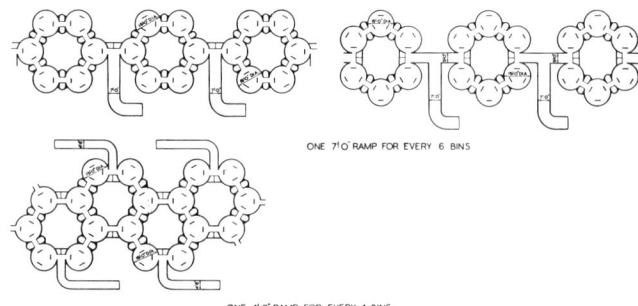

ONE 7'0" RAMP FOR EVERY 6 BINS

ONE 4'6" RAMP FOR EVERY 4 BINS

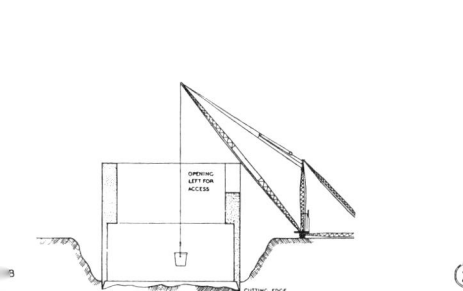
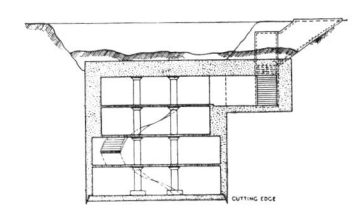
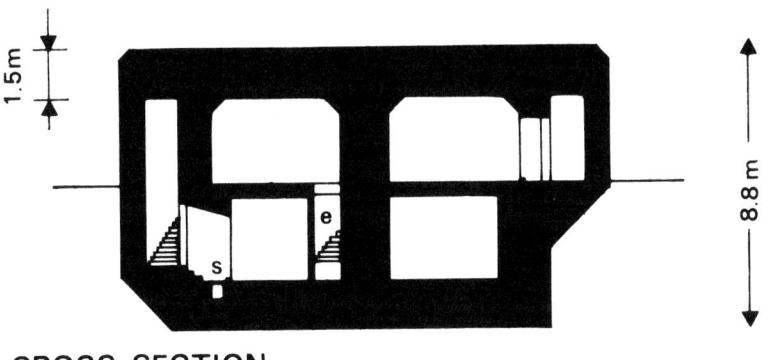

CROSS SECTION

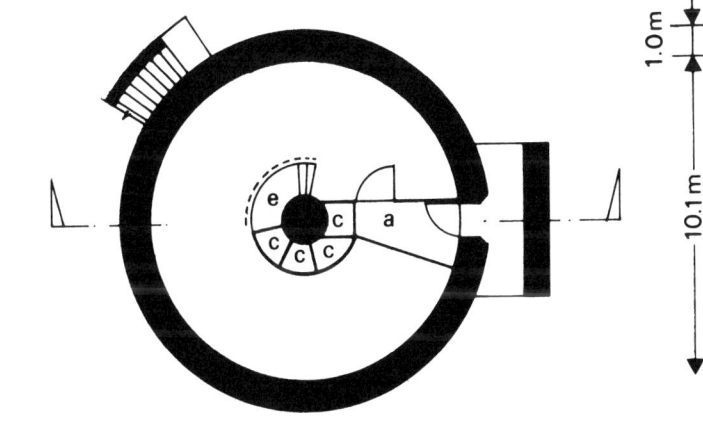

GROUND LEVEL

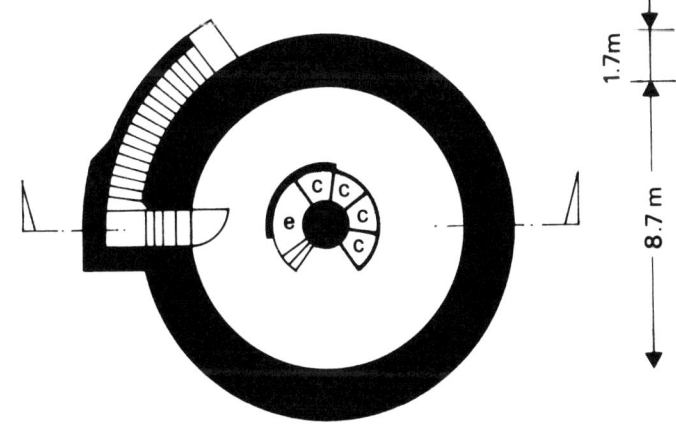

BASEMENT LEVEL

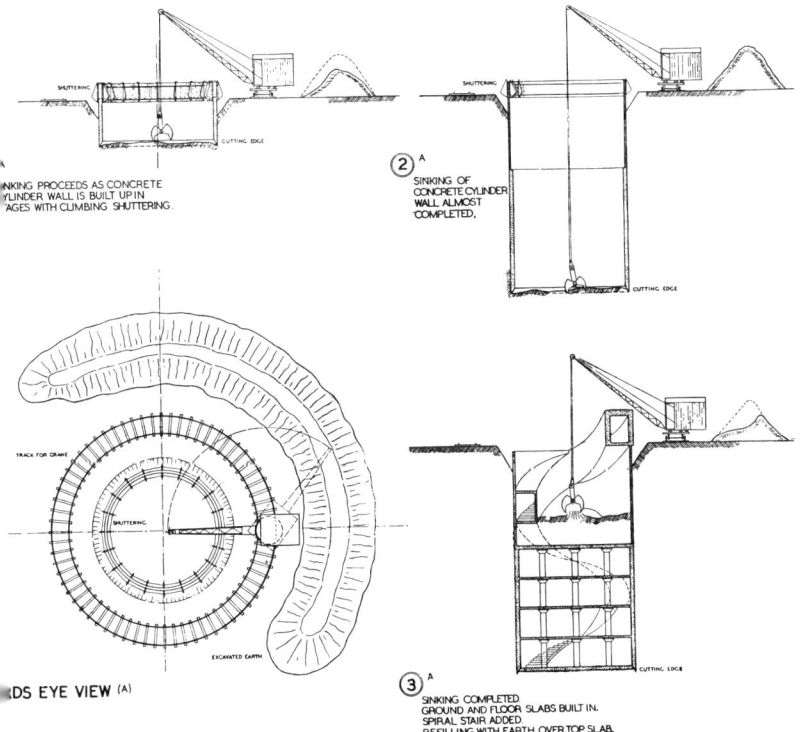

7. Some of the bomb-proof and high density trench shelter designs prepared by Tecton and Ove Arup for the Finsbury Borough Council during the winter of 1938–39.

The 200 person circular shelter proposed by the Design Panel of the Engineering Precautions (Air Raid) Committee.
c   chemical closet
e   emergency stairs
a   airlock
s   sump hole

that sense 'deep', because they were to be constructed downwards and not upwards, so as not to obstruct the open space for which they were planned. They had to be multi-storied for reasons of economy of space and money. It is obvious, that if it is decided to provide such an expensive roof protection, as 10 feet of reinforced concrete, then it would be highly extravagant to use it for the protection of only one storey, when it can just as well protect several stories. It comes to this, the heavily protected shelters are not bombproof because they are deep, but they have to be deep because they are bombproof." (13).

In an attempt to resolve the issue of deep shelters, the Government had set up an independent committee under Lord Hailey, but it was not until April 1939, a month after the result of the Finsbury report, that its findings were published. Its general conclusions were that bombproof shelter (14), either deep or heavily concreted, should not be provided for the general public and in doing so backed up the current Government policy, although by now a few underground shelters had already been built in certain suburbs such as Harrow. The Committee had gone back to basic principles and their reasons for a decision against strong protection were: the huge amount of labour and materials which would be needed, the time factor (it took two years to construct 26 km of underground tunnel in peace time, and this would only provide shelters for about 160 000 persons), poor accessibility and the danger of panic, danger of a 'deep shelter mentality' developing, and perhaps quite validly the calculated catchment areas of strong shelters. On the basis of a seven-minute warning the Committee calculated a catchment area of only about 300 metres radius at night which they stated would not fill the huge shelters being proposed. Though some of the reasoning proved faulty later, at least the danger of panic in relation to access proved to be a very real factor, even if overrated. At Bethnal Green tube station, in March 1943, a woman fell on the way down the staircase. The people following her panicked, and in the resulting chaos 173 people were killed in the crush (15). For certain vital services such as hospitals, however, the Hailey Committee did recommend that deep shelters should be provided. The Government consequently undertook to pay grants for the protection of vital industrial plant in May 1939, a policy leading to the construction of underground factories and other provisions, but on a minute scale.

Following the report by the Committee the Government were able to feel reasonably complacent. It confirmed their decision to channel resources into active rather than passive defences. The Official History records that on April 20th, 1939:

"The Lord Privy Seal told Parliament that the Government saw no grounds for departing from their general policy; though they would select some establishments to be given heavier than normal protection." (16).

Finsbury Borough Council did continue with their own deep-shelter plans despite Government disapproval, but work had scarcely started before war was declared and in September 1939 the contractors were able to invoke the war clause in their contract, and work stopped. The Hailey Report and the failure of the Finsbury scheme were not the end of the deep shelter controversy, however, for the subject was going to be a recurrent one.

Almost no strong or 'bomb-proof' shelters had thus been provided when the war began, and the Government had even re-stated, in Spring 1939, that the London Underground would not be available for public shelter (17). They claimed that the Tubes would be needed for communication and the evacuation of casualties, although there were undertones in this of the continual fear of 'deep shelter mentality', that a public once safe underground would be unwilling to emerge again, a theory similar to that held by the British on the Western Front in World War I. In fact it eventually proved impossible to prevent the public using the Tubes as shelters. But moreover, even if the Government had decided to provide bombproof shelters they would hardly have known how to construct them, for although the problems of bomb defence had been considered since the First World War, the secrecy surrounding the subject in England had stopped any practical experiments being undertaken.

In Germany and France such tests had been carried out in the early 30's but it was not until June 1939 that the first handbook of technical data on bomb resistance, based on actual bombing trials was published. A few weeks later the Home Office issued a revised version, 'Structural Handbook 5A', recommending a roof thickness of 1.52 m for concrete designed to resist a medium case 500-lb bomb, and 2.29 m to resist a heavy case bomb (18). The figures in 5A were determined by a panel of engineers and architects, formed by the Institute of Civil Engineers as a Design Panel of the Engineering Precautions (Air Raid)

◀ 9. Constructed by Manchester Corporation in October 1939, these brick and concrete shelters, with only 50 person capacity, reflected Government policy at this time. Their design was soon proved inadequate.

EVOLUTION OF THE CELLULAR SHELTER

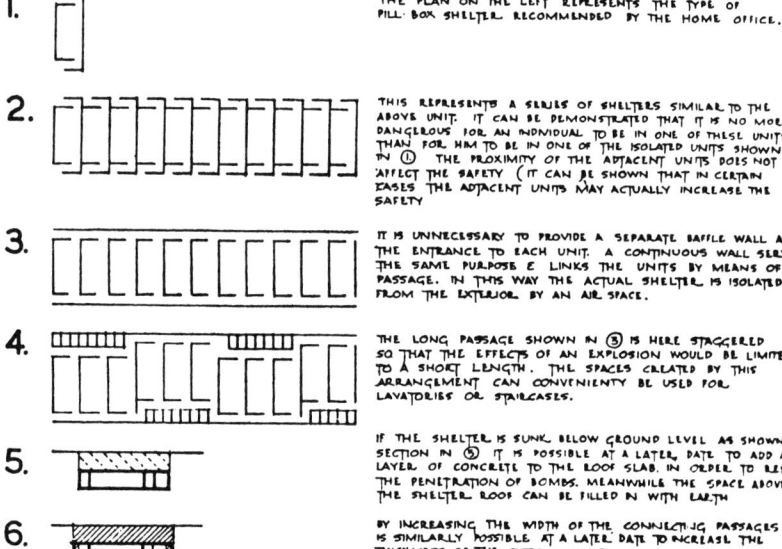

1. THE PLAN ON THE LEFT REPRESENTS THE TYPE OF PILL-BOX SHELTER RECOMMENDED BY THE HOME OFFICE.

2. THIS REPRESENTS A SERIES OF SHELTERS SIMILAR TO THE ABOVE UNIT. IT CAN BE DEMONSTRATED THAT IT IS NO MORE DANGEROUS FOR AN INDIVIDUAL TO BE IN ONE OF THESE UNITS THAN FOR HIM TO BE IN ONE OF THE ISOLATED UNITS SHOWN IN ①. THE PROXIMITY OF THE ADJACENT UNITS DOES NOT AFFECT THE SAFETY (IT CAN BE SHOWN THAT IN CERTAIN CASES THE ADJACENT UNITS MAY ACTUALLY INCREASE THE SAFETY

3. IT IS UNNECESSARY TO PROVIDE A SEPARATE BAFFLE WALL AT THE ENTRANCE TO EACH UNIT. A CONTINUOUS WALL SERVES THE SAME PURPOSE & LINKS THE UNITS BY MEANS OF A PASSAGE. IN THIS WAY THE ACTUAL SHELTER IS ISOLATED FROM THE EXTERIOR BY AN AIR SPACE.

4. THE LONG PASSAGE SHOWN IN ③ IS HERE STAGGERED SO THAT THE EFFECTS OF AN EXPLOSION WOULD BE LIMITED TO A SHORT LENGTH. THE SPACES CREATED BY THIS ARRANGEMENT CAN CONVENIENTLY BE USED FOR LAVATORIES OR STAIRCASES.

5. IF THE SHELTER IS SUNK BELOW GROUND LEVEL AS SHOWN IN SECTION IN ⑤ IT IS POSSIBLE AT A LATER DATE TO ADD A LAYER OF CONCRETE TO THE ROOF SLAB, IN ORDER TO RESIST THE PENETRATION OF BOMBS. MEANWHILE THE SPACE ABOVE THE SHELTER ROOF CAN BE FILLED IN WITH EARTH.

6. BY INCREASING THE WIDTH OF THE CONNECTING PASSAGES IT IS SIMILARLY POSSIBLE AT A LATER DATE TO INCREASE THE THICKNESS OF THE OUTER WALLS.

CAPACITY

STAGE 1. SHELTER NATURALLY VENTILATED — CAPACITY IS 576 PERSONS.

STAGE 2. SHELTER MECHANICALLY VENTILATED AT A RATE NOT LESS THAN 450 CU. FT. PER PERSON PER HOUR. — CAPACITY IS 770 PERSONS.

KEY TO MATERIALS, ETC.

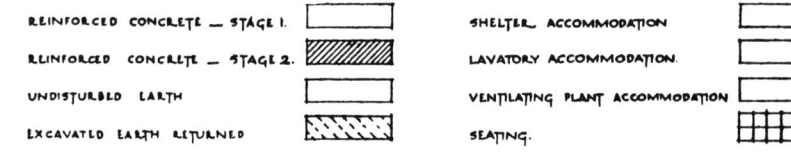

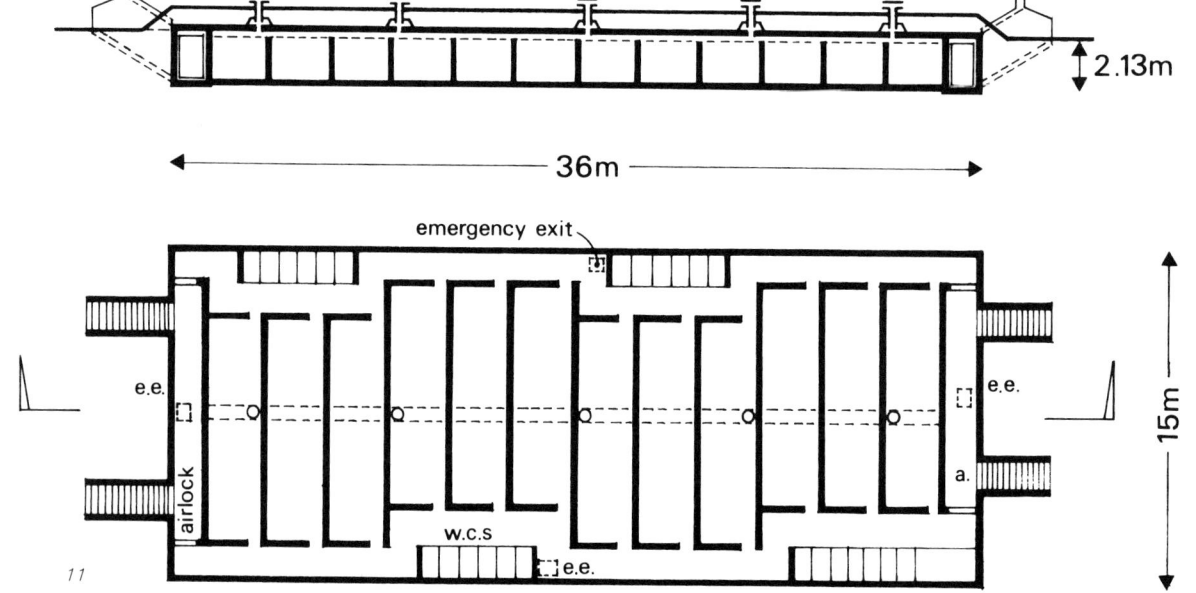

10. The cellular design by the ARP Co-ordinating Committee as published in 'The Builder' in December 1939.

11. The compartmental shelter constructed at Mandeville Houses, Finsbury, in 1940.

Committee (19). Formed at the special request of the Home Office this panel prepared about 30 provisional designs based on these figures, essentially two-storey structures of rectangular or circular plan form. Bomb-proof and large (up to 1 200 persons), these played no real part in future Government policy. The designs are however interesting on two accounts. Firstly they show that the Government had finally acquired the technical knowledge to prepare bomb-proof shelters where it saw fit. Secondly, and perhaps of only detailed interest, the circular plan shelters put forward by the Design Panel had certain superficial similarities with Tecton/Arup's proposals for Finsbury a few months earlier.

In the last few days before the war a number of steel-framed buildings (20) had been labelled as Public Shelters, but the Government was now forced to recognize that, like the Anderson in the domestic field, the trenches and existing buildings would not form sufficient shelter in cetain areas, where strong building or land suitable for trenches were not available. For the first time the Government asked the Local Authorities to build structures specially designed for public shelter. They were to be 'brick and concrete surface shelters', similar to the domestic shelters authorized earlier in the year. Designed for a maximum of 50 persons these shelters were still basically within the context of the dispersal policy. While the previous provisions were officially aimed at people caught in the street, and the poor, this does not mean that people elsewhere were all right. Industries, offices and shops had been asked to provide shelter for their employees (21), and the population as a whole had been encouraged to buy commercially manufactured shelters in steel or concrete or to strengthen part of their house. On the industrial side the provision of shelters was very incomplete, and only after four months of war had anything like a majority of shelters been produced. In the case of offices and shops the position was much worse. In the private sector, it can safely be said that the population was not spending large amounts of their income on shelters. Instead they preferred to rely on the Government provisions.

Much of the concern given to the shelter provision prior to the war had been due to the fear of a 'knock out blow'. British Chiefs of Staff as well as several members of the public—and the press— had assumed that Germany would, in the event of war, launch an immediate bomb attack on London. Minutes after war was declared in 1939, therefore, the sirens called out all over Britain, people filed into shelters and the traffic stopped. The Official History states:

"The fact that this threat did not materialize at the time or in the form in which it was expected may too easily obscure its historical reality." (22).

This failure of the expected 'knock out blow' to materialize caused no halt in the public demand for 'deep shelters'. In late 1939 the unofficial ARP Co-ordinating Committee submitted a memorandum to Sir John Anderson, Minister of Home Security . . .

"in which they stated it to be their view that the first big raid would reveal the terrible inadequacy of the whole Government scheme, and that consequently a national scheme of heavily protected shelters should be begun." (23).

As a result of this memorandum Anderson asked the ARP Co-ordinating Committee to submit to him designs for communal air-raid shelters for urban populations. On December 14th (1939) the Committee submitted a design for a cellular shelter based on multiples of the standard pill-box type shelter recommended by the Home Office. Subsequently each compartment contained only 50 to 80 people, depending on whether artificial or natural ventilation was used, and localized the effect of bomb damage. The total shelter size suggested was about 576 to 770 persons in 11 compartments. The main feature of the design was that it could be constructed in two stages. The first stage was the provision of a reinforced concrete shelter about 1.5 m below ground level with only a minimum slab thickness. When covered over again with earth this would give a rather better protection than the standard covered and lined trench. When the opportunity arose the earth fill could be re-excavated and a second stage of reinforced concrete protection could be added. It was a solution which overcame the problem of providing a quick partial protection, and would allow the eventual development of a heavily protected shelter (24). In Parliament on June 12th, 1940, however, Anderson claimed that there was no time left for the 'two stage shelters' proposed, thus continuing the now traditional difference between the 'deep shelter' extreme and Government policy. But despite the Government's official rejection of the cellular shelters, some were actually built in Finsbury.

The failure of the deep shelter scheme in Finsbury had left the Borough without proper shelter provision. Between the outbreak of the war and October 1940, when Ove Arup made his report 'London's Shelter

Problem', 27 shelters (some of them two storied) were constructed, catering for 11 750 people at a cost of £8 per head. The shelters were started in the first rush after war was declared in September and before Home Office approval had been obtained. Shortly afterwards the Home Office intervened, preventing the construction of further shelters of this type by refusing to allow waterproofing of walls or roofs, reinforcement in the floors, or the installation of proper w.c.'s or water supply in future shelters. But this attitude of the Home Office proved to be rather short-sighted. The compartmental shelters with reinforced concrete walls and floors offered a much better protection than the brick and concrete shelters being recommended at this time (25). They also had the advantage of size when the problem of providing amenities arose in the late 1940, making such provision for smaller shelters expensive. On an initial cost basis, before the question of amenities arose, the compartmental shelters may have cost more, but the Home Office, in their attempt to preserve the dispersal policy, chose what was in principle an inherently expensive solution. By attempting to save money and effort for active defence, the Government opted for the ineffective brick shelters, which eventually had to be strengthened and improved.

In early 1940, shortage of steel slowed down production of the Anderson shelter and caused the length of the shelters being provided to be shorter. By April 1940 production had to be suspended. Some 2 300 000 shelters had been produced, (shelter for an estimated 12.5 million, and as all but a thousand of these had been supplied free (26), this represented a cost to the Government of something well over £10 000 000. Despite the huge numbers which had been produced the shelter problem was far from solved, and the Government was forced to put more emphasis on the brick and concrete domestic surface shelter. In March 'communal domestic surface shelters' were introduced but, designed for a maximum of only 48 persons, they were still very much a part of the dispersal policy. Intended for use in areas of adjoining housing with no suitable areas for small domestic shelters, they were usually situated in the street. They were built directly on the road or pavement with weak brick walls and a loose concrete roof, and sometimes proved to be decidedly more dangerous than the houses they were intended to replace for protection. In any case they were a great deal more uncomfortable, and trivial though this may sound it was to become an important factor. The major initial fault was, however, to result from a statement made by the Ministry concerning the use of cement.

The demands on cement being made at this time from all quarters, including coastal fortifications, now caused a shortage of cement as well as steel, although the shortage was later attributed by Communists and others to a manufacturers' 'Cement Ring' (27). To try and counter this the Ministry attempted, in a statement during April 1940, to reduce the amount of cement being used in the brick and concrete shelters. This statement was misinterpreted as permission to use a mortar composed of sand and lime with no cement. As the brick walls literally relied on the tensile strength of the mortar for the horizontal pressures of near blasts, the damage caused was considerable. Before the statement was amended in July 1940, and the mistake rectified, a large number of shelters had been built without any cement at all in their mortar.

The first real test of the shelter provisions came in September 1940, exactly two years after the Munich crisis. Civilians had seen no action in the first eight months of war, in the so-called 'Phoney War', and it was only in May 1940 that the first bombs were dropped on England, marking the start of a series of small raids on airfields and towns. During July–October 1940, this escalated into the 'Battle of Britain' as the Germans attempted to destroy RAF resistance in preparation for the invasion 'Sea Lion'. Only in September did the Luftwaffe switch its attention to the wholesale attack of the British population.

Structurally the Anderson shelters had stood up well to the small, quick raids of the summer of 1940 and thus already justified the public's confidence in them. Their tensile properties allowed them to spring back into shape easily and they had a good impact resistance against blast for this reason. They proved to be usually undamaged by 50-kg bombs dropping up to 2 metres away and 250-kg bombs dropping 6 metres away, although unprotected exits in some cases proved lethal. The structural efficiency of the other forms of shelter which had been provided was much poorer and the brick and concrete surface shelters proved to be particularly poor. Roofs were lifted by blast, brick walls penetrated by splinters and even cases of collapse without the aid of the German bombs were recorded. The trench shelter with timber or pre-cast concrete linings also proved to be vulnerable: their roofs lifted

12. Surface shelters built for a school at Southgate: total capacity 750.

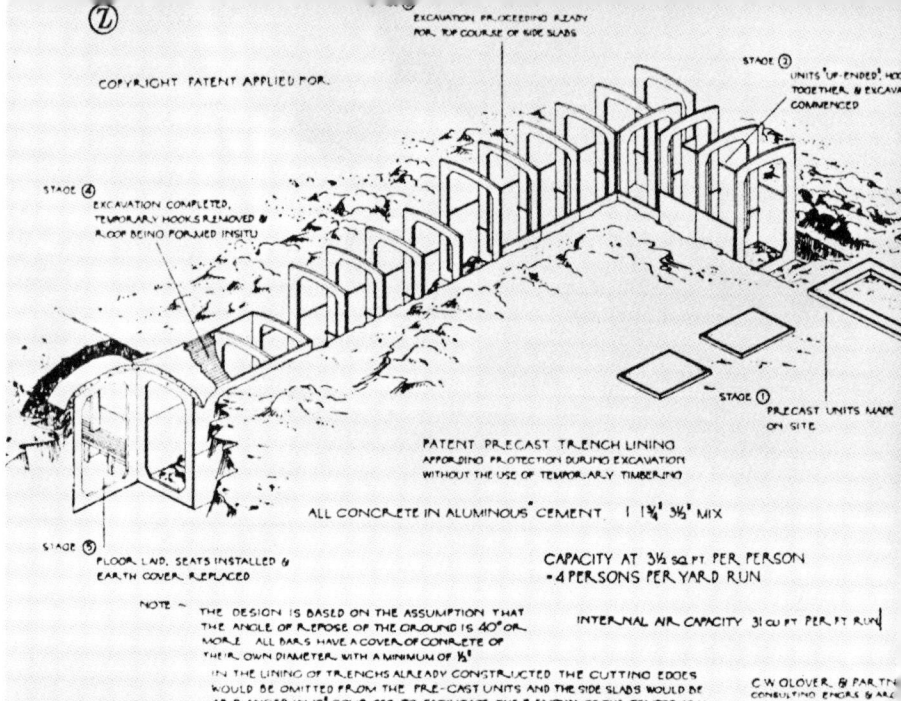

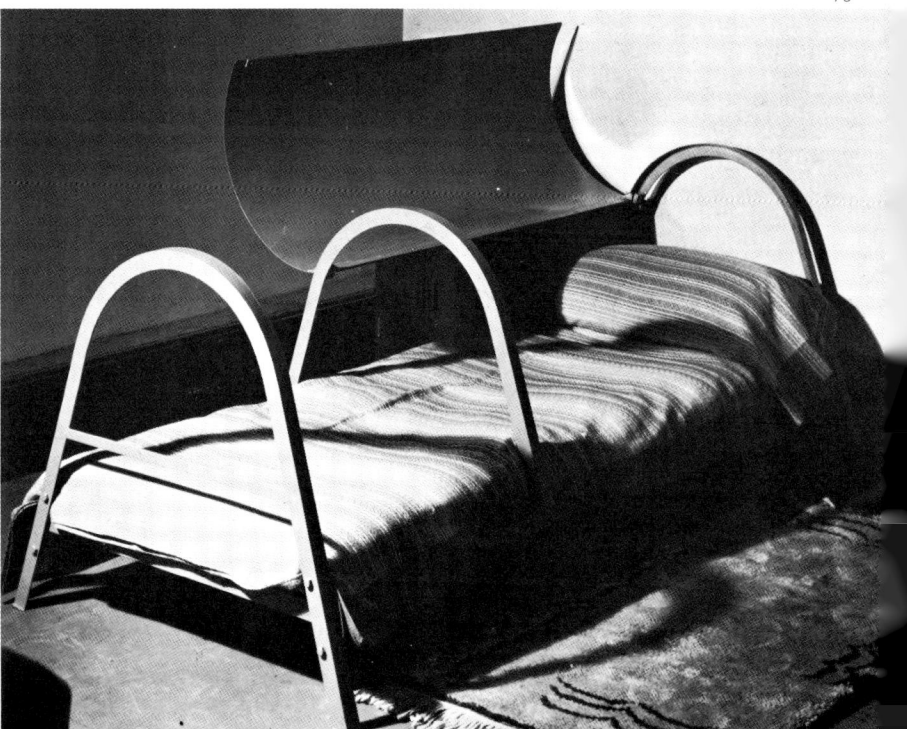

13. The long raids of the Blitz emphasised the weaknesses of existing shelters. The lack of strength and facilities soon discouraged people from using shelters such as this one at Park Royal.
14. Diagram of trench reinforcement being carried out using 'Glover' precast concrete elements.
15. The Heal's shelter bed: middle class version of the Morrison shelter.

and the sides pushed in easily. During heavy bombardment, the fact that both the Anderson and the brick/concrete shelters were only splinter proof became only too obvious. Even near misses blasted shelters into oblivion and often such incidents would leave no trace of the occupants, such was the limited protection provided. In some of the provincial towns such as Hull, Merseyside, Clydebank, Bristol and Southampton, people took to 'trekking' to the surrounding countryside for the night. In London people flocked into the tubes like they had in the First World War, despite Government instructions to the contrary.

Another serious fault which was found was not directly connected with strength but with amenities; the prolonged night raids of the first week of the Blitz led to people having to spend much of the night in their shelters. The brick and concrete shelters, the trench shelters, and the Anderson, were damp and cold places designed only for the short daylight raids which were expected. When, thanks to the effectiveness of the RAF, the raids switched to night, people were obliged to eat, sleep, excrete and continue their private lives in the shelters night after night. Shelters designed for short periods became dormitories as people, especially in London, developed the habit of going into the shelters at dusk and staying there until morning. The Official History suggests one cause for the lack of amenities in the shelters, although the prime one must of course be Government policy:

"The authorities had been so preoccupied with the primary task of making shelters safe that they had paid little attention to amenities. This tendency had been intensified by the fact that shelter responsibility had been borne in Regions mainly by technical staffs and in local councils by borough engineers." (28).

Not surprisingly between 1939 and 1941 there was a 10 per cent rise in tuberculosis (29). The dangers to health from cold and damp in the winter of 1940–41 were in some cases greater than the dangers from high explosives. Some shelters became more popular than others, and in London the underground stations were particularly popular. They at least were warm and reasonably dry. The racket of anti-aircraft guns and exploding bombs was inaudible at that depth, giving some chance of a night's sleep. The flimsy and damp surface shelters became emptier and emptier, the lack of public confidence in their strength making the situation even worse.

In November 1940, at the height of the London Blitz, it was shown by a shelter census that only 40 per cent of the city's population were using any form of shelter. Apart from the fact that the more prosperous citizens had largely been left to fend for themselves, this figure reflected the fact that many people preferred to risk it in the comfort of their homes. Although a few more Andersons were produced during the Blitz (30), the rejection of the surface shelters subsequently prompted the Government to start work on a shelter which people could use in their own homes. The replacement of Anderson by Herbert Morrison as Minister of Home Security in October 1940, was the start of this and other attempts to improve conditions in the shelters (31), prompted of course by public opinion and agitation.

The design of a shelter to be used inside the home owed its origin to Blitz experience which showed it to be a feasible propostition. People had survived bombing successfully under staircases and furniture. In January 1941 the Government was able to approve a shelter design which fulfilled the requirements, the 'Morrison' shelter, and an order for 400 000 was placed (32). The Morrison was a steel framework with mesh sides, steel mattress base, and 3 mm steel plate top. The main variety used was 1.98 m long, by 1.22 m wide and 840 mm high, which could be conveniently used as a table as well as a shelter. A further contract for another million of these Morrisons was made later in 1941, but none of them were available in any number before the Blitz ended. A later version of the shelter, which was not produced in any large numbers, was larger with double bunks (33). Like the Anderson, both were distributed on the low income basis (now £350 p.a.) and put on sale to the public.

To make the surface shelters more attractive to the public the Government started a scheme of strengthening and improving both existing shelters and new ones. Some of the weak ones built with lime mortar were pulled down altogether, and others were strengthened by adding an inner or outer skin of reinforced brickwork or concrete keyed into the existing walls. A new design for public and communal shelters was prepared in December 1940 using reinforced concrete, reinforced brickwork or reinforced concrete blockwork. The reinforcement rods of the stronger wall constructions were to be carried through and tied in the roof and floor. The roof had an overhang to improve weather protection and the new individual brick surface shelters were to be constructed of reinforced brickwork. None of these improvements were

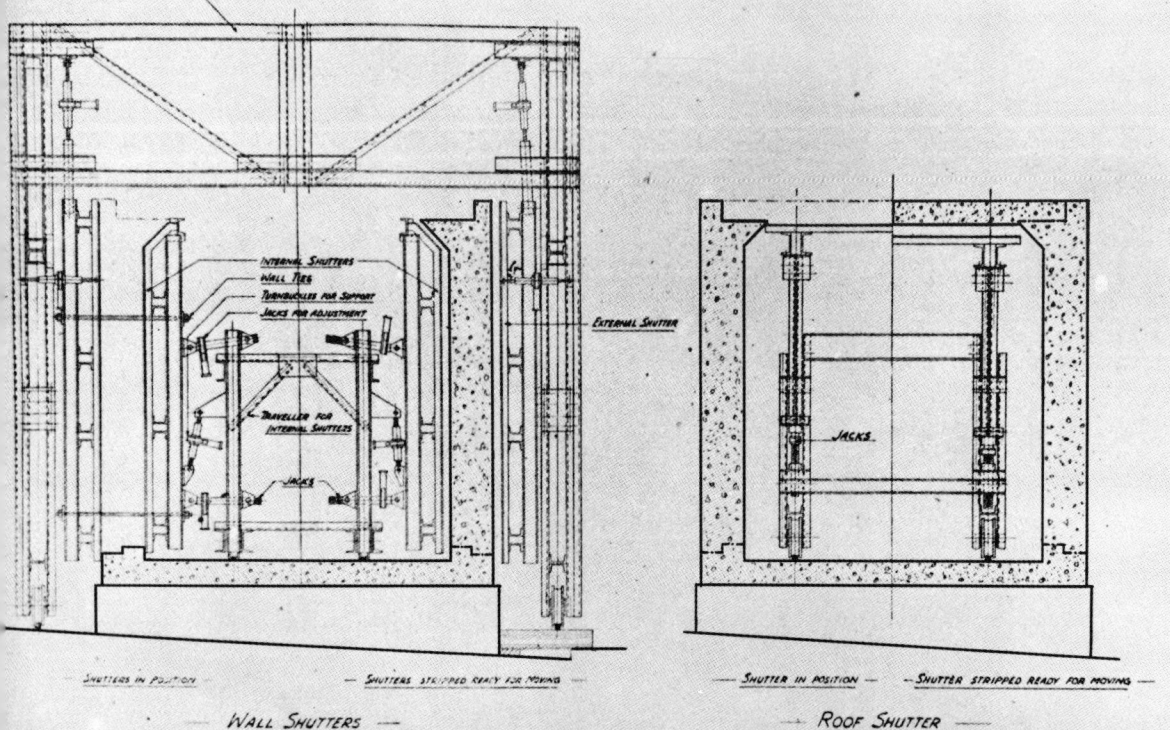

16, 17. Construction of shelters in St. Pancras using the mobile shuttering method.

18. The 'Admiralty Citadel' in Horse Guards Parade, London: constructed in 1940, primarily as protection against air attack, for service and civilian staff of the Admiralty.

19. A London tube station being used as shelter in the winter of 1940–41.

really a fundamental change in thinking, though, and a more interesting advance on the principles behind the improvements was developed at borough engineer level.

In the Borough of St. Pancras, where the emergency committee decided to switch over from the construction of brickwork communal shelters to the construction of reinforced concrete ones, a remarkable piece of initiative was shown. To speed up production, a system of steel shuttering was hired which cut down production time to a figure 30 per cent that of the original time with traditional shuttering. The steel shuttering had been originally designed for the construction of reinforced concrete culverts and, working on a travelling basis, it allowed the sort of progressive construction suitable for culvert work. It proved ideally adaptable to the shelter problem and sections of rectangular 'concrete tube' were literally laid in the streets of the borough in suitable lengths to meet Home Office requirements, although the shelter, because of its method of construction, was not produced with the overhangs required in the new standards of December 1940 (34).

The Home Office structural measures in the new designs were a little belated. The engineer Ove Arup, who had consistently been ignored or over-ruled by the central body of control had already observed that if shelters were not designed to resist direct hits then the main protection was the wall, not the roof. This was logical for the force which was being resisted was mainly the horizontal one of the blast and splinters. The brickwork used in the Home Office designs before December 1940, with no reinforcement, had only negligible resistance against horizontal movement: any horizontal thrust imposed a tensile stress in the brickwork which was resisted literally by the adhesive strength of the mortar.

Arup's designs for shelters, which he called 'Wall Shelters', based on this principle, were built in Finsbury during 1940, but the principle of concentrating on wall strength was used even earlier than this. 'The Builder' of November 24th, 1939 (35), published a design for a small domestic size surface shelter with 380 mm reinforced concrete, yet only a 130 mm roof slab. When the Research and Experiments Dept. undertook to test and advise on commercially produced shelters in April 1941, they must therefore have received a mixed reception from engineers and architects involved in shelter work. One engineer at least had grown a little doubtful of the Home Office's attitude. In discussing some of the problems involved in shelter design in October 1940, two months before the Home Office revised designs were issued, Ove Arup wrote rather sadly:

"It seems to be very difficult for the layman to appreciate . . . (the logic of multi-storey shelters) . . . and this applies even more to the Home Office, who apparently deal with these matters mainly from a layman's point of view." (36).

As already mentioned, the other reason for the unpopularity of the Home Office shelters was their lack of amenities. There was no heating, poor ventilation, lack of privacy, and only improvised sanitary arrangements, although the lack of some of these amenities was just as critical in the underground 'shelters' which were always crowded. The crowded, unhealthy situation in the underground at night caused considerable concern amongst the medical authorities, and improvement of shelter conditions became as much if not more important than structural improvements. The Government set up the Horder Committee to investigate the situation.

After only two months of the Blitz caterers and voluntary bodies were supplying food, and this was one aspect which the Government had to adopt and improve upon. By the end of April 1941 food was supplied in most large shelters being used all night. In doing this, the Government was having to accept the fact that people were using the Underground as shelters whether the Government liked it or not. So in addition to trying to improve conditions in Home Office surface shelters by providing bunks, heating and chemical closets, they were also obliged to improve the conditions in the Underground Stations where people were congregating. In some stations during the Blitz, the lines were pulled up, tarmac laid, and bunks for 22 000 installed. In other stations, by drawing white lines on the platforms, and by issuing place tickets, the authorities managed to keep traffic running. Shelterers were allowed to take shelter only after 4 p.m. although in one case where this rule was not applied, Liverpool St. Station, something like the 'deep shelter mentality' which the Government feared began to appear, as some people stopped in shelter for a week or more at a time.

The provision of bunks in the communal shelters, incidentally, shows just how unprepared the Home Office had been for the all night raids. The provision of bunks meant that communal shelters had to be en-

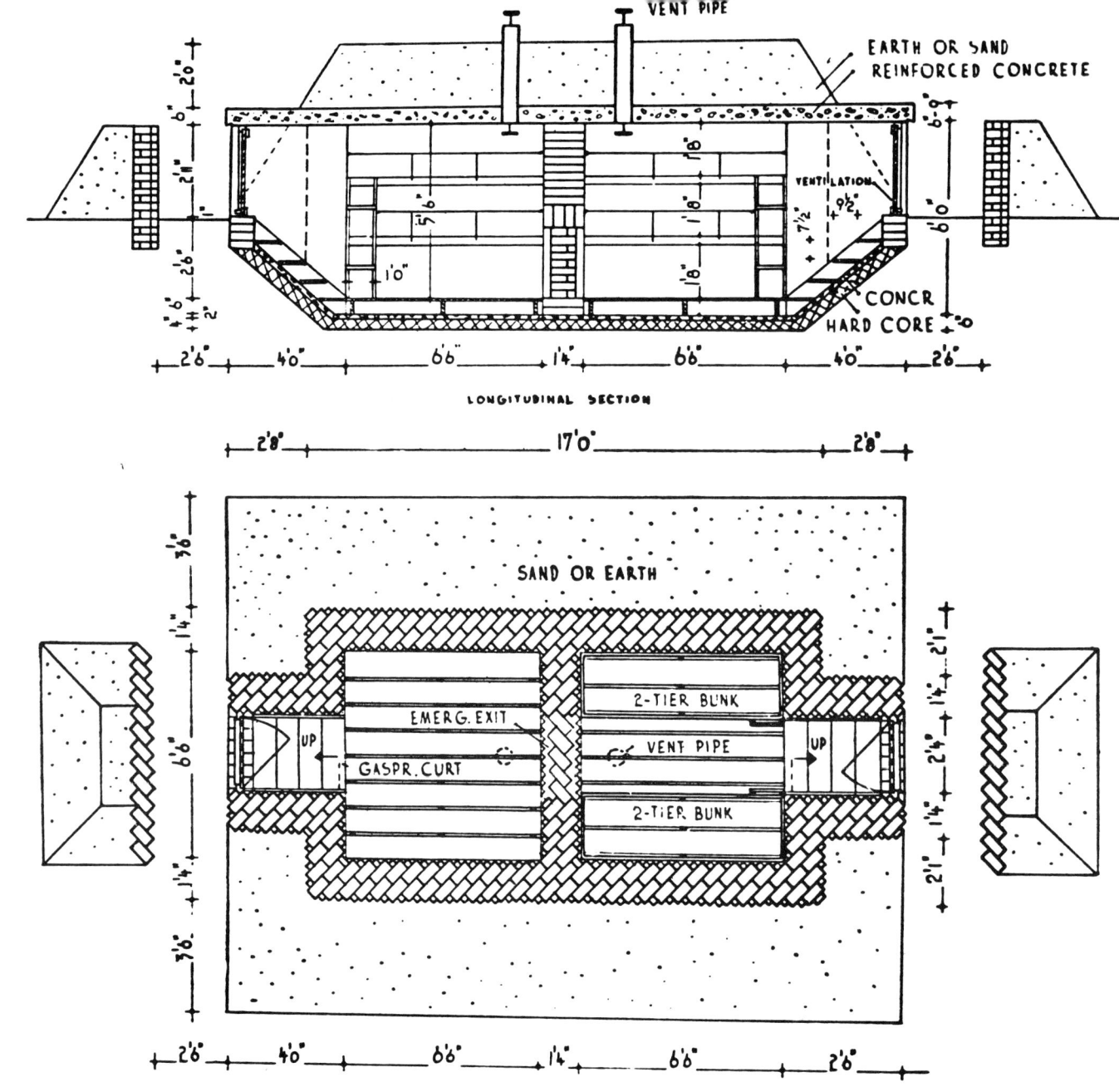

20. Walter Segal's proposals for 'semi-detached' family shelters, as published in 1941.

larged in size as from October 1941 to take bunks. After the spring of 1941 water borne sanitation was put into new shelters, and on the Underground the chemical closets proved completely inadequate for the huge numbers of people. Water borne sanitation could not be provided there as most of the platforms were below sewer level. The difficult situation of large numbers of people in conditions of close proximity and primitive sanitary arrangements posed an urgent problem, and for a time chemical closets had to be manually carried up to sewer level. Eventually a series of 'sewage ejector plants' was devised using the compressed air available at stations to force the sewage up to sewer level.

The improvements made in the conditions in the Tubes perhaps deserves special mention, when one remembers the Government opposition to their use even in the early months of war. By February 1941 bunks were installed in most of the stations and some stations were equipped with catering equipment, distributing food and drink to the other shelters by special train. Bunks led to a system of place reservation, and the welfare facilities gradually improved. The Official History describes how this was allowed to come about:

"The Ministry of Home Security approached welfare activities with some apprehension lest shelters should be made too attractive and the dispersal policy endangered. It soon, however, became clear that since large numbers were determined to congregate in shelters, it was better to acccpt the situation and help sustain morale by furnishing entertainment and education." (37).

The use of shelters all night and the way the Tubes were resorted to brought a renewal of the 'deep shelter' controversy yet again. The Communists still agitated in the hope of using the issue as 'a potential revolutionary flashpoint' (38) while the general public simply requisitioned the Tubes as deep shelters. Nor was faith in the Tubes broken by the fact that they proved to be not entirely safe. At Bank Station in January 1941 111 people were killed by a direct hit, and at Balham Tube Station a direct hit during the Blitz fractured water and sewage pipes, drowning 64 people (39). Protection in fact proved to be partly psychological, and despite warnings of their weakness, railway arches became a favourite form of shelter.

One of the main criticisms of deep shelters had been that people would not be able to get to them in time, but with the night raids of the Blitz people were going to the shelters before any warnings were given and automatically spent the night there. The problem of access was now of only secondary importance. At first the Government claimed that it was not known how long this type of raid would last, and that the Luftwaffe might soon change its tactics. The failings in both the structural and amenity sense, of the Home Office shelters, made agitation easy and as this agitation came from not only the political left, but from the moderate press, the Government was forced to review the situation.

In October 1940 the Cabinet decided that although deep shelters could not be anywhere extensively provided, a new system of tunnels linked to the London Tubes should be bored and tunnelling authorized in the provinces where economically possible. But the cry for deep shelter was not a universal one. In the technical magazine 'Building' the architect Walter Segal wrote in early 1941:

"The provision of deep shelters, so desirable in principle, should reasonably not be started to any great extent as long as thousands of civilians are without protection at all and as long as thousands are sheltered under the most unhealthy conditions. Money spent to improve the health and comfort conditions in present shelters (among other items including the reduction of the number of shelterers and providing more and better shelter room for those who are without shelter at all) seems to be much better spent than upon the construction of bomb-proof shelters which would benefit only a small part of the population." (40).

The Government's short-lived softening towards the provision of shelters must have been to some extent a political expedient, part of the attempt to cool agitation. The decision to build shelter extension tunnels for 10 London tube stations was not a technically sound one as time proved. None of the extensions were ready until one year after the Blitz ended and even then each extension took only 8,000 persons. The cost, as Arup had predicted, was astronomical. The original estimate of £15 per person, quite expensive anyway, worked out in practice at something approaching £35 a head, yet in January 1941 the Government had decided not to construct heavily protected surface shelters for the public because of the estimated cost of £25 per head.

Obviously there was some confusion over shelter decisions. Having been pushed by public opinion in late 1940, with the Tubes being used

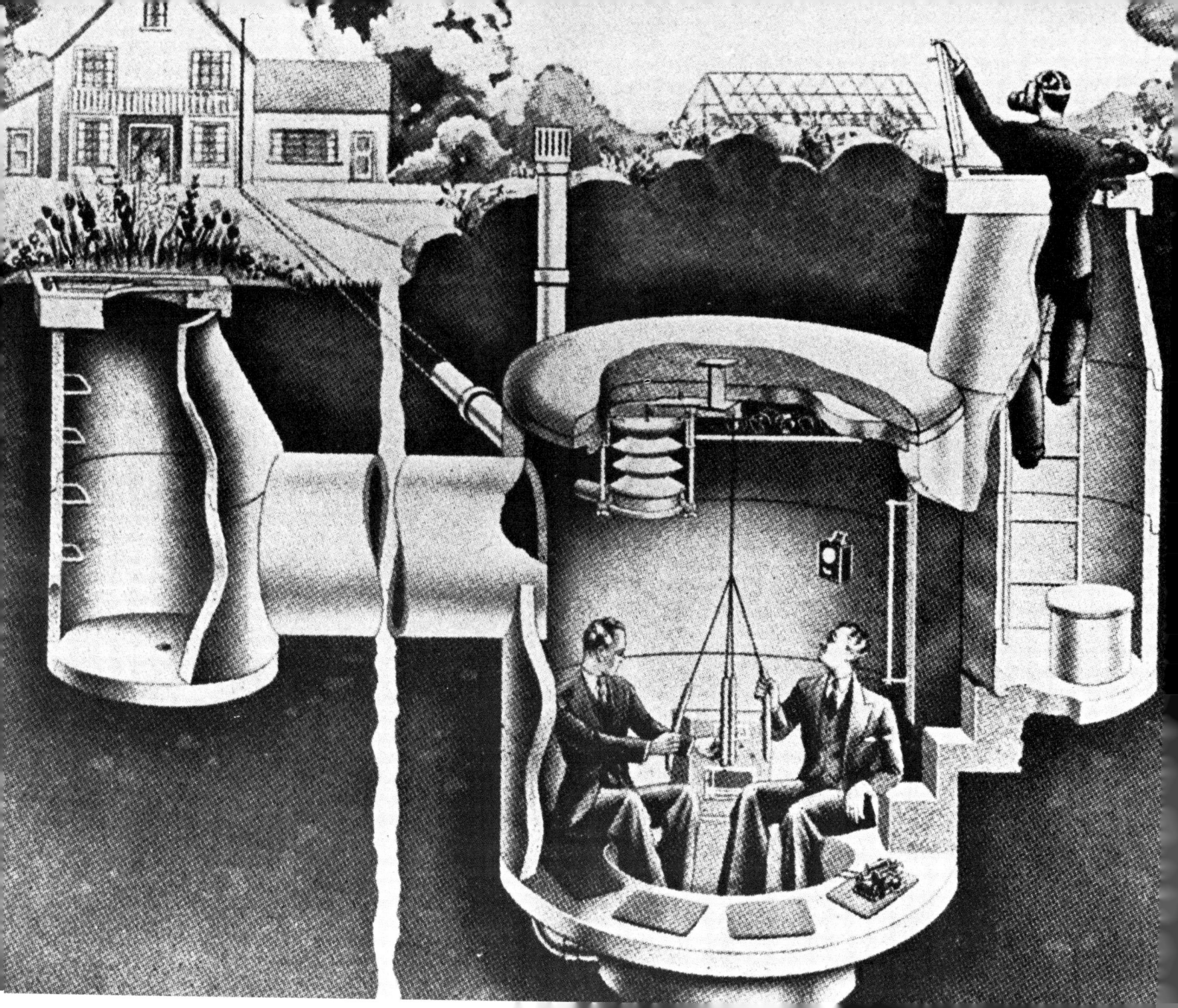

extensively, the Government had allowed further tunnelling to take place. Yet their indecision is shown by the fact that in February 1941 the Government was already changing its mind about the deep shelter projects and in January 1942 the Ministry of Home Security announced that it was most unlikely that further deep shelter projects would be approved. In fact of the 10 Tube shelters which were built in London, only a few were used for public shelter, and even then not until 1944. They were unsuitable for the intermittent raids of 1942–43 and used mainly for a variety of military purposes. Part of Goodge St. extensions was made available for General Eisenhower's London HQ in late 1942, two others were adapted for operational staff of Government departments, and several more were used for week-end accommodation of British and American troops. At the end of 1943 it was decided that all the extended Tube shelters should be reserved as extra 'citadel' accommodation.

In the rest of the country some tunnels had been constructed when the Government momentarily changed its mind. Dover had the best tunnel protection for her population as she was not only bombed regularly, but also shelled from the Pas de Calais, and the Government gave permission for the caves in the chalk cliffs to be extended. With the extensions this gave underground accommodation for 14 000 and as the wartime population was only 17 to 18 000 (1941) this gave a high percentage of them protection (41). The rest of the British population were not as fortunate and in assessing overall British civilian shelter provision it must be concluded that deep or bomb-proof shelter played little part in it. Ove Arup, constantly opposed to the Government's policy, stated:

"The main reason for the failure of the government policy is to my mind that the authorities have been obsessed with this idea of dispersion, to the exclusion of everything else. They have thereby from the start chosen the most expensive solution, and the need for economy has made it impossible to 'deliver the goods'. What lies behind this obsession I do not know—I hesitate to believe that it can really be due to inability to understand a simple mathematical truth. . . ." (42).

This statement seems to sum up the shelter situation. The dispersal theory was not unrealistic for quick, short raids, but during the Blitz the strengthening and improving of amenities in small (50 person) shelters made this system more expensive than if larger sized, stronger shelters, where amenities could be provided more economically, had been provided from the start. Yet the politicians put a blind faith in the principle of dispersal. Churchill himself said in August 1940, to the cries of 'Hear, hear,' that 'Dispersal is the sovereign remedy against heavy casualties'. On October 10th, 1940, when the Blitz was well under way, the Home Secretary put similar faith in the principle:

"Dispersal and the smaller domestic shelters are still the best safeguard against heavy casualties."

It was fortunate that the Blitz was the major aerial offensive. In eight months it caused over 65 per cent of the total civilian deaths over 1939–45 in Britain. If the Luftwaffe had been able to unleash the sort of tonnage which the RAF and USAAF dropped on Germany the results might have been terrifying. Dresden, with ARP measures and shelter provision much better than an equivalent British city, suffered 135 000 deaths in 24 hours of bombing in 1945 . . . twice the total number killed over the entire period of war in Britain. Britain was lucky. Her poor shelter programme, brightened only by the technical genius of the mass-produced 'Morrison' and 'Anderson' shelters, would not have stood up to much heavier attacks than those of the Blitz. Added to which her shelter programme was drastically cut 12 months after the Blitz ended as energy was channelled into other directions. When the Blitz ended in May 1941, the Luftwaffe had switched its attention to Russia. The British reaction was not to strengthen the defences but a gradual winding down of the shelter programme, a decrease in production of fighters, and a massive switch to the production of bombers.

◀ *21. Cut away view of a man-hole type domestic shelter as published in 1938.*

12 GERMAN BOMB-PROOF MANIA

Protection of civilian morale: the public shelters and flak towers. Protection of military projects: the underground and bomb-proof factories, submarine pens, the V sites and the 'National Redoubt'

Events

1914 Oct. British Naval aircraft raid Zeppelin sheds at Dusseldorf and Cologne.

1918 June Independent Air Force formed under Trenchard and organized for attack on enemy's 'home front'.

1940 May 10 Germans invade Low Countries. Churchill takes over as PM and starts air attacks on Ruhr rail targets.
Aug. 25 First air attack on Berlin.

1941 Aug. Butt Report criticizes accuracy of RAF raids on industrial targets: starts policy of 'area bombing' of civilians.

1943 Jan. Casablanca directive gives heavy bombing of submarine pens priority over area bombing until May.
May Wuppertal raid. First to cause large civilian casualties.
July–Aug. Hamburg raids. The first 'fire storm'. 50 000 killed.
Oct. Kassel raid. The second 'fire storm' but followed by many more on other cities through 1943–44.

1944 June 6 D-Day.
June 13 V-weapon attack on Britain begins.
July 'Crossbow' bombing offensive on the V-weapon bunkers reaches its climax.

1945 Feb. Dresden 'fire storm' raid. 135 000 killed in 24 hours.
March Allies pushing towards Berlin halted by fear of 'National Redoubt' in Alps.
April Bombing offensive and ground victory finally bring German war effort to a halt.
April 30 3.30 p.m. Adolf Hitler commits suicide in the Chancellery bunker.

---

1. *The Harburg flak tower near Hamburg.*

● Germany's war-time strategy, at least up to 1943, is generally associated with a policy of all-out attack. It is curious to find, therefore, that bomb-proof protection on a huge and sometimes indiscriminate scale was, even in the beginning, one of the main features of her wartime building programme. This fact is suitably reflected in Speer's description of the Führer's own bunker beneath the Chancellery in Berlin:

"His study (Hitler's) roofed with more than 16 feet of concrete, then topped with six feet of earth, was undoubtedly the safest place in Berlin." (1).

As far as the general population was concerned, shelter provision was hardly less massive in construction; and it was made available on a far wider scale than in Britain. The US Strategic Bombing Survey on 'Public Air Raid Shelters in Germany' stated that:

"By the end of the war, bunker facilities in Germany on a design basis could accommodate roughly 15 per cent of the population in all principal towns and cities. By overcrowding bunkers could accommodate up to 75 per cent of these populations." (2).

This policy was to some extent a reflection of Allied bombing policy; while Germany had conceived her bombing force as part of 'blitzkrieg' attack, Britain had planned heavy bombers to form a strategic bombing force for attack on the heart of an enemy country. As Churchill stated on September 3rd, 1940, 'The bombers alone provide the means of victory' (3), a strategy which the Allies attempted to apply right up to 1945. Yet German shelter policy was in a way not only a result of this strategy but based on the memory of 1918, when the collapse of German morale on the Home Front had supposedly led to her defeat. Her regard for troop protection on the Western Front during 1914–18 might be seen as a further factor. On the British side too, the attitude towards the protection of the civilian population was a hangover from the First World War. The reluctance to provide bomb-proof shelter on the ground that it might lead to 'Shelter Mentality' echoed her attitude on the Western Front.

In the pre-war years the German ARP laws of 1935 and 1937 had already been much more advanced in practical detail than their British counterpart although little scientific development of shelter design had taken place before 1937. In that year special testing laboratories were set up to test various types of construction and offices were set up for studying and co-ordinating construction. The first shelters built before

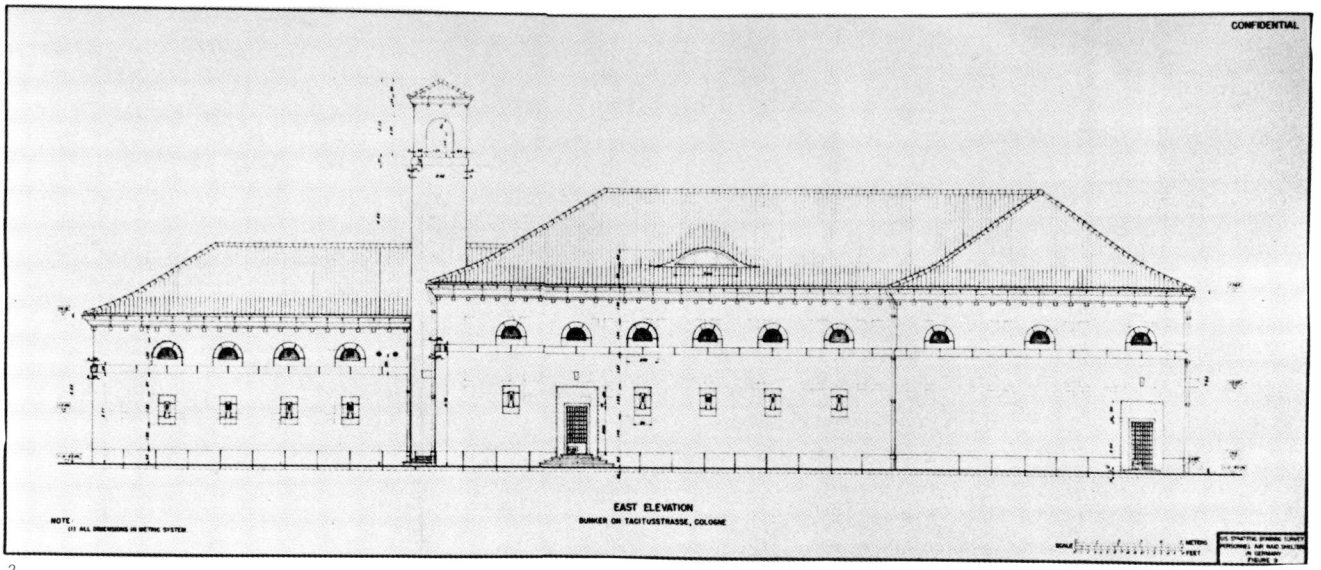

2, 3. Elevation and plan of an air raid bunker in Tacitusstrasse, Cologne. This type of massive concrete construction, with its architectural veneer, had no British counterpart.

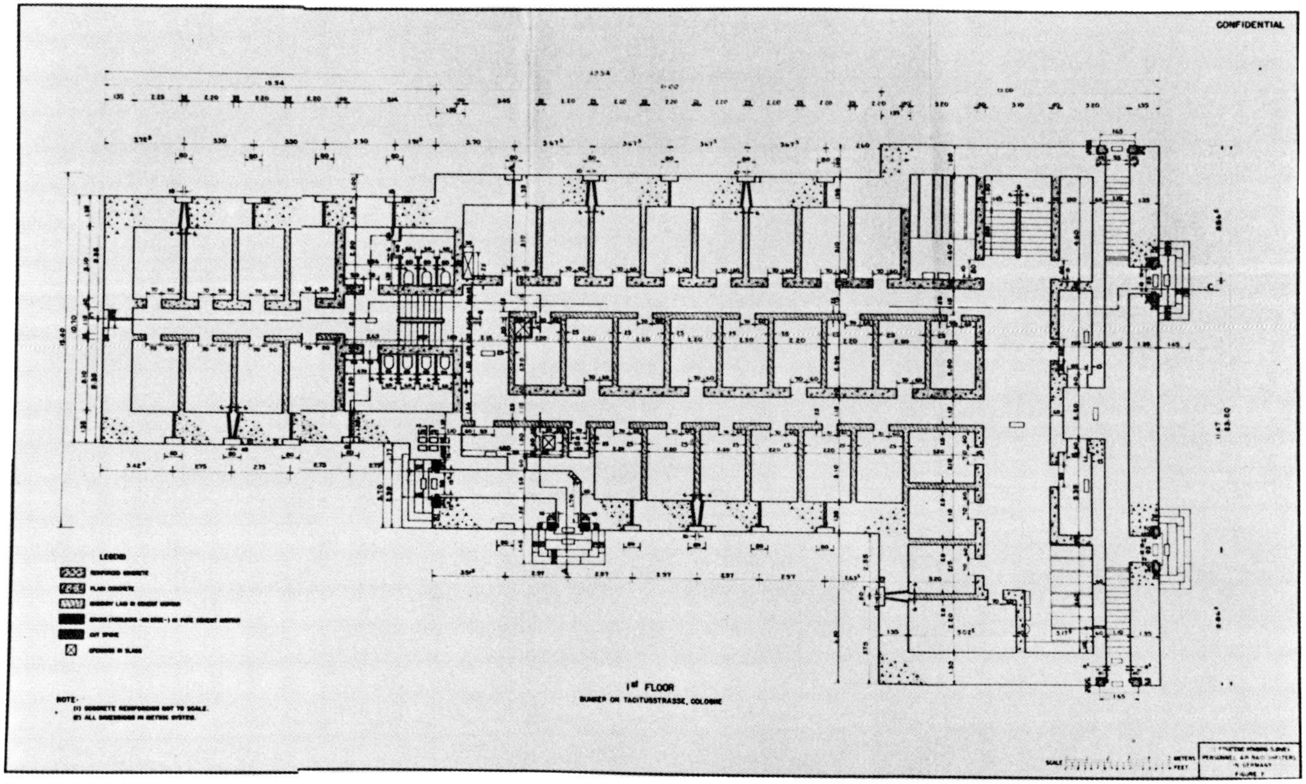

1940 appear to have been of two types: the tunnel shelter and the cellar shelter (modified building basement) (4). Similar to British shelter provision at this time, they lacked amenities and only gave poor protection. After 1940, however, almost all the shelters constructed were of the bunker type and it is these which stand out as the outstanding feature of German shelter provision.

While large numbers of bunkers were built, this was not achieved by any reduction in standards. The German bunkers were inevitably heavily protected concrete structures and, unlike British shelters, were designed from the start for use during raids of long duration, having very complete services, air conditioning and sanitary arrangements. The bunkers were designed for large capacities, usually 500 to 4 000 each and sometimes as many as 12 000 to 18 000 persons, but were subdivided internally into many compartments and were often multi-storey. Their interiors were luxurious compared with their British contemporaries. German specifications called for special entries, gas locks, warden's room, fresh air and halls. They also required separate rooms, toilets and washrooms for each sex. Mechanical ventilation was the general standard with gas locks and filter systems protecting against gas attacks and in some cases air cooling was provided. Separate water supplies (wells), sewage and power systems were usually provided for emergencies and in some of the larger bunkers even electric elevators were installed (5).

The first planned programme for bunker construction began in 1940 when approval was given for the erection of bunkers to provide for five per cent of the population of some 70 cities. Based on the provision of protection against 500-lb bombs these shelters had roof slabs of 1.4 metres but a later scheme for additional bunkers in the 70 cities of the original programme and for similar provision in other cities called for much heavier protection. This later programme provided protection against 1,000-lb bombs by the design of a roof slab of two metres but some cities constructed ceiling slabs of 2.5 metres to protect against 2,000-lb bombs and a few bunkers were designed to resist even larger bombs (6). By 1941 the bunker programme was well developed. A set of eight pamphlets called 'Codes of Practice for Building (Planning) Air Raid Protection Shelters' was published in July by the Reichsminister of Airways and Chief of the Luftwaffe, who was charged with the co-ordination and inspection of shelter construction throughout Germany (7). These pamphlets set down minimum standards for shelters, particularly bunkers, and the general principles contained in them remained essentially unaltered throughout the war (8). The main variations on the standards in these pamphlets were created by the effect of war-time demand on critical building materials which called for a more economical use of cement and steel. Towards the end of the war concrete standards subsequently often fell below specification.

Bunker design throughout Germany developed certain basic standards because of this co-ordination but there were specific regional and local variations. At vulnerable targets such as Hamburg even greater protection than normal might be provided with less concern over appearance. Foundations might vary with sub-soil conditions (9). The detailed design of individual bunkers was often determined by use. For example, bunkers at factories had only few beds while bunkers in residential areas had sleeping facilities for 90 per cent of their design capacity. As with any building project, the views of the individual architects and engineers involved also affected the result despite any co-ordination of specification which occurred (10). Such remarkable futuristic ideas as the rocket shaped 'bomb deflecting' or 'ant-hill' bunker (300 persons) were put forward and presumably built (11). In searching for a common term to describe German bunkers for the civilian population the only true generalization which can be made therefore is that, unlike British shelters, they were 'bomb-proof'.

One outstanding feature of German bunker design during the early years of the shelter programme, the 1940–41 period, was the consideration taken in making the external appearance of a bunker match the architectural character of its neighbours, although no consideration of the post-war use of bunkers was made. Bunkers entirely underground were not unknown but as the commonest form of bunker was the type which was mostly above ground, with only a small part below ground level, external appearance had to be considered. The amount of external decoration varied from area to area, Cologne being apparently one of the more active in this context. Building lines were related to surrounding structures and the bomb-proof roofs of bunkers in the residential districts of Cologne had tiled roofs added to conform with their neighbours. Later in the war, when labour and materials became more difficult to obtain, bunkers were built with plain concrete finishes which

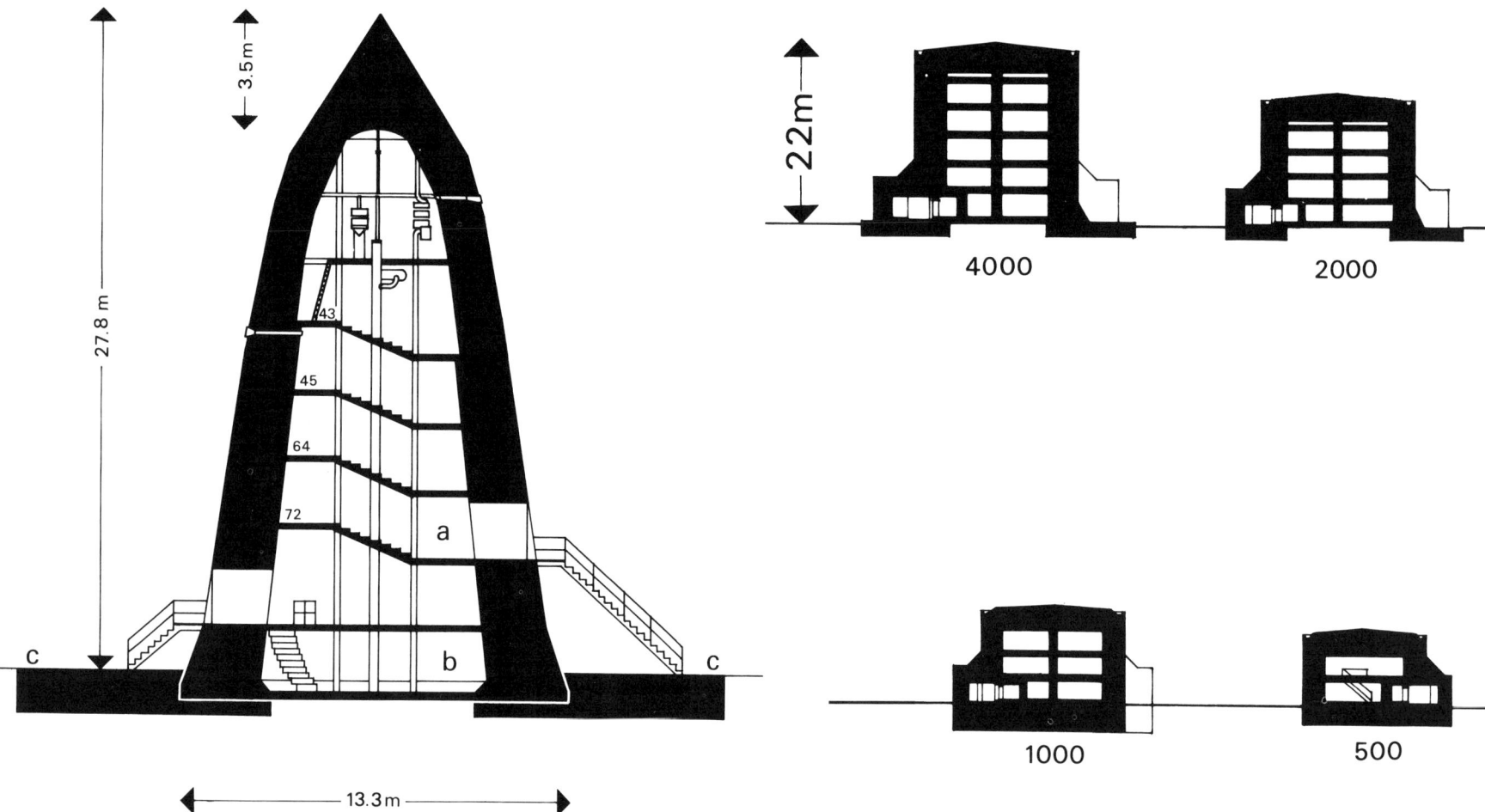

4. The 'ant-hill' or bomb deflecting shelter. A similar form of roof construction was used by the French army on the Western Front in 1914–18. The numbers indicate floor capacities, a the switchboard, b the air raid command post, c the bomb activating apron.

5. Standard designs for air raid bunkers as issued by the Speer Ministry in 1944. Designed for capacities of 4000, 2000, 1000, and 500 persons respectively, all had a roof thickness of 2.5 metres.

usually allowed for the addition of a complete veneer of brick or stone and decorative roofs at a later date. This was the primary method of camouflage but other methods were developed, such as the painting of false windows or the disguising of bunkers as bombed and burned out buildings (12).

Although not all bunkers were treated in this way, there is some comparison to be made with the German use of buildings as bunker camouflage on the Western Front in 1914–18. However, the primary reason for the application of 'architectural veneers' to German civilian bunkers in the Second World War probably owed as much to Nazi architectural ideology as to considerations of camouflage. Official specifications issued concerning the appearance of bunkers stated a desire for some architectural conformity within neighbourhoods where bunkers were to be erected but the bunkers which were completed in the early part of the war, with their decorative veneer, in fact add stylistic evidence to the conclusion that bunkers were used as another extension of Nazi architectural propaganda—almost as remarkable in their contrast between use and appearance as the Weather Service Broadcasting Station, constructed for the Luftwaffe, with its 'folk style' thatched roof and 60 metre lattice steel aerial (13). Confirmation of this projection of Nazi ideology is made in the US Strategic Bombing Survey which illustrates, amongst other examples, a bunker on Tacitus-strasse, Cologne. The bunker is not merely 'camouflaged' but rather, with its neo-Romanesque exterior, another example of the divergent stylistic preferences of the Third Reich (14).

With the attempts being made in 1940–41 to fit the bunker into both local architectural context and Nazi ideology the architects and engineers involved had much more freedom than later. It was not until December 1944, following the inclusion of the Reich within the OT's area of control, that specifications for standard unit type designs were issued by the Reichsminister for Armament and War Production, and Head of OT, Albert Speer. The designs were for bunkers of 500, 1000, 2000 and 4000 persons capacity and of two, three, four and five storeys respectively. As Ove Arup had concluded in London, the large multi-storey shelters were much more economical. In fact the 500 person shelters required a volume of 3.0 cubic metres of concrete per person whilst the 4000 person shelter required only 1.8 cubic metres. The main feature of the standard design was that the different sized bunkers were dimensionally co-ordinated with each other. Thus certain sections having common dimensions could be prefabricated and precast beams could be used for roof sections, eliminating timber formwork, although precast beam construction was common for public bunkers before then and there was already a tendency towards re-usable prefabricated formwork before the December 1944 order. Speer's order for the standardized bunkers was issued too late for these new types to be widely used. In any case by the end of 1944 much of the bunker programme had been carried out and the US Strategic Bombing Survey team in 1945 found no bunkers which had been built on these standards. Their main significance is that they were the official standard at the end of the war (15).

Speer's order of December 1944 illustrates that the emphasis on heavy concrete as a passive defence measure remained throughout the war. But the huge public shelters were not the only part of the German civil defence programme which relied on heavy concrete construction. Germany, like Britain, had her fighters to intercept and destroy enemy bombers but both countries put emphasis on the use of anti-aircraft batteries. But whereas British flak installations were mostly little more than temporary dug-outs, within the Reich huge monolithic structures, 50 metres or more in height, were built. These 'Flaktürme', with their heavy bomb-proof construction, appear to have been unique to Germany and in Britain the only comparable equivalent were the Sea Forts in the Thames and Mersey estuaries.

The Luftwaffe, who were in charge of flak installations in Germany, were far more likely to build huge and costly structures at the least excuse, perhaps reflecting Goering's obsession with the political status symbol. However, as the German air arm was itself inclined towards the use of low-level planes such as the Stukas, there may have been a more reasonable explanation for the construction of high bomb-proof flak towers. Rising above surrounding buildings and vegetation their height gave an unobstructed field of low angle fire. The flak towers might be as many as ten storeys high for this reason. The lower floors housed ammunition and quarters for the AA crews but the height needed for the tower to function left space which could be used for air raid shelters and hospital purposes. Thus in a way the towers were bomb-proof shelters and an extension of the bunker programme, providing both passive and active defence. In the latter category, apart

6. One of the early fortress flak towers built as part of the West Wall's 'Air Zone West'.

7. Four of the huge flak towers which guarded the German Marine Barracks outside the French town of Angers. All of the towers were put out of action by Allied bombing and one (right foreground) was reduced to rubble.

8. View from the Berlin Zoo flak tower looking towards the adjacent radar tower.

9. Diagrammatic ground floor plan of the Berlin Zoo flak tower. Known as the Tiergarten Tower, it was built in 1941–2 using 190 000 tons of monolithic concrete. The tower was 37 metres high.

2.5m

71m

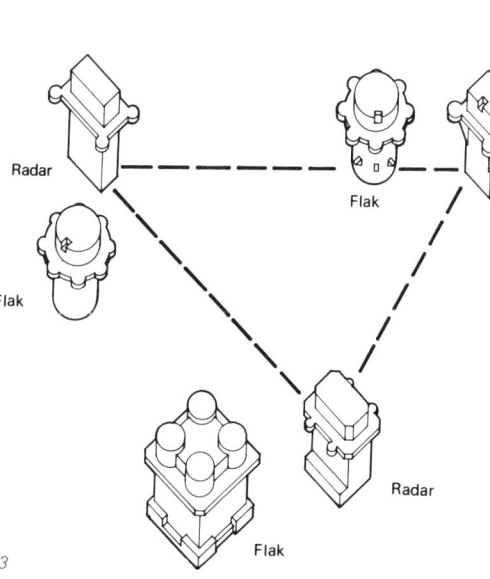

10, 11, 12. Two types of flak tower (10/11) and one of the radar towers (12) in Vienna.

13. Diagrammatic layout of the Vienna towers.

from their use in AA defence, some of the Berlin towers were also used as fortresses by SS troops.

Flak towers had been built by the Organization Todt as early as 1936–40 on the West Wall in what was known as 'Luftzone West'. These and later flak towers constructed by the OT in France were essentially economic in their use of building materials and blended in with existing watertowers, windmills and silos. Between 1939 and June 1944, however, OT work was confined to the German occupied territories. The majority of flak towers were constructed inside the Reich and it can therefore be assumed that these were mainly built by civilian contractors operating directly under the Luftwaffe. Only in 1943 was Amtbau, a central organisation for inner Reich construction, established (16). The flak towers built in the Reich (including Austria) were huge monumental structures and they differed strongly from those built by the OT in the occupied territories. Most of them were built in cities, coinciding with Allied targets; their monumental construction might therefore have had the additional objective of providing the German people with some form of psychological and political reassurance.

The design of individual flak towers within the Reich varied, but there were certain similarities. There was provision for 88 mm anti-aircraft guns on the roof and a gallery level below this from which 20 mm or 40 mm guns could be fired at low-flying aircraft. The overall shape and style of the tower varied to some extent with time and place. The large Berlin flak towers were fortress-like with steel shuttered windows (17). Designed in the monumental style of the Ordensburgen —quasi-military schools for a 'heroic' party leadership—they roman-ticized mediaeval military tradition. This neo-Romanesque appearance suggests that the Berlin towers were built early in the war, corresponding with the tendency towards 'applied style' of the 1940–41 civilian bunkers. Their appearance is probably further explained in that being designed in a 'heavy' monumental style, they were comparable to the neo-classicism which Hitler himself favoured for his capital (18). Other flak towers such as those at Harburg, represented the same basic function but were more identifiable with the progressive style of the modern movement'. Although the Nazis had used the controversy of the new style in the early 30's as propaganda material and denounced it, they soon adapted it in this way as the other extreme of Nazi stylistic ideology, as 'evidence of the revolutionary and modern character of the Nazi regime' (19).

One feature which was common to most flak towers was the close presence of another tower, very similar in design. This was the radar tower housing the 'Würzburg' gun-laying equipment. In the Berlin zoo the radar tower, though not identical to its flak tower, was in the same modernized romanticism and complementary. Likewise in Vienna the radar towers were in complementary designs. Vienna had three pairs of towers (20), each pair consisting of a flak and radar tower. These pairs were distributed across the city, equidistant from each other, forming an equilateral triangle in layout. Two of the flak towers were circular in plan with a gallery running around their walls containing eight gun positions each for low-flying aircraft, in addition to their main roof-mounted artillery. Next to each of these towers was a rectangular radar tower, again with a gallery and a gun position on each corner of the gallery. The third pair of towers were quite different. In this case the flak tower was square in plan, not round, with a gallery level and four small circular towers above this. The radar tower was again rectangular in plan but with different features to the others. Thus of the six towers in Vienna there were four distinctly different types, and all were individual in detail. Despite the logic of the overall city layout, the conclusion must be made that standardization was non-existent where flak towers were concerned.

Massive as both the public shelters and the flak towers were, they proved to be of only limited value when both targets and the weaponry of aerial bombing began to change after 1942. During 1940 and 1941 RAF Bomber Command had concentrated on pin-point attacks on industrial targets but an official report in August 1941, the Butt Report, discovered that only between 7 and 22 per cent of the planes were getting within 8 kilometres of the target (21). The obvious conclusion was to switch to targets which did not require precision navigation and this report marked the birth of the 'area bombing' policy. When Churchill's adviser, Professor F. A. Lindemann, was asked to suggest a bombing policy to assist the Russians he was naturally preoccupied with the implications of the Butt Report. His final report in Spring 1942 was the confirmation of a ruthless and 'systematic' attack on the German civil population. Over 15 months he calculated that it could make one-third of the German population homeless (22).

As the size of the Allied bombing fleet increased and its effectiveness

14, 15, 16, 17. Tower for land defence at the centre of the Brest submarine pen's north west face (14) and the right hand corner of this face (15)—both showing the burster slab which was added to the main roof. The south west face of the pen (16) and detail of the flak tower on the southern corner (17), showing the blockwork used to thicken the original roof slab.

grew with the use of improved navigational aids, the civilian population flocked to the bunkers in large numbers. These became overcrowded by five or six times their design capacity. A bunker in Hamburg designed for the use of 18 000 people was in practice used by up to 60 000 persons and at times even more. The population, as in Britain, felt safer in bomb-proof shelters. Such overcrowding made the service provision inadequate. There were insufficient bunks, the sanitary arrangements were overstretched and the mechanical ventilation became ineffective. These faults were not in themselves fatal and the US Strategic Bombing Survey stated in 1945 that:

"Allied aerial attacks were not generally effective against bunker types of shelters. Incendiaries and smaller sized bombs did not damage bunkers, and effects of large size near misses were confined to exterior superficial damage." (23).

This survey did not however take into consideration the complete changes brought about by the 'fire storms', which were the direct result of the area bombing policy when it found full effect in mid-1943. Apart from incredible heat, they produced vast air suction—creating the 'fire typhoon', a meteorological phenomenon. People and objects were sucked into the fire, masonry was observed to reach temperatures of 1000 degrees centigrade. Under these conditions bomb-proof protection became of secondary importance. The first fire-storm of the war at Hamburg in July 1943 was in a city which had the most advanced air-raid precautions in Germany. Despite these precautions the casualties during the five days of the Battle of Hamburg were astronomically high, a conservative estimate being 43 000 dead—more than the number killed by the Luftwaffe in the eight months of the Blitz.

David Irving describes the effect of this first fire storm on the Hamburg bunkers:

"When rescue teams finally cleared their way into hermetically sealed bunkers and shelters, after several weeks, the heat generated inside them had been so intense that nothing remained of their occupants: only a soft undulating layer of grey ash was left in one bunker, from which the number of victims could only be estimated as between 250 and 300' by the doctors.... The uncommon temperature in these bunkers was further testified to by the pools of molten metal which had formerly been pots, pans and cooking utensils taken into the shelters." (24).

The next fire storm, at Kassel, had similar results. Of the 5000 deaths some 70 per cent were caused by asphyxiation. The Police President in his report put this down to the commonly held belief that bunkers were the safest place to be during an air raid. In fact a bunker was only safe for the first few minutes of a raid. After this it should have been evacuated, before the fire storms developed, but people were reluctant to leave the apparent safety of their bunkers and this tendency was reinforced by the fact that most German bunkers were designed for use over long periods. Any lessons which might have been learnt from these early raids about the danger of too passive an attitude amongst shelterers were not publicized. Goebbels thought that the truth about the nature of the 'fire typhoon' could cause widespread panic and so no realistic ARP instructions, particularly with regard to bunkers, were issued. Despite the failure of bunkers under fire storm conditions, however, it would be wrong to conclude that German civilian bunkers were completely ineffective. Dresden, with comparatively poor shelter provision, though better than most English cities of its size, testifies to this. The number of deaths during the RAF and USAAF raid on Dresden in February 1945 is estimated at 135 000—a figure three times as many as at Hamburg, with its extensive shelter provision, during the first fire storm of July 1943 (25).

The Casablanca Conference of January 1943 had affirmed the policy of 'area bombing' but before this policy was implemented it required that the submarine pens be the first target. The success of the U boats against Allied shipping in the Atlantic during 1942 had made this necessary. Although the bombing of German cities tended to make most of the news during the war and has been the focus of most writers subsequently, there were in fact several times when such attacks on military installations connected with the German war effort assumed vital strategic importance.

The U-boat had established itself in the First War as 'a potentially war winning weapon' (36) and under the energetic and efficient Doenitz the German submarine fleet was again built up to a force of 400 submarines by 1943. Having decided to put so much energy into a submarine war the German navy naturally put a corresponding amount of effort into protecting their U-boats in harbour from Allied bombing. The success of the resulting protection, largely designed on the

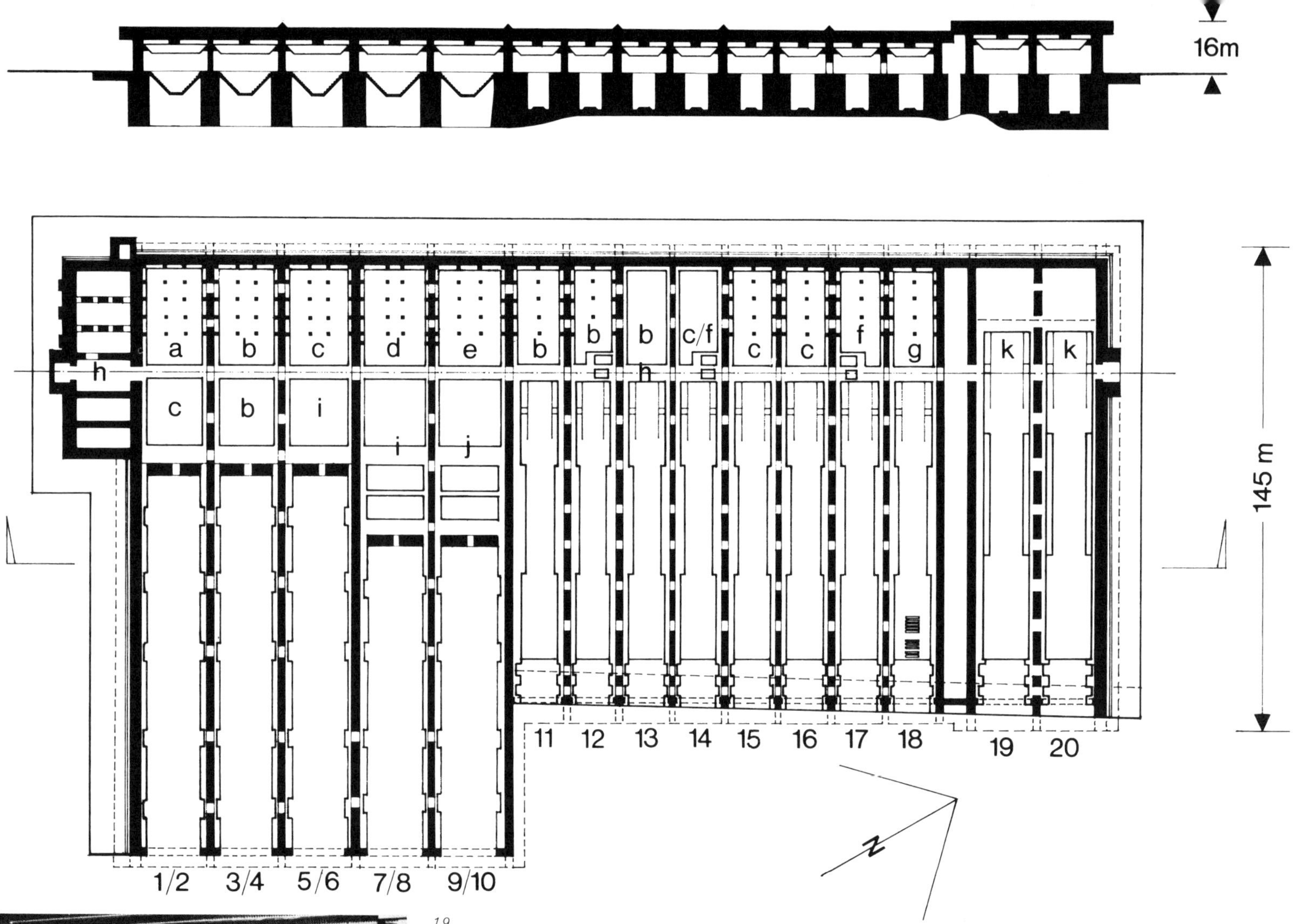

18. Plan and section through the main hangars of the Brest pen. Roof thickening and additions to the north west face are not shown.
   a. ordnance repair
   b. machine shop
   c. ship repair
   d. generator
   e. electrical repair shop
   f. pump room
   g. fire department
   h. corridor
   i. torpedo room
   j. ordnance room
   k. Pumps
   1–10 wet pens
   11–20 dry pens
19. Interior of the Brest pen.

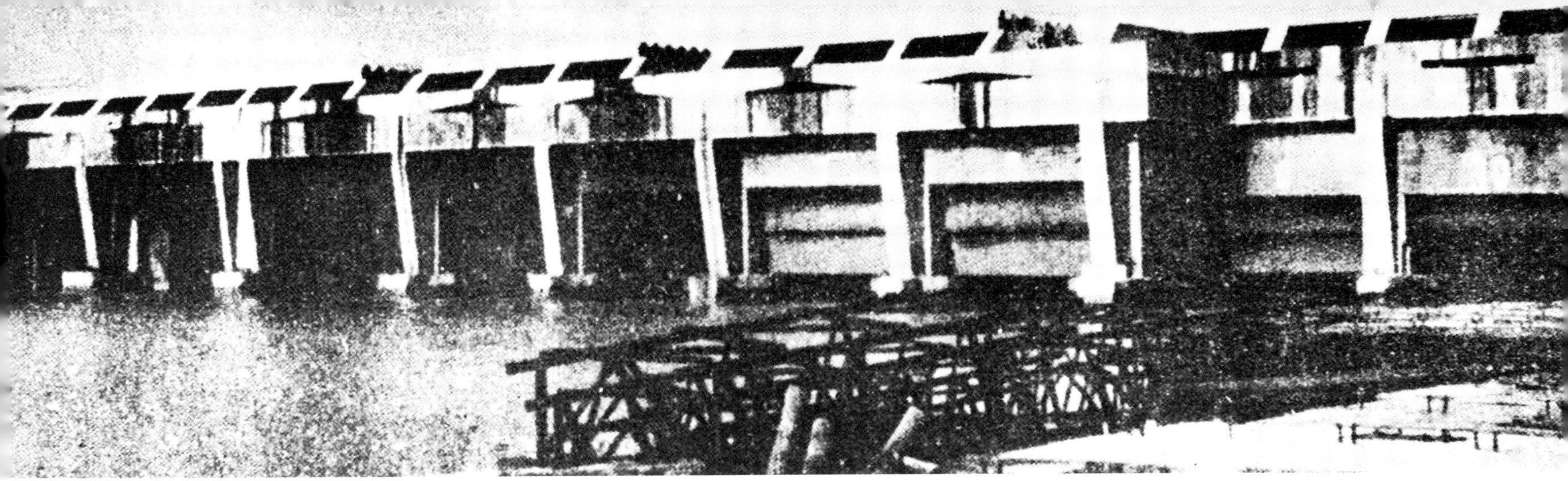

20

21

22

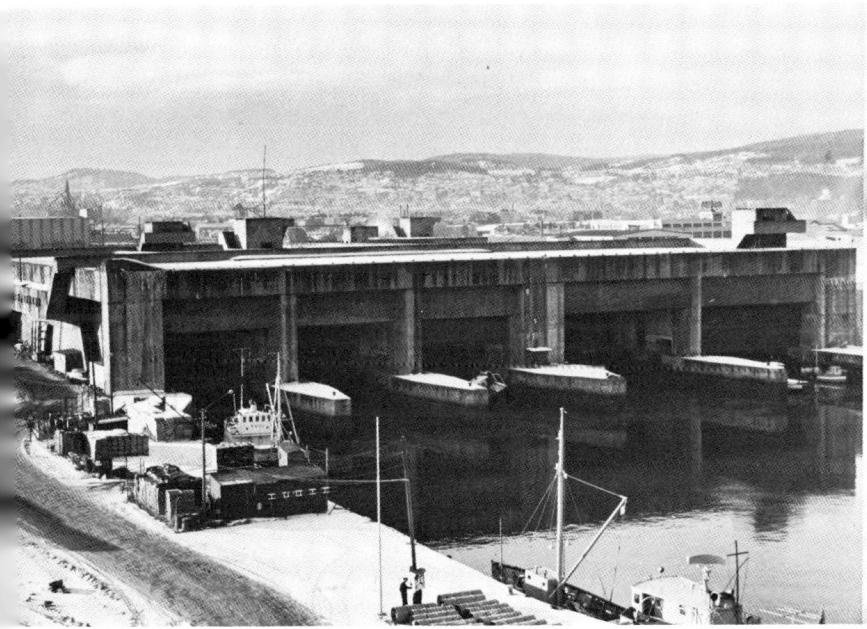

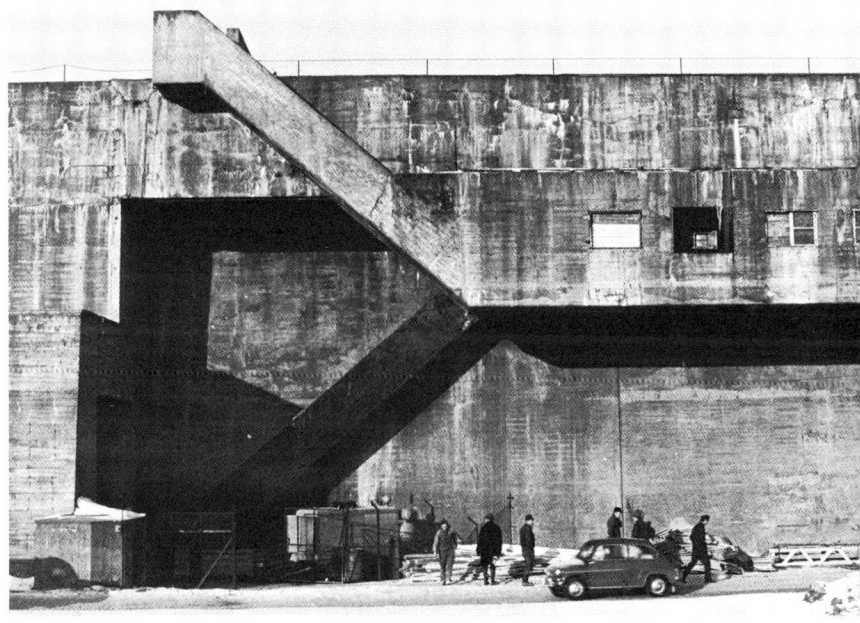

20. The large Bordeaux pen was similar in design to the Brest pen. It also received additional roof protection in the form of a burster slab.

21, 22. Photographs of 'Dora I': the submarine pen constructed by the OT in Trondheim at a cost (to the Bank of Norway) of some £20M.

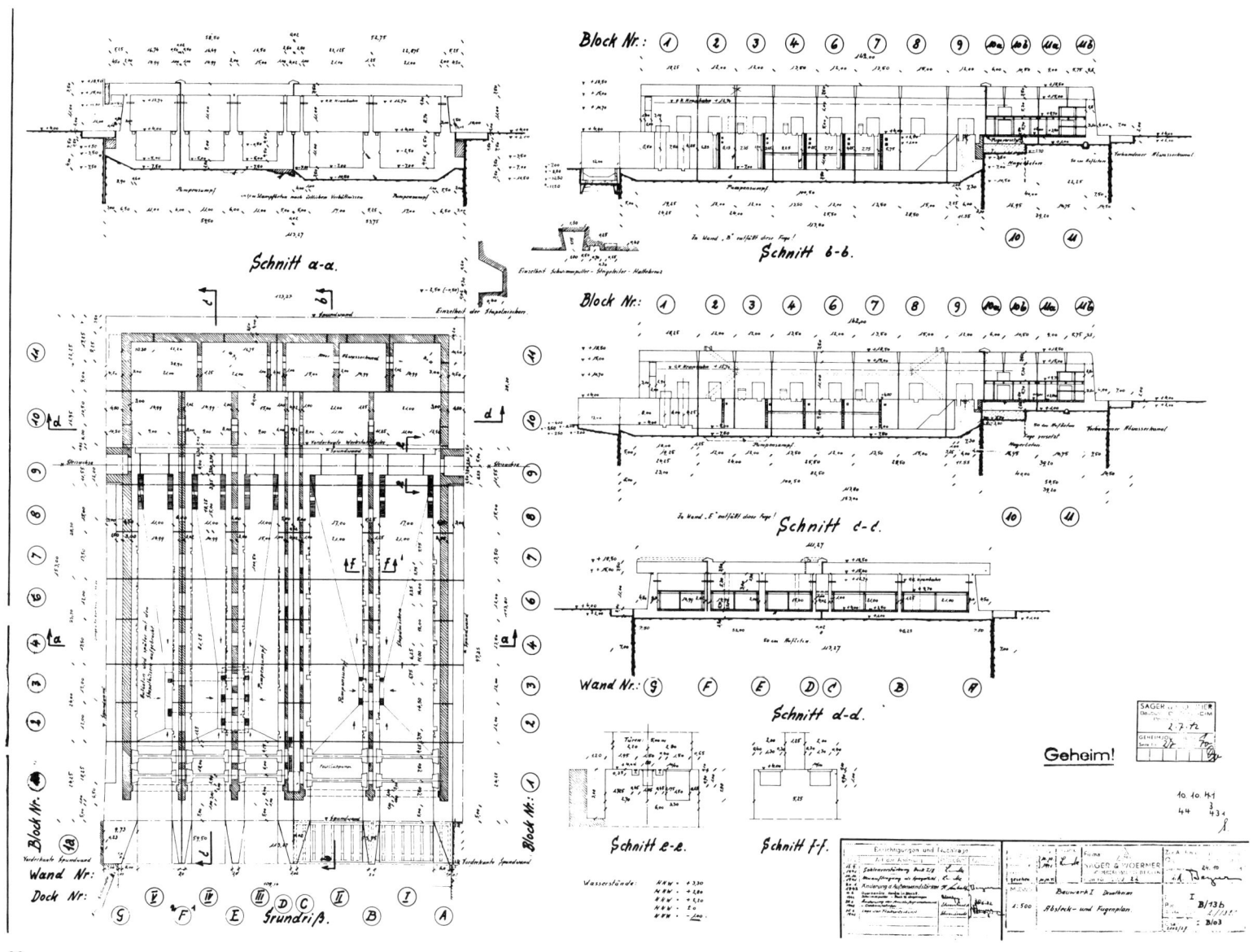

23. The original working drawing for the pen by the Berlin firm Sager and Woerner. Signed by the OT Oberbauleitung for Trondheim in October 1941.

precedent of the pens built at Bruges in 1914–18 (27), but with five times the thickness of concrete, is best shown by the fact that of the 785 U-boats lost during 1939–45, only 7.6 per cent were lost due to bombing raids (28).

In the Reich itself, construction of submarine pens started early. At Hamburg, for example, the Deutsche-Werft pens had been built using pre-cast pre-stressed concrete trusses (29) manufactured by Wayss and Freytag at their beam factory at Heilbronn (30), but transportation problems probably caused the use of such beams to be limited mainly within Germany. The pens inside the Reich soon assumed secondary importance, both numerically and strategically, however. On June 23rd, 1940, when France fell, no time was lost in moving U-boat bases to the Atlantic coast. Only one month later Lorient was opened as a base and pens were built there as well as at Cherbourg, Brest, St. Nazaire, La Pallice and Bordeaux. These enabled a shorter, safer route into the Atlantic, allowing 25 per cent more boats to be kept in the operational area attacking the convoys (31). The fact that some of the smaller U-boats could now be used west of the British Isles and French dock-yards used for repair, relieving pressure on the German yards, were additional advantages. The French Atlantic ports became the main base of German naval effort now, not the Reich, and this fact was emphasized by Admiral Doenitz setting up his shore HQ at Brest. From there he directed his wolf packs in the 'Battle of the Atlantic' for nearly two years (32).

Nor were the French Atlantic ports the only submarine bases outside the Reich. Pens were built for example at Trondheim in Norway, allowing access to the most northerly parts of the Atlantic, at Marseilles in southern France, allowing direct access to the Mediterranean. Some of the other pens such as Le Havre, Boulogne, Ijmuiden, Rotterdam, Bergen, Poortershaven give some impression of the considerable number of installations which were built overall. Added to which, most of the bases were extremely heavily fortified and it is obvious that the construction of pens and ancillary fortification for the U-boat war comprised a major part of German war construction. Outside the Reich the OT was supposed to take a large share of the responsibility for their construction, but the navy often contracted directly to local firms in order to avoid OT red tape. With this situation there was a limit to the amount of standardization which was possible. Not only were there problems of individual sites and soil conditions to consider but the method of control sub-division, the OT Einsatzgruppen, tended to substantiate geographical differences. Thus, while the squat pens built at Bergen had little similarity to the Brest pens, within the Atlantic coast section of Einsatzgruppe West the pens at Bordeaux and Brest were almost identical.

With Brest as Doenitz's HQ, the pens there deserve special description. Started in 1941 the submarine base consisted of not only the concrete pens but also underground workshops, shops and accommodation for the crews, in addition to large numbers of gun emplacements (33). The pens themselves were mostly above ground but the depth of water necessary for access by the U-boats meant that internally a large part of the pens were below water level. The overall size of the pens was about 180 by 300 metres. Within this space 10 submarines could shelter in the five wet pens and a further 10 in the 10 dry pens. The wet pens had only one leaf doors extending to 1.8 metres above the working floor although the dry pens, where repair work was carried out, had double overlapping leaves. The problem of access, as usual, created the weak link. In addition to this weakness the maximum external wall thickness at Brest was less than the 3.5 metres maximum at the Deutsche Werft Pen, Hamburg. The roof protection was also initially less. The roof slab was however being continually added to, as was the case with many other 'bomb-proof' structures built by the Germans.

The original roof slab was about 3.5 metres deep concrete strengthened by steel trusses fastened to 2mm corrugated metal which acted as permanent formwork and ceiling finish. The trusses were closely spaced and the entire slab was densely reinforced with steel bars—but even this amount of protection was insufficient to resist the heavier bombs. The original slab was thus added to later, bringing the thickness of the roof up to 5.5 metres over 60 to 80 per cent of the roof area. About 15 to 20 per cent of the roof had a quite different method of thickening, in some cases built over the original slab, in others on the thickened slab. This was the 'burster slab'—a layer of pre-cast slabs about 1.5 metres deep and laid on thick supporting walls at six metre centres leaving a two metre air space. It is the external appearance of these slabs which give the Brest and Bordeaux pens their distinctive character. Nor were these continual additions to the roof slab the only protection of the pens against air attack. Truncated pyramids of solid concrete formed

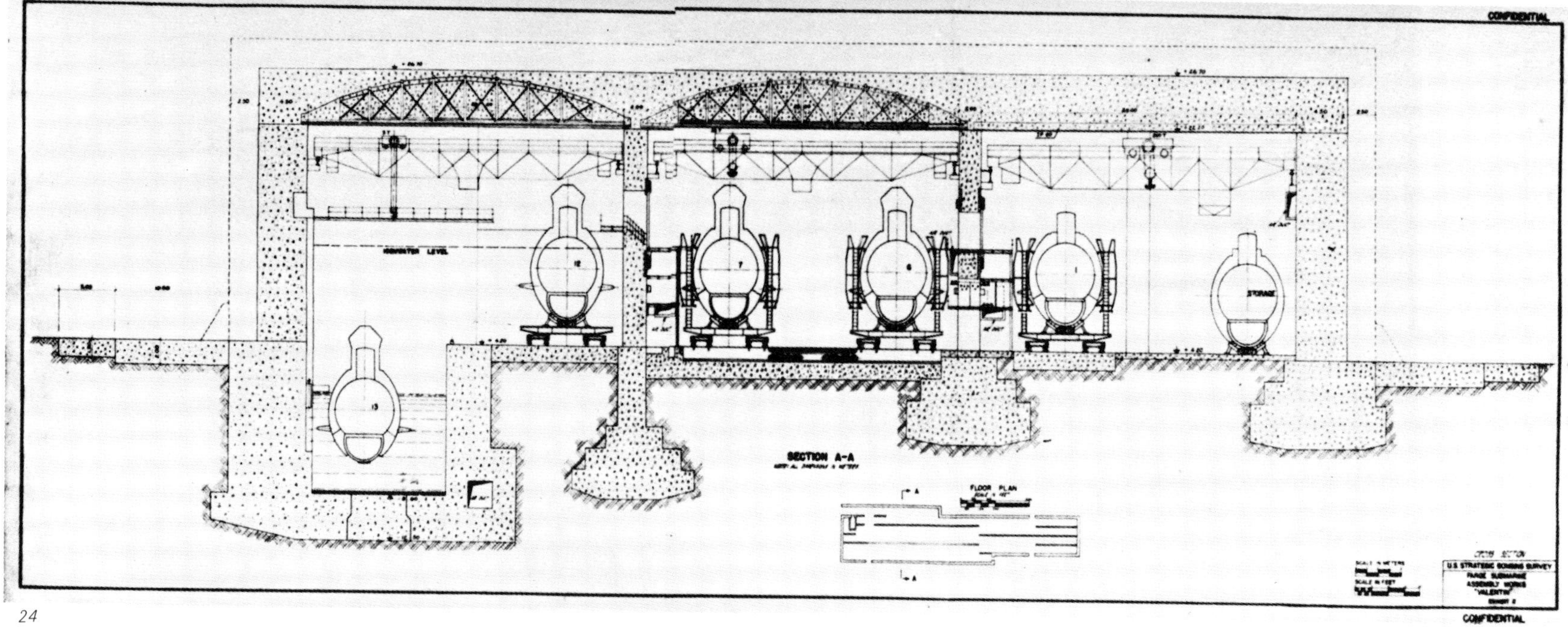

24

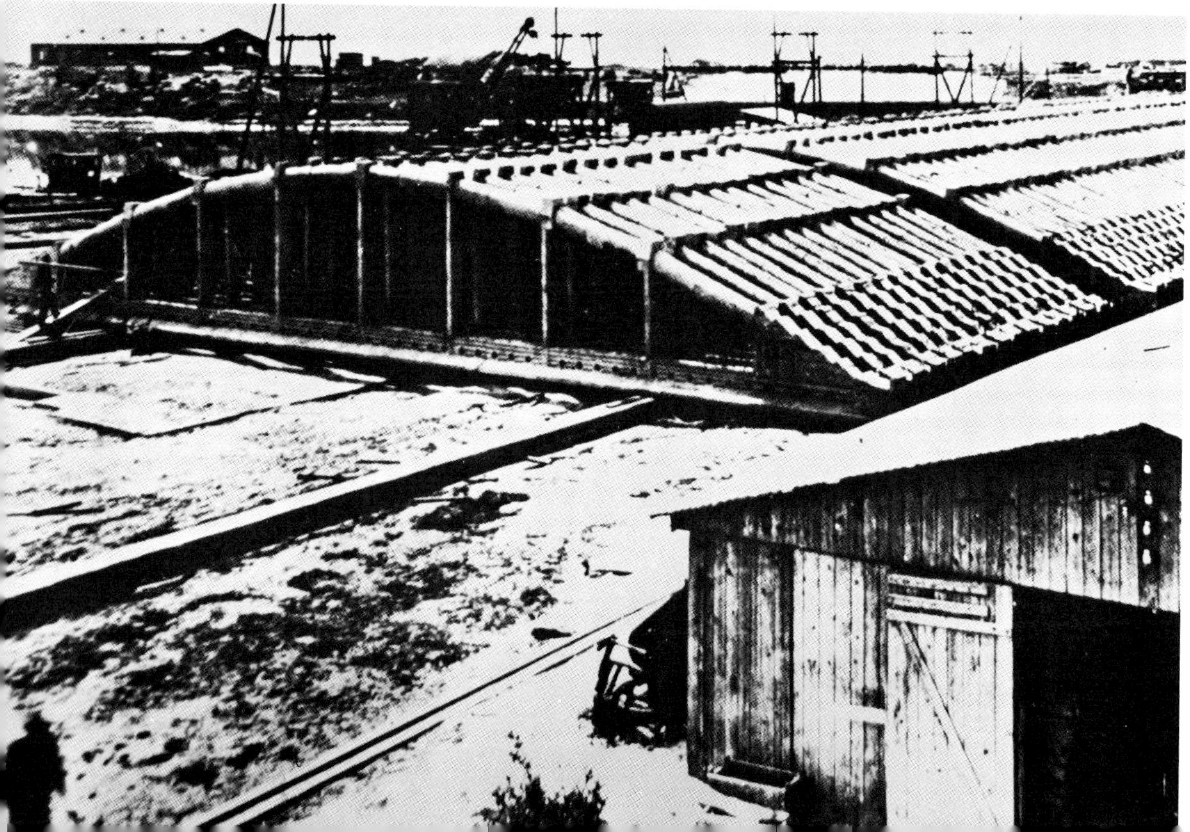

25

24, 25. Section through the Farge Assembly Plant (24) and photograph of the reinforced concrete trusses used in its construction (25).

the pens' personal flak towers and added to the peculiar three-dimensional qualities brought about by providing air protection for submarines (34). Not all the pens were as visually exciting as the Brest pens, however. Some were merely huge monolithic blocks—impressive only by their sheer size. Bergen, much smaller, was quite provincial by comparison.

The success of the German submarine campaign against commerce in the Atlantic was considerable. Following the decision by Roosevelt and Churchill, at Casablanca in January 1943, to divert their major bombing effort on to the submarine pens in the Atlantic ports, over four months, some 11 000 tons of high explosive and 8000 tons of incendiaries were dropped on the pens at Brest, Lorient, St. Nazaire, La Pallice, and Bordeaux with little or no effect. Bombs gouged out craters 11 metres in diameter without penetrating the slab or damaging the underside (35). It was not until Barnes Wallis's invention was used, the Tallboy Bomb, that any of the Atlantic pens were pierced. Even then an Allied Intelligence report on Brest remarked of the six heavy bombs which pierced the roof slab:

"Surprisingly little damage was observed in the interior of the pens. Evidently the blasts had spent themselves in tearing through the slab. Interior walls, which were less than 30 feet from the blast were not damaged. The strongest blast did not damage more than six steel trusses." (36).

In any case by the time the Tallboy bombs were being used on the pens, the Allies had already won the 'Battle of the Atlantic' at sea. By May 1943 the U-boats were only sinking one merchant ship per U-boat lost, compared with an exchange rate of over 40 ships per U-boat in the winter of 1940–41 (37). The U-boats were withdrawn from the North Atlantic and the campaign against shipping collapsed.

Admiral Doenitz's reaction was to improve the U-boat. As it happens, however, like so many other German 'super-weapons', these new submarines were only just available in quantity when the war ended. The success of the U-boat in the first three years of the war was never repeated. However the production of priority armaments such as these new U-boats raises another aspect of bomb-proof shelter: the protection of industry.

The construction of an assembly plant for one of these new submarines at Farge, 15 km north of Bremen, is a case in point. Constructed by the Weser River, the plant assembled prefabricated and sectionalized components which had been produced inland. The assembly plant offered similar protection to some of the submarine pens themselves and was in any case similar in design to the pens, using the same pre-stressed concrete trusses used at Deutsche Werft. With a roof thickness of 7 metres in parts the amount of protection was in fact much greater than in other pens (38). Most of the pens had been started in the period 1940–42 and the shelter at Farge probably represented the latest requirements for protection in 1945; it was only 90 per cent complete when the war ended and no production was ever begun in it. The main point about this shelter, however, is not the incredible thickness of concrete involved nor its similarity to the pens (its function was not much different). It is that it represents the increasing attention given to the protection of war industry in the Reich, by dispersal as well as by bomb-proof construction, a tendency which increased with the accelerating activity of the RAF and USAAF.

In Britain the protection of vital industrial plant was first recommended by the Hailey Report, in May 1939, and this policy was implemented to a limited extent by a reluctant government (39). When the bombing began to take effect in 1940 the Ministry of Aircraft Production (MAP), under Lord Beaverbrook, was enthusiastic about the idea of building factories underground. The twin tunnels of the central line between Wanstead and Ilford were converted into a 28 000 square metre factory in which the Plessey Company produced aircraft components (40). The largest British underground factory was set up in some caves near Bath, producing Bristol aero-engines and gun-barrels for fighter planes. This factory had cost £12M to construct before it was completed in 1942 and because of this high cost there were only three other large underground schemes (41). The construction of bomb-proof shelters for the British war industry was therefore negligible (42).

In Germany outline plans for the dispersal of important plants had been drawn up as early as 1939 but the German dispersal of industry also meant in many cases the construction of bomb-proof shelters for important war plants even though they were dispersed in rural areas. But, whatever the causes of British hesitation to divert money and manpower to protecting industry from air attack, these misgivings were echoed in the attitude of Albert Speer, the Reichsminister for Armament

26

27

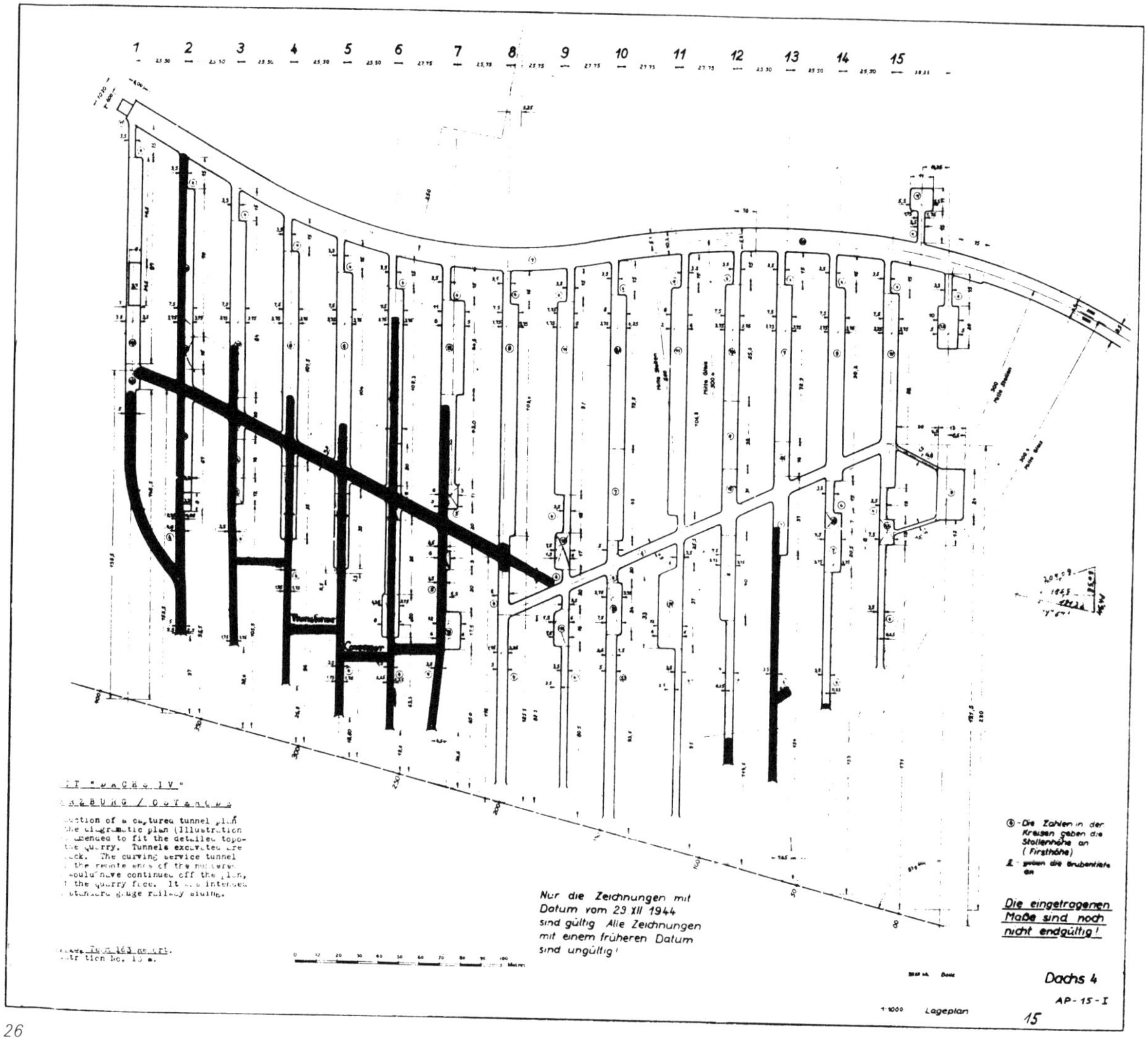

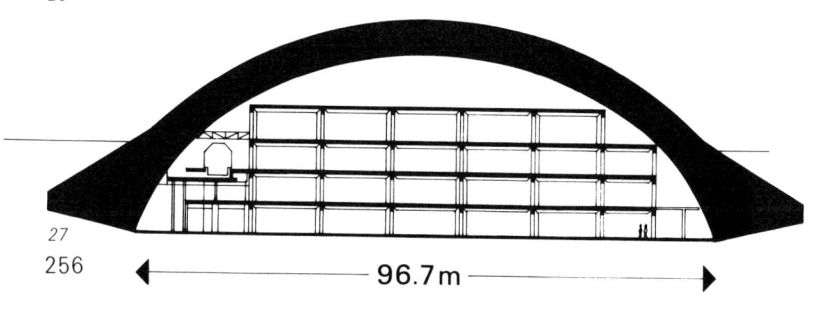

26. The 'Dachs IV' underground factory. The tunnels actually excavated by the time of Allied capture are shown in black.

27. Section through the bomb-proof roof of the Landsberg plant.

and War Production. In his opinion bombers could not be combated with concrete 'and in any case it would have taken many years of work before our plants could be placed underground or behind massive concrete' (43).

His scepticism in this respect had some interesting political repercussions. Both Himmler and Goering were jealous of Speer's power and influence; here was an issue on which Speer was in disagreement with Hitler who, by 1943, was demanding a general transfer of industry to caves and huge shelters to prevent bombing cutting back production. Thus, when Speer fell ill in 1943, Goering used the opportunity to undermine his position. He summoned Xaver Dorsch (44), the ambitious Ministerial Director of OTZ, the OT's Central Headquarters in Berlin, and told him of Hitler's demand for bomb-proof factory shelters. Dorsch, it turned out, did have some suitable designs ready but the problem was that the OT's field of operations had hitherto been confined to the occupied territories. That same evening Hitler authorized Dorsch 'to carry out the building of such major structures inside the Reich as well as outside' (45). On April 29th, Amt Bau, the government agency within Germany, was put under Dorsch as head of OTZ. Hitler made one of his 'dreaded impulsive decrees' making Dorsch his direct subordinate and giving the big shelters priority over all other construction projects. About the same time Goering also put Dorsch in charge of a project for the construction of underground hangars for Luftwaffe fighter bases in the Reich (46).

The political implications of all this were obvious. Indirectly Goering would have control over OT via the grateful Dorsch—something he had wanted since, if not before, Todt's death. However, when Speer recovered from his illness and returned to Berlin in 1944, he quietly re-established his influence with Hitler. On August 24th, 1944 Hitler decreed that Speer, as Chief of OT, was now in command of all official construction agencies within the Reich (47). Thus far from decreasing Speer's power the intrigue increased it at Goering's expense, ultimately putting Luftwaffe and naval construction agencies in the Reich under OT control through Dorsch's renewed subordination to Speer. The construction of underground factories, ironically, was to continue under Speer's control.

In October 1944 a State department was founded under Speer, the 'Industriekontor', which was hopefully to end the administrative confusion so typical of the bomb-proof dispersion of German industry. But the majority of projects had been started before the foundation of this department. Many private dispersals, including most of the known Messerschmitt projects, did not in fact appear on the Industriekontor files at all (48).

One of the Messerschmitt bomb-proof assembly plants was constructed in a densely wooded area 4 km north-west of Landsberg. Designated 'Weingut II', both the main plant and subsidiary facilities were well camouflaged and an Allied intelligence report on the plant in June 1945 discovered that future plans were to improve this further as the necessity for a tell-tale air strip was to be eliminated by the proposal to catapult completed aircraft from the assembly building for delivery to existing airfields' (49). The main interest of this plant was not however, its camouflage but its roof construction.

The main bomb-proof roof construction at Weingut II rose about 15 metres above the ground and dropped about the same distance below, the concrete roof having an arch span of 96.7 metres and a crown thickness of five metres. By 1944 there were definite plans to double this thickness, the surface being keyed with this in mind. The arch-type roof was constructed completely independent of the internal structure which sub-divided the space into four storeys of factory space, complete with overhead cranes, air-conditioning and a railway platform allowing trains to enter and deliver under protection before leaving the plant. This type of arch construction was used several times by the Germans for large bomb-proof constructions. Instead of using conventional timber methods the formwork was built up in the shape of a huge gravel mound, graded with the finer gravel at the top, finished with a smooth thin coat of concrete. This was painted with camouflage paint on top of which the main slab was laid. After setting and gaining strength the gravel was excavated, allowing the thin 'forming' coat of concrete to fall away, leaving a smooth internal finish to the complete slab. Excavation was aided by building a double track railway in a box culvert running through the factory along its centre-line. The technique was similar to the 'Verbunkerung' method used later for flying bomb shelters such as Siracourt, and the massive A4 rocket launching site at Wizernes (50). Such forms of construction were relatively rare, however, when compared with the underground factories which were intended to compose a major part of the German bomb-

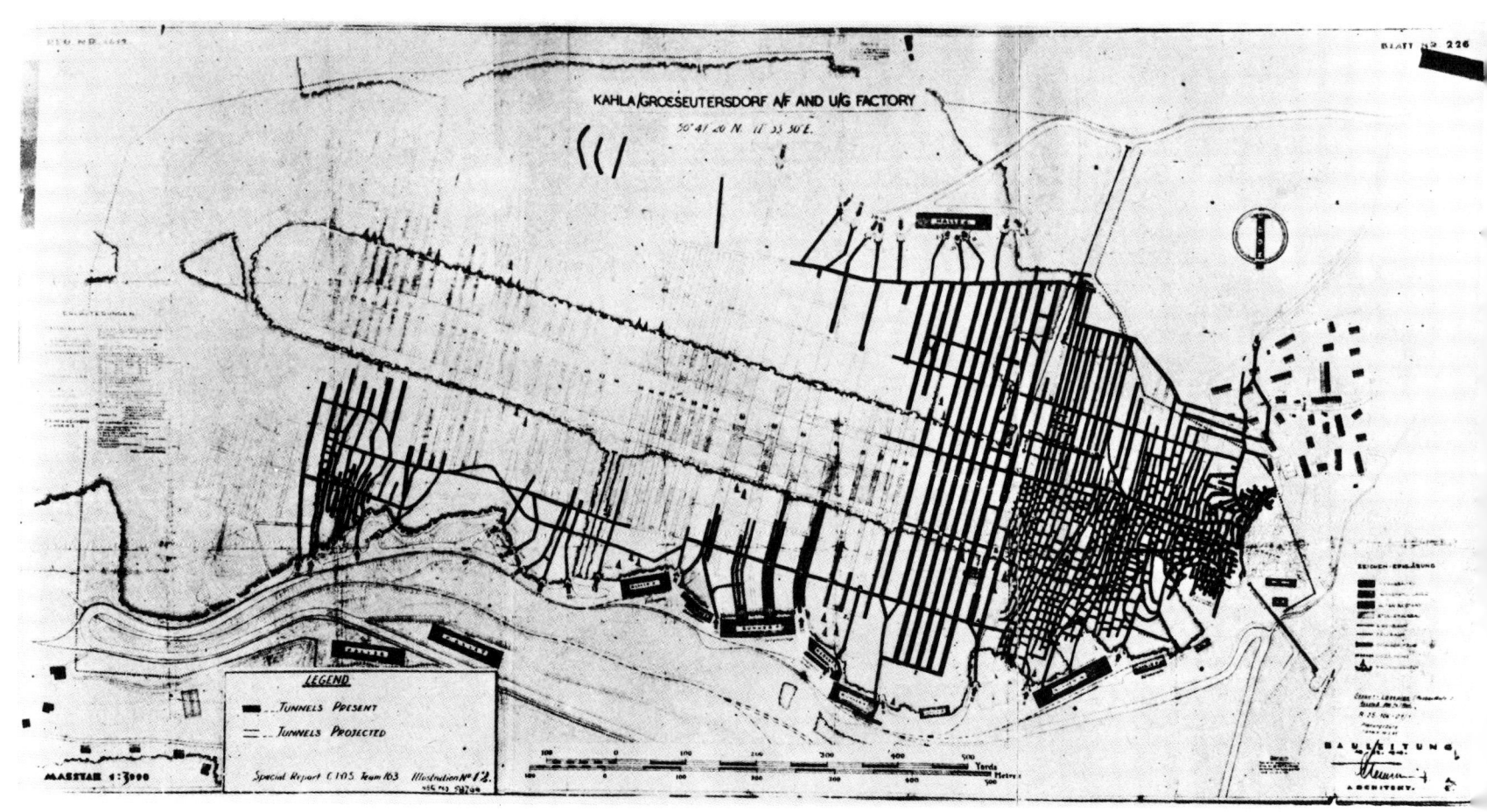

28. The Kahla/Grosseutersdorf A/F and U/G Factory.

proof protection of industry.

Plans in the 'Industriekontor' files, at the end of the war, show that tunnelled, underground factories had been prepared for a variety of production processes. Some of the larger ones being planned were complexes with floor areas of over 95 000 square metres. These factories were built under direction from the factory production firm. Consultant geologists such as Prof. Schriel, and Prof. Deubel had the main initiative in determining the tunnel plan. A report on 'Underground Factories in Central Germany' ascertains that:

". . . it is clear that most of the responsibility for an underground project devolved on the geologist. Briefed with the floor space required, and with details of processes and machines requiring minimum height and width allowances, he had to find a suitable site, having due regard to thickness of overburden (200–300 feet was considered safe in softer rocks like gypsum), to the proximity of a railway, and to the availability of industrial water supplies. Once found, the site was tested for overall suitability of rock by means of vertical and horizontal trial borings. . . . With the information derived from these borings the geologist would draw a tunnel plan in consultation with the architect and representatives of the tenant firm, if known." (51).

From this point the main responsibility for the layout and co-ordination of the tunnel work was with individual architects such as Fleming, Alfred Hesse, and Prof. Herbert Rimpl.

By 1945, however, a major part of the plans for such underground facilities was as yet not completed and some, such as the Dachs IV project, were hardly begun (52). Though these projects had started late, this was not the only reason for their incompletion. The provision of underground or otherwise bomb-proof factories never received the unanimous party support which had been given to the public shelter and submarine pen projects; nor did the constant power struggle revolving around the subject improve matters.

Speer was not altogether successful in preventing the SS from infiltrating his ministry. The underground and bomb-proof factories relied increasingly on slave labour controlled by the SS for both production and initial construction. The OT particularly had become dependent on forced labour supplies to increase the manpower of the individual nucleus firms already during the early stages of the Atlantic Wall construction. Through this means Himmler was able to ultimately establish a powerful grip on armaments and production, using the Generalbevollmächtigter für den Arbeitseinsatz (Plenipotentiary-General for Manpower Allocation), SS-Obergruppenführer Fritz Sauckel.

The most successful SS infiltration, however, was to take place within the secret weapon programme: the construction of the firing sites and factories for the Fi 103 flying bomb, better known as the V1, the A4 rocket or V2, and the little known multiple charge gun the V3. The V-weapon programme was subsequently to absorb more than its proper share of bomb-proof protection.

In the last two years of the war Hitler became more and more interested in the production of a 'secret weapon' that would once more turn the tide in his favour. When the Baltic testing station of Peenemünde was bombed in 1943 construction work had thus to be increased to offset the setback caused. Himmler saw this as the ideal opportunity to inject the SS into a position of influence in the programme. To do this he offered Speer assistance with post-bombing construction. Speer unwittingly accepted and assistance arrived in the shape of an SS construction engineer, designer of concentration camps and gas chambers, Major-General Dr. Ing. Hans Kammler. The initial job given to him was the direction of minor construction works connected with the A4 programme. By 1945 Kammler was the supreme tactical commander of all German secret weapons.

When, in August 1943, the Führer ordered that production of these weapons be resumed in bomb-proof and underground factories, Kammler planned to divide Peenemünde's functions into three parts and disperse them across the Reich. Firstly an overland firing range was to be set up as part of the SS training camp at Blitzna in Poland. Secondly a development works was to be set up in an underground cavern to be excavated by convict labour in the cliffs at Traunsee in Austria. Called the 'Cement project', work on this excavation was to be finished in February 1945 but was amended on July 6th, 1944, when Hitler decided to cut back on some long-term projects and directed Speer to convert the whole tunnel complex into a tank-gear factory (53). Thirdly, a main underground production works was to be set up.

Initially the underground production works was itself to be divided into three underground production works: a 'Central Works' was to be constructed in the heart of Germany, an 'Eastern Works' was planned

near Riga, and the Vienna-Friedrichshafen group of factories were designated 'Southern Works'. Plans for these latter two 'Works' were eventually abandoned, however, and effort concentrated on the 'Central Works', which grew to be the largest underground factory, as far as is known, in the world. Gerhard Degenkolb, appointed Director of the Special A4 Committee in Speer's ministry, selected a site for the 'Central Works' or 'Mittelwerk' under the Kohnstein mountain in the Harz region of central Germany, little more than 160 km south-west of Berlin (54). Before the war the Government's Industrial Research Association had adapted a number of tunnels close by the town of Nordhausen for the storage of critical chemicals but this was now to be greatly enlarged.

The scheme submitted by Central Works to the German War Office in October 1943 was for two main tunnels running nearly 2 km on an S-shaped parallel course about 210 metres apart and linked by 43 parallel galleries, 'like the rungs of a ladder' (55), 27 of which were to be for A4 production and the remaining for jet-engine assembly. The total works would have an estimated floor space of about 96 000 square metres, which included 5000 for convict sleeping quarters. The ladder-like plan was almost ideal from the production line point of view. The A4 could follow a definite course through the tunnel system, travelling down one main railway tunnel and having components and sub-assemblies from each side of the galleries fitted as it passed. Constructed by Waffen SS labour battalions the Works employed 16 000 slave labourers, enabling Himmler and Kammler to exert considerable control. During Speer's illness in early 1944 the SS subsequently attempted to eliminate one of the A4's chief designers, Wernher von Braun, from the secret programme as part of their general bid for Speer's Ministry. Xaver Dorsch, head of the OTZ was—as already mentioned—very much involved in this. Despite the continual power struggles however, from January 1st, 1944, when the first rockets were produced at Nordhausen, until the plant's capture in April 1945, nearly 6000 A4 rockets were assembled there.

Apart from the actual production of the army's A4 ('Aggregat') rocket and the Luftwaffe's Fi 103 flying bomb, the preparations for the use of these two secret weapons also provided the German High Command with large, and to some extent exaggerated, bomb-proof shelter problems. The two weapons were later named the V2 and V1 respectively as the main 'Vergeltungswaffen' (weapons of reprisal) of Hitler's revenge policy but although the army's V2 was in fact the second of the V-weapons to be used against England it had always had more attention lavished on it, much to the Luftwaffe's annoyance.

Despite the many doubts of the value of the A4 Hitler approved Speer's proposals for large 'bunkers' in the Cap Gris Nez area of Northern France and at Obersalzberg on February 29th, 1943 he saw and approved plans by the OT for the first of these huge 'bunkers' at Watten near Calais. Code-named 'North West Power Station' this site enabled the bunker to command a large part of south-east England. The OT were to construct the bunkers using 120 000 cubic metres of concrete, after which von Braun's engineers were to install the firing gear (56). Walter Dornberger, an artillery officer in charge of A4 research, in fact preferred mobile launching sites and was strongly opposed to the Watten idea but Hitler saw large bunkers as a means of attracting Allied bombers away from German cities, as the impregnable U-boat shelters had done, an opinion strengthened by the fact that at this time there was no bomb powerful enough to pierce the pens.

However, the Allied civil engineers observing the construction of the Watten bunker, notably Sir Malcolm MacAlpine, advised that the structure would be vulnerable when the shuttering was still up but the concrete incomplete. On this advice an attack by the Flying Fortresses of the US 8th Air Force on August 27th, 1943 effectively stopped construction.

By this time a number of concrete shelters had been prepared in the Le Havre and Cherbourg areas from which rockets could be taken to the mobile projectors, simple steel launching tables. An underground shelter had already been chosen at Wizernes, quite near the Watten bunker, and Xaver Dorsch now suggested that this be made into a large launching bunker, leaving the badly damaged Watten for other purposes. Dorsch proposed to build a million ton dome of solid concrete using the 'Verbunkerung' method (57), the same method employed in the bomb-proof factory at Landsberg and the flying bomb shelter at Siracourt. The Wizernes dome was to be a refinement on these. Constructed by the edge of a 30 metre quarry face, the dome would house workshops, barracks, rocket storage, excavated in the chalk, from which two tunnels would run into the quarry. Along these tunnels Dorsch proposed that rockets could be hauled from the main preparation

◀ *29. Interior of the military underground hospital built on Jersey by the OT using slave labour: the doctor's rest room.*

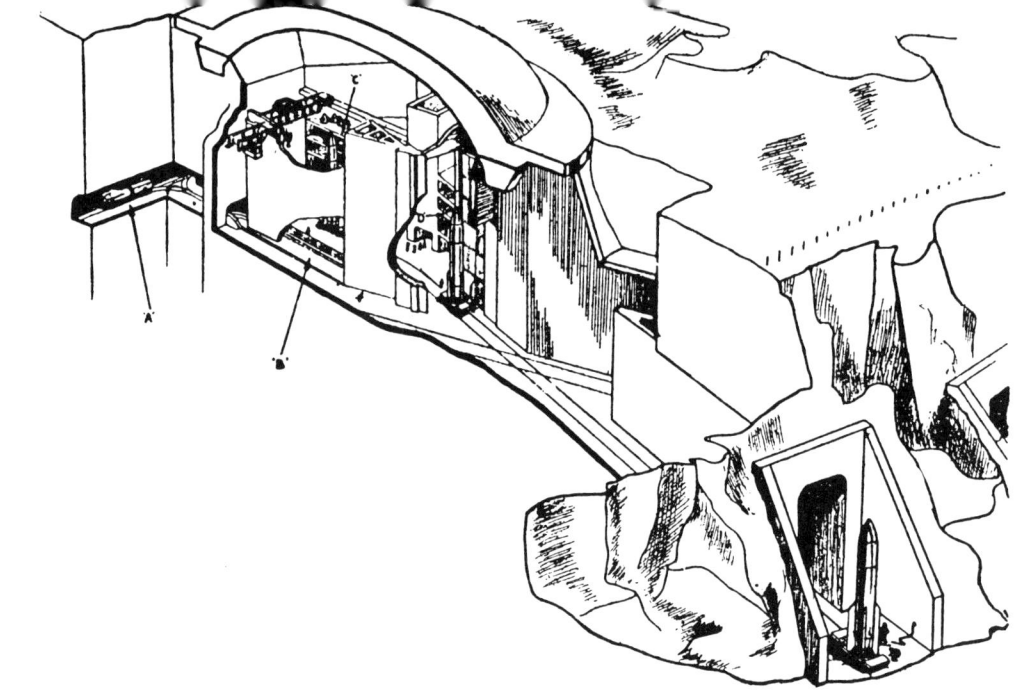

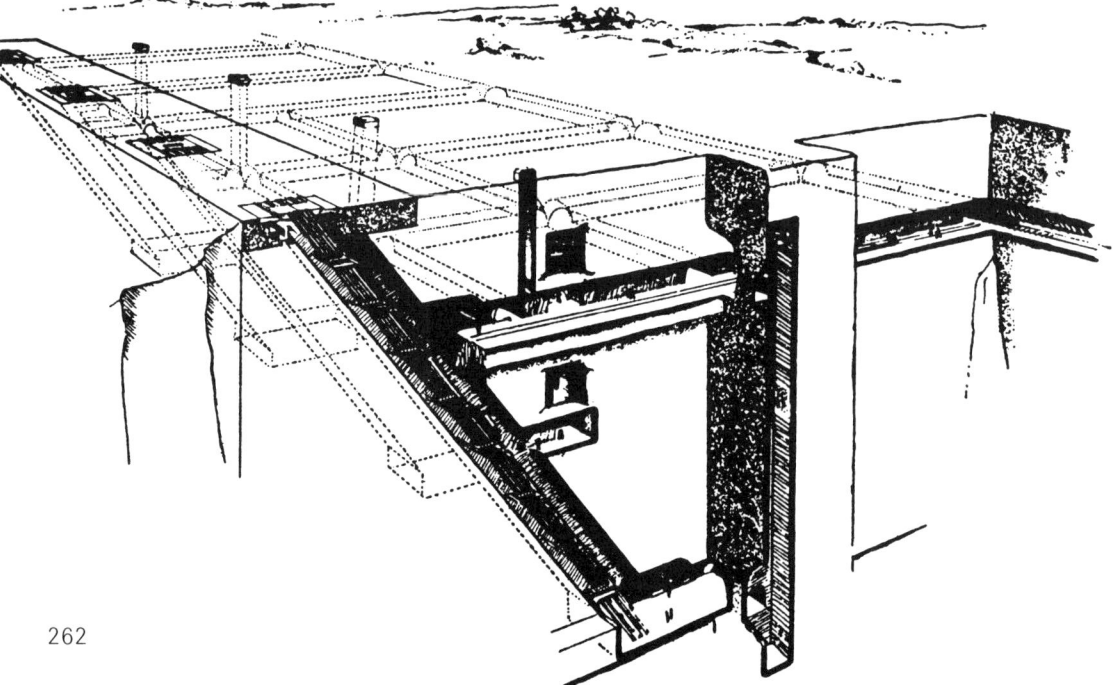

30. Isometric sketch of the planned A4 (V2) rocket launching bunker at Wizernes (as illustrated by Irving).

31. Isometric sketch of the V3 installation (as illustrated by Irving).

bay and launched at a rate of 50 per day. This huge project predictably became one of Hitler's favourite schemes but the spectacular structure was never completed. The site was made unapproachable by the dropping of the RAF six-ton 'earthquake' bombs in July 1944 (58), leaving the eventual launching of the A4 to mobile batteries under Kammler which, much belated, caused only insignificant damage.

The A4 itself was a drastic mistake. During the war Dr. R. V. Jones of British Intelligence saw the A4 not as part of a military strategy but a result of 'the innate German thirst for romanticism' (59). This opinion is applicable not only to the A4 rocket but also to a great deal of the German military construction programme. But whilst the £6 500 A4s were of little use in bombarding England the £125 Fi 103, an expendable pilotless aircraft launched from a 70 metre ramp, was extremely effective. Developed by a Luftwaffe fearing the attention being lavished on the A4 rocket, the flying bomb was ready for use long before its competitor. Like the A4 however, the problem of shelter for the missile was allowed to assume gigantic proportions and by late 1943 some 40 000 OT personnel were employed on the construction programme in France. The large flying bomb bunkers, the unfinished A4 bunkers at Watten and Wizernes near Calais, and Sottevast and Martinvast near Cherbourg, caused huge amounts of labour and materials to be diverted from the Atlantic Wall programme.

In January 1943 Lt.-Gen. Walter von Axthelm, C-in-C of German Anti-Aircraft Artillery, had visited Peenemünde and seen a flying bomb launched. Impressed by this he discussed the use of the weapon against England with Field Marshal Milch, only to find that Milch already had strong opinions. Milch proposed that eight giant launching bunkers be built on the Channel coast, providing storage for several thousand missiles and facilities to enable launching even whilst under enemy bombardment. This was the exact opposite to von Axthelm's conclusions. He thought the bunkers Milch proposed would be easily spotted and bombed before completion. Instead he suggested 100 small sites which could not possibly all be destroyed but as neither would agree they decided to leave the final decision to Goering. Not until nearly six months later did Goering make his decision and his conclusion was in any case a compromise. Ninety-six small sites and four large ones were to be constructed and Hitler affirmed the decision to build the four large bunkers, including them in the Atlantic Wall programme, although with some reservations about the amount of concrete required, which in itself gives some indication of their scale.

One of the largest bunkers was built using the 'Verbunkerung' method at Siracourt and another at Lottinghem, the former being almost complete by D-Day. By October 1943 nearly 60 per cent of the smaller sites had been constructed in Cherbourg and the Pas de Calais but about this time British Intelligence became aware of their existence. Shaped like skis on their sides, with a central bearing on London, these sites were easily identifiable and became the subject of heavy bombing throughout the winter of 1943–44. By early February the Germans were unconcerned however. Having prepared a new prefabricated launching system, which took only a week to erect, they happily allowed the 'ski' sites to be bombed while foundations were quietly prepared for the new launching sites. At the same time the bomb-proof supply and distribution system for the flying bombs was completed. Eight large forward supply sites were constructed just behind the line of ski sites and about 30 km apart. One was constructed at Valognes on the Cherbourg peninsula. Each of these supply sites held 250 bombs and they were in turn supplied from a total of 5000 bombs stored in three large dumps in caves at Creil, Le Mans and Chartres. The first bombs were launched from the new catapult rigs using this complex bomb-proof distribution system, one week after D-Day on June 13th, 1944.

During the next few weeks the RAF mounted heavy attacks on both supply sites and bunkers, smashing the Siracourt bunker with one of the six-ton 'Tallboy' earthquake bombs. Gradually as the Allied advance into Europe proceeded the sites were overrun but from June 13th, to September 1st, it is estimated that the flying bomb, at a total German expenditure of £13M including manufacture and shelter construction, cost the Allies £50M in loss of production, extra defence and the 'Crossbow' bombing offensive on the V-weapon sites. As this figure does not include the cost of permanent repairs, the effort expended on the construction of bomb-proof sites for the V1 can be seen, unlike the V2, to have been more than justified.

Another project, being constructed only 8 km from the French coast, would have been similarly justified if it had been completed. This was the little known V3, Hitler's third revenge weapon, given the go-ahead after the bombing of Peenemünde in the summer of 1943.

The V3 was the 'multiple charge' or 'high-pressure pump'. By using a

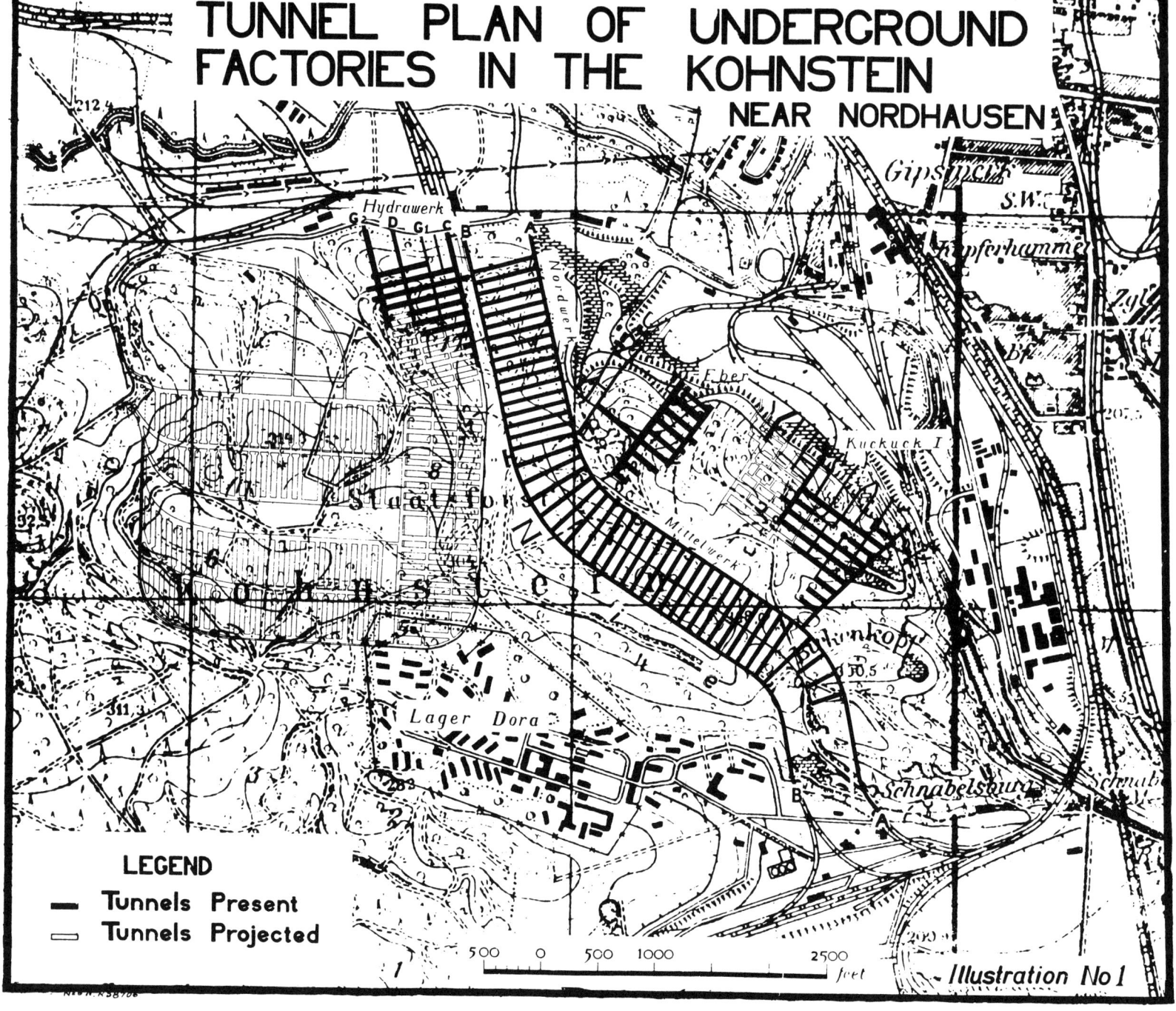

32. Plan of the underground complex in the Kohnstein.

series of synchronized electronically detonated charges along a 126.8 m gun barrel, a finned dart-like projectile, 2.7 m long and 150 mm in diameter, was accelerated up to speeds which would enable it to cover the 150 km to central London. The V3 bunkers were situated under a hill at Mimoyecques near Calais and originally two sites had been planned there, each with five batteries of five guns. One site was badly damaged by the 9th Air Force in November 1943 but even so it was calculated that the remaining 25 barrels under construction would be able to fire 300 shells per hour on London. The guns were almost complete when the Allies overran the site in August 1944, although the testing of the barrel design at the Baltic station had still not reached a satisfactory conclusion (60). No doubt given time the subterranean workings of the V3, returning to the offensive character of the early Atlantic Wall of 1940–41, would have been more of a threat than the V1 or V2.

As the British and American armies gradually advanced across France into Southern Germany and the Russians pushed down towards Berlin from the north, the V-weapon sites were replaced by a new threat: the 'National Redoubt'. Eisenhower was certain that the main problem after making contact with the Russians was to push south-east towards Bavaria and Western Austria, where Allied intelligence had warned him that the Nazis were preparing an impregnable fortress in the Alpine mountains. The German obsession with bomb-proof construction made this seem credible. Under criticism from Churchill, Eisenhower thus allowed the Russians to beat him to Berlin. His reason is explained by an intelligence report of March 11th, 1945, which was typical of the information he was acting on. It stated:

"Here, defended by nature and by the most efficient secret weapons yet invented, the powers that have hitherto guided Germany will survive to reorganize her resurrection; here armaments will be manufactured in bombproof factories. Food and equipment will be stored in vast underground caverns and a specially selected corps of young men will be trained in guerilla warfare, so that a whole underground army can be fitted and directed to liberate Germany from the occupying forces." (61).

This idea of a National Redoubt, backed by the OT, did not exist in reality. It was a huge bluff, perhaps the last great 'Propaganda Wall' which Goebbels devised. The concept of a last stand in some great subterranean fortress was not completely fictional however. At the Central Works in the Harz mountains, additional underground factories were planned and started with just this objective in mind. Equipment from Peenemünde was evacuated there and work on the Nordwerk, at the northern end of Mittelwerk, producing Junkers jet engines, was pushed ahead. A large liquid oxygen plant, the Hydrawerk, and a second Junkers factory were planned and tunnelling started. An underground synthetic oil plant was also well under way, Kuckuck I (Cuckoo Project), and at nearby Woffleben a further network of tunnels had been blasted out to house a Henschel missile factory. In April 1945 Kammler of the Waffen SS was entrusted with the defence of this enlarged underground complex. The fact that the Harz mountains were overrun uneventfully soon afterwards does not alter the historical fact that a last stand was prepared for, although not in Bavaria, but only 160 km south-west of Berlin (62).

PART FOUR CONTEXT

13 CONTEXT

Trends in military architecture since 1945; the relationship between military and civil architecture.

The basic characteristic of military architecture has always been, as we have shown, its adaptability to changes in technology and to the consequent changes in military tactics and strategy. The result has been a process of constant evolution. Since 1945, under the pressure of the technological acceleration of the second half of the twentieth century, this process has continued but in a way which, paradoxically, is both traditional and innovative in character.

During 1900–45, one can identify the emergence of two strands of military building activity: firstly an evolutionary strand which countered new developments like the tank simply with further variations on the theme of fortress design; and secondly a strand which responded to real quantum shifts in the availability of technology, in an altogether more innovative way. In Europe, France with her Maginot Line, a direct development from the girdle fortress and the experience of 1914–18, and Germany, with her vast programme of bomb-proof construction— both deriving from traditional fortress architecture—tended towards the first grouping; while Britain tended much more towards the second, with Mulberry Harbour and the hutting programme. These two groupings were also apparent in military architecture outside North West Europe in this period. Both concrete and underground installations of an evolutionary nature were constructed by the Japanese and Americans during World War II (1). At the same time, shortage of steel in wartime USA led to innovations such as laminated timber arches, plywood box beams, and concrete shell construction (2).

Since 1945 the rate of technological advance in weaponry has developed so rapidly that, to counteract it, the group of military architecture that can be characterized as 'innovative' has become almost as typical as the fortress was in 1914. This marks an interesting change in the military attitude to technology. Before 1914 Zeppelin had found it extremely hard to get army support for his projects (3) and Allied generals had stated that the machine gun was of little military consequence (4): it was the number of men bayonet to bayonet that would dictate the ultimate result. By 1945 it was research and production which were the keys to military success.

Since the war, the military client has thus proved to be a willing patron for new ideas. In Britain, for example, military research into pneumatic structures has been carried out by MINTECH, the Research and Development Establishment at Cardington (5). In the USA, under the pressure of constant military activity, this tendency has been even more pronounced. Buckminster Fuller's radomes and geodesic domes, initially unacceptable to a conservative civilian building industry, were developed under military interest. The Space Race, with its undeniably military and political objectives, has also resulted in constructional innovation, both in the form of the huge structures required for rocket assembly and launching, and of the new materials and methods involved in the rocket programme itself.

Nevertheless, while the rate of innovation in military construction has increased it is still possible to identify a reliance on traditional forms of defence—on fortification. China, fearing Russian attack, has constructed what has recently been described as 'a world of underground tunnels and shelters'. Sophisticated military hardware is still accompanied by heavy concrete installations. Those recently reported to have been built by Israel near the Suez Canal are a case in point. 'The Guardian' newspaper stated:

"Officers claim that bunkers on the Bar-Lev line have been made proof against attack even from the Soviet made 203-mm guns on the Egyptian side, a construction project, covering logistic bases and installations in the rear as well as bunkers in the front line has been completed just on time . . ." (6).

These fortifications are not primitive however, as can be seen from the amount of concern given to their environmental design:

"For the comfort of the men within immediate air support an ice factory has been transferred to the heart of Sinai, to operate water coolers in the bunkers." (7).

Such sophistication does not disguise the fact that the Bar-Lev line is almost traditional in design, which suggests that new concepts such as the atomic bomb will not put an end to traditional methods of fortification on a local scale. Simultaneously, however, the nature of modern defence has changed so rapidly since the advent of Hitler's V-weapons that it is no longer evolutionary. Innovation, as opposed to the more gradual process of evolution, has been the dominant feature of post-1945 military construction, including defence.

America's DEW (Distant Early Warning) line is a characteristic example, consisting of a line of radar stations stretching round the North American continent. The Texas Towers constructed in the Atlantic as part of this line are not unlike the Sea Forts built in the 1940s

◀ *1. Fylingdales early warning station.*

2. Cluster of Buckminster Fuller's light portable 'Dymaxion Deployment Units' installed at the head of the Persian Gulf during WWII: used as dormitories for aviators and mechanics assembling US pursuit planes for delivery to Russia.

3. Converted Butler grain bin redesigned by Fuller, while steel was still available in the USA, as the Dymaxion Deployment Unit: test assembly at Kansas City in August 1940. War emergency thus prompted acceptance of ideas Fuller had developed in 1928.

4. US Navy hangar using welded steel arch ribs was developed to house coastal patrol blimps early in WWII when steel was still plentiful: span 100 metres, rise 56 metres.

5. Later steel shortage and scarcity of large solid timbers in wartime USA led to the introduction of plywood girders and laminated beams for long span military structures: crescent type hangar truss with glued laminated chords of 50 metres.

6. In Italy the Second World War gave Pier Luigi Nervi the opportunity to construct and develop innovative reinforced concrete structures: construction of military hangars of prefabricated components 1939–41.

7. Hangar of pre-cast concrete components by Luigi Nervi 1939–41: preliminary trial erection.

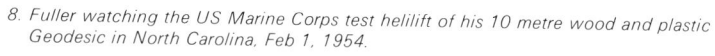

8. Fuller watching the US Marine Corps test helilift of his 10 metre wood and plastic Geodesic in North Carolina, Feb 1, 1954.

9. Mock US Marines landing 1956: Geodesic on carrier flight deck before helicopter lift off at 60 knots.

10. Ten metre diameter plastic radome being tested on Mount Washington in 1954: similar domes now in use on the US Arctic radar line.

11, 12. Vehicle Assembly Building at the Kennedy Space Centre: the world's largest building, with a volume of 130 million cubic feet.

13. Texas tower radar station.

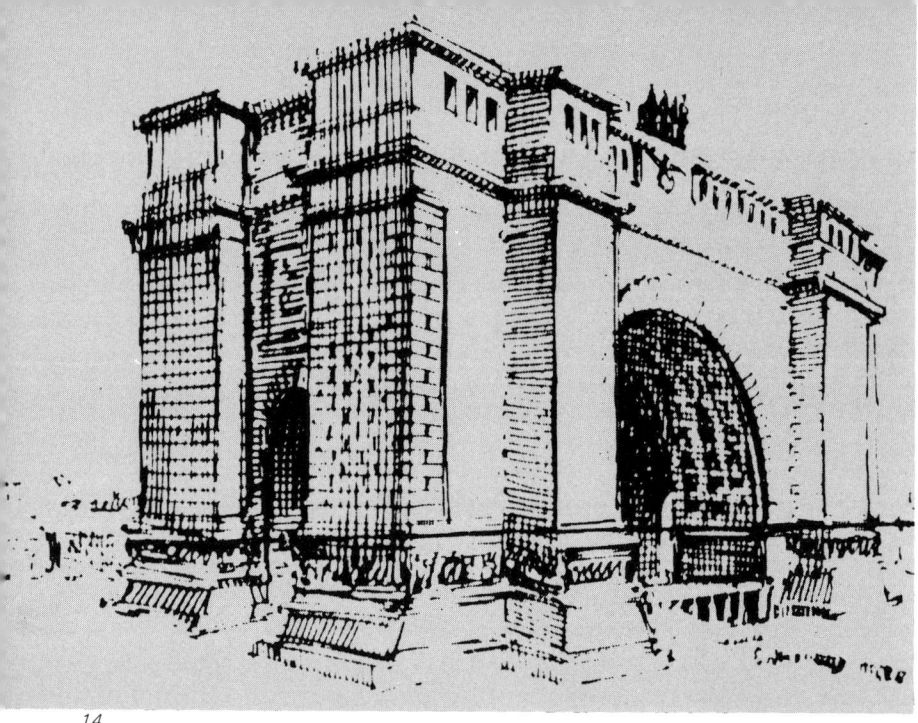

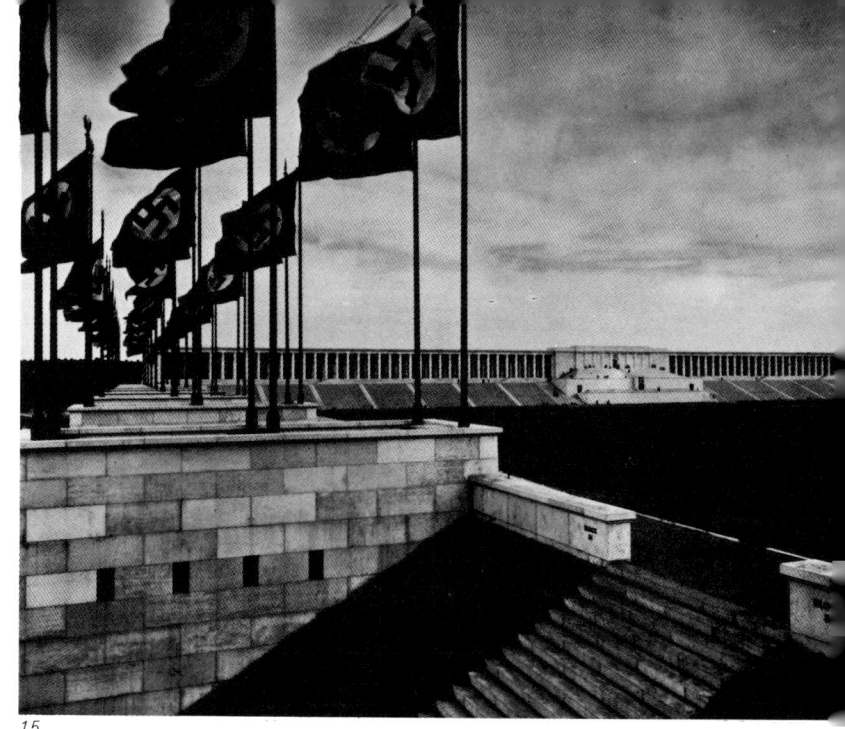

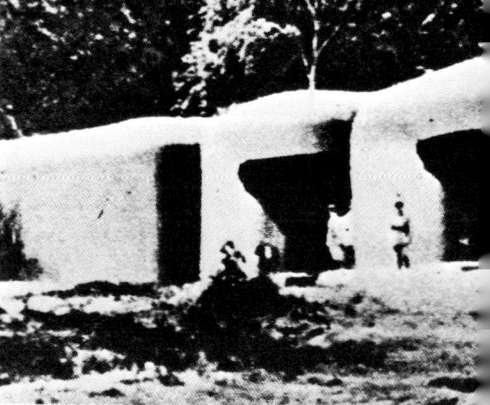

14

15

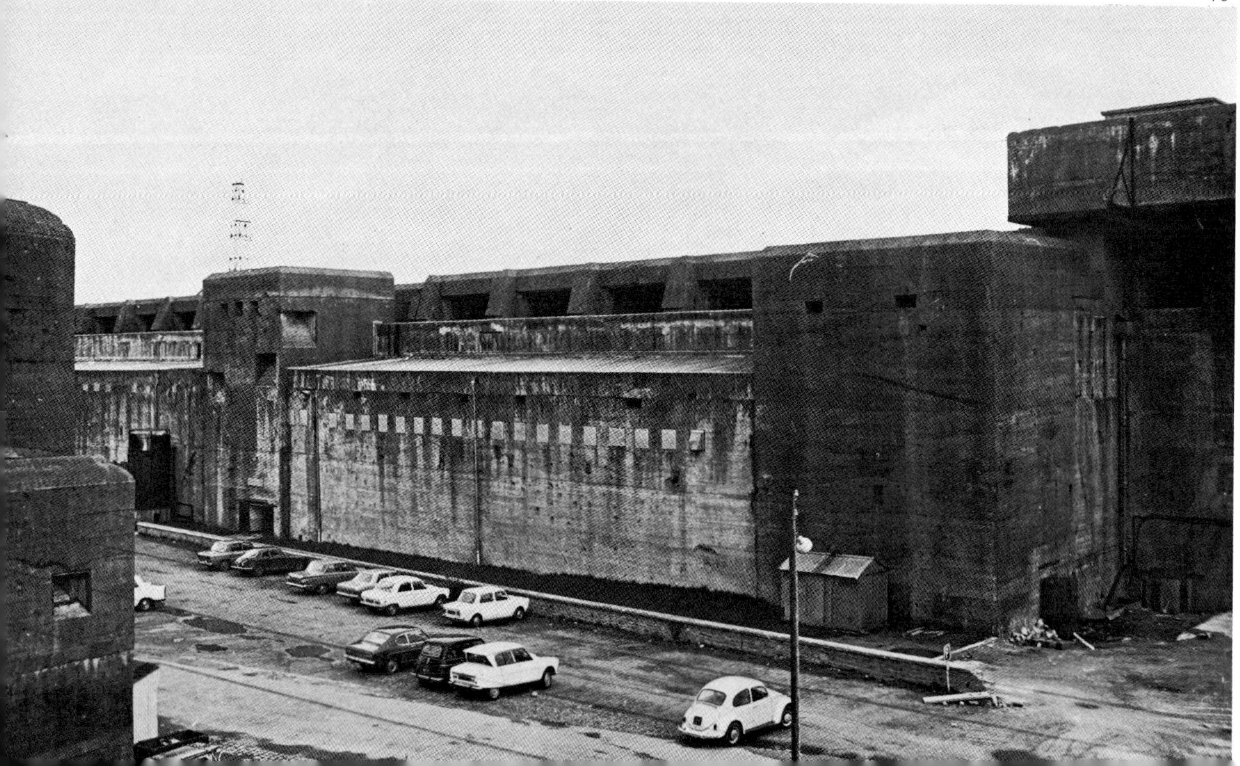

16

17

Hitler's own sketch for a triumphal arch in Berlin.

15. *The Zeppelinfeld in Nürnberg 1937: architect Albert Speer.*

16. *Recent photograph of the German submarine pens at Brest.*

17. *Casemated artillery block on the Maginot Line: its fluent form symbolically French in character.*

in their basic conception but being unarmed, with pneumatic 'radomes' protecting the radar equipment, these structures can hardly be seen as fortifications in the traditional sense. Egon Eis concedes in his book 'Forts of Folly':

"The radar island, the unarmed bastion of the new age whose only weapon is the radar screen and the radio tower, can attempt to do what no classical fortress could have achieved: to keep bombers and rockets at bay." (8).

In Britain the equally impressive installation at Fylingdales in North Yorkshire, built for a similar purpose, confirms the opinion that on an international scale such defence construction is no longer evolutionary. Concrete has been replaced by the invisible and rapidly changing shield of advanced technology.

Indicative of the innovative nature of modern defence development and even more spectacular than the DEW line is a new system of land defence which has recently been 'successfully' tested by US troops in Vietnam: the 'Electronic Battlefield'. 'The Observer' stated that:

"... already at least £1500 million has been secretly expended on the system. It employs an awesome array of sensing devices, laser beams, night-seeing automata and computers to create an electromagnetic environment in which, say the boffins of CDC (Combat Development Command). 'Nothing hostile can survive.'" (9).

Instead of concrete fortifications, the defence zone consists of what has been described as 'a gaint pinball machine': sensors based on sound, vision and smell. Heat detectors respond to body heat; 'people-sniffers' respond to human sweat and urine. A central computer analyses the information and presents the defender with the enemy's possible intentions and the possible military counter-measures. General Westmoreland, the Army's Chief of Staff is quoted as saying:

"I see combat areas under 24 hours surveillance, battlefields on which we can destroy anything we locate through instant communications and almost instantaneous application of firepower ..." (10).

It is implied that an electronic defence system stretching from the Alps to the North Sea may be possible: an idea which puts the reinforced concrete of Hitler's Atlantic Wall almost light years away.

Sufficient has been said and shown in this book to make the point that military construction has formed a very large proportion of the total constructional activity of this century. The interesting question now arises whether its relative neglect by architectural historians is paralleled in a corresponding disregard for military construction by designers—'the givers of form' as the Germans call them—and whether, for its part, military construction shows any sign of having been influenced by non-functional, stylistic considerations.

Certainly, if there has been a lack of interaction, it has not been because military and civil engineers or architects were different groups of people. In all countries engineers and architects of repute were drawn on for expert help in military construction projects; and in Germany men like Speer, Todt and Kammler reached extremely influential positions in the political hierarchy—a fact which to some extent reflected the relatively high political status of architecture and engineering in Nazi Germany. Similarly it was primarily in Germany that military architecture developed a style which corresponded, at least in spirit, to civilian building of the same period by reflecting the image-making that was always present in the physical manifestations of Nazi ideology.

Speer himself likens the design he made for Hitler's new Chancellery to a 'Cecil B. de Mille set ... it had been the very expression of tyranny' (11) an example which gives some impression of the political involvement of the National Socialists with architectural style. Like the radio or the press, visual style could be and was used to project the political and military images which were required, and in the case of architectural style the divergent tastes of the Nazi hierarchy were only too obvious. The most striking examples of this use of style were the flak towers and public shelters within the Reich, where a definite and intentional stylistic junction between civil and military architecture was apparent, yet other aspects of German military construction showed more subtle suggestions of architectural style. The naval observation posts on the Atlantic Wall were a case in point—with their apparently expressionist imagery. But actual intentions are more difficult to determine. There are even very good reasons why, in this particular instance, the influence of architectural style could be discounted. Designed by the navy, mainly on headlands, these structures can be identified fairly directly with naval architecture—the warship bridge translated into concrete. At the same time national characteristics were also involved and these can be applied more generally to Nazi construc-

18. Observation tower at La Corbière Jersey: an example of the high standard of German finishes.

19. Recent photograph of Atlantic Wall casemate near Brest: camouflage using rough textured render to match surrounding landscape.

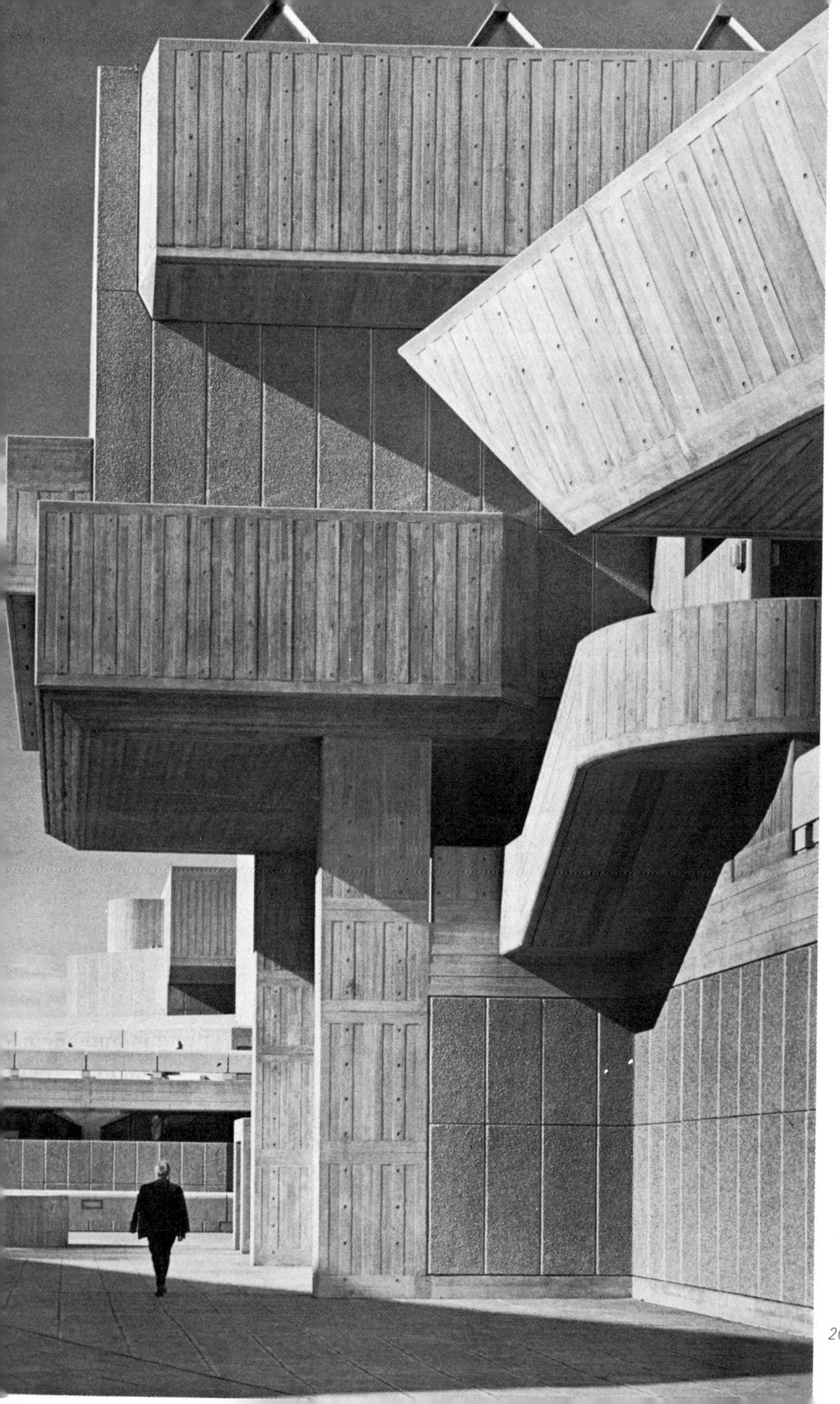

20

21

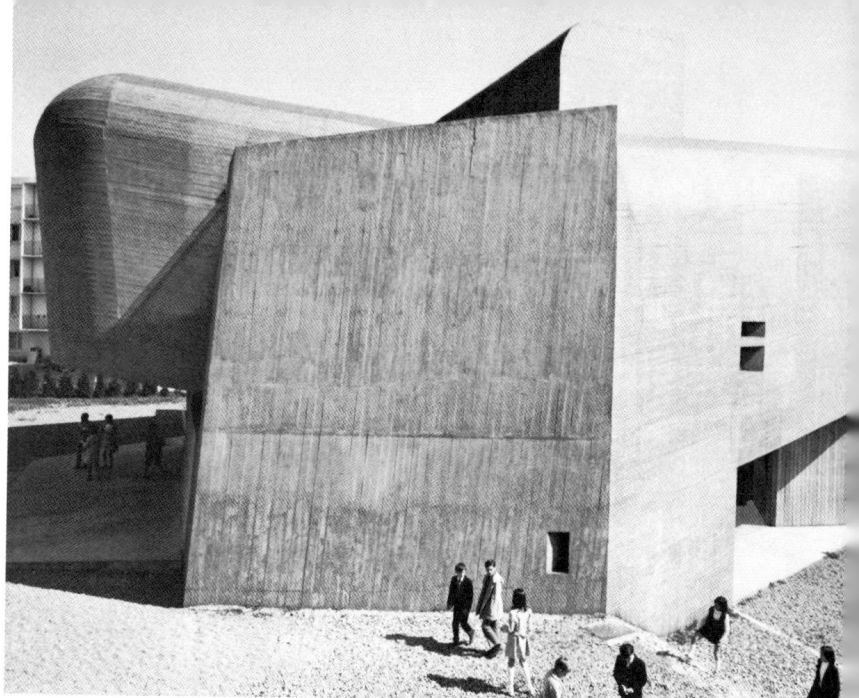

20. Hayward Gallery and Queen Elizabeth Hall, London

21. Parish centre of Sainte Bernadette: Nevers

tion as a stylistic factor.

The first of these characteristics was the tendency in German military construction towards the monumental approach, a tendency which had been typical German military design since 1914. Despite the implementation of prestressing, new shuttering techniques, prefabrication and other aspects of concrete technology it was a tendency which continued during 1939–45. The second national characteristic was the high standard of German craftsmanship, coupled with a regard for detail and finishes. The rounded corners of the Atlantic Wall bunkers may well have owed their origins to the influence of ballistics on fortification design, but on a cost performance basis the return for extra time, labour and shuttering was certainly deemed unnecessary both in Britain during 1940 and among the Allies along the larger part of the Western Front 1914–18. Similarly camouflage, in Britain achieved by fancy dress, was also generally effected on the Atlantic Wall by the use of such fluent forms, by the application of long earth banks, textured concrete, the subtle use of camouflage paint, and other more aesthetically and craft-orientated means.

The acceptance of both of these characteristics as national ones certainly tends to reduce the stylistic significance of certain military structures, such as the naval observation posts, and post-war events would appear to substantiate their validity. High standards of workmanship have continued in post-war Germany (12) and, according to Kidder Smith, in his 'New Architecture in Europe', a 'monument complex' has continued for a considerable time as a feature of her post-war civil architecture (13).

While the evolutionary aspects of military architecture merged with civil architectural style in this way and many otherwise functional structures, such as the flak towers or submarine pens, became monumental, it has sometimes been implied that style was a two-way traffic: that certain modern civil architecture has consciously evoked a 'bunker image'. Such suggestions can be seen on further examination to be based on doubtful evidence. A case in point is the work of Adolf Loos. His rectilinear and stark designs are often jokingly compared with the pill-boxes of the First World War, but such a specific comparison or any other general comparison with the 'functionalism' of the modern movement has no justification. Loos' rejection of ornament saw its roots in the work of Sullivan in the United States and Otto Wagner in Vienna during the 1890s, and found expression as early as 1908 in his treatise 'Ornament und Verbrechen', Ornament and Crime. But while the Loos comparison has never been a serious one, it is interesting how this theme of comparison between fortification design and twentieth century architecture has been a recurrent one, and not only amongst a critical public. An article on the Channel Island fortifications by Michel Santiago, published in the 'Architectural Review' in 1963, stated:

"... the most disquieting of all is the fire-control tower at Mannez Quarry (on Alderney), for its blind forms with their snubbed-off corners seem to belong unmistakably to the plastic minded sixties." (14).

A common comparison is that between the heavy German bunkers of the 1940s and the reinforced concrete monoliths which have become such a feature of post-war architecture. A typical article in this field was published in 'Architectural Design' in 1967 under the title 'Culture Bunkers':

"The form and finishes of military installations are being used for the most hallowed of new buildings—cultural and civic centres. Throughout Europe and even in America architects are setting up their culture bunkers. The Queen Elizabeth Hall in London is not an isolated example." (15).

Another architectural critic has termed the Queen Elizabeth Hall 'quasi fortified' (16), but in this case the heavy concrete construction was quite definitely for reasons of sound insulation (17) and any stylistic link up with World War II bunkers must be discounted. If anything, the Queen Elizabeth Hall is Corbusian in style (18).

Yet although the post-war use of exposed reinforced concrete finishes does not appear to originate directly from experience in wartime construction, it does seem to stem from one man, and, significantly, one building: Le Corbusier's 'Unité d'Habitation' at Marseilles— started in 1948. In 'The New Brutalism', Banham cites this as:

"... the first genuinely post-war building, in the sense that its innovations separated it definitively from Modern Architecture before 1939." (19).

By 'innovations' Banham refers to the single fact that Le Corbusier had abandoned the 'pre-war fiction that reinforced concrete was a precise "machine-age" material' and exploited rough wooden formwork to form what he called 'béton brut', a crude surface which reflected the grain and defects of the timber. The events which prompted Le Cor-

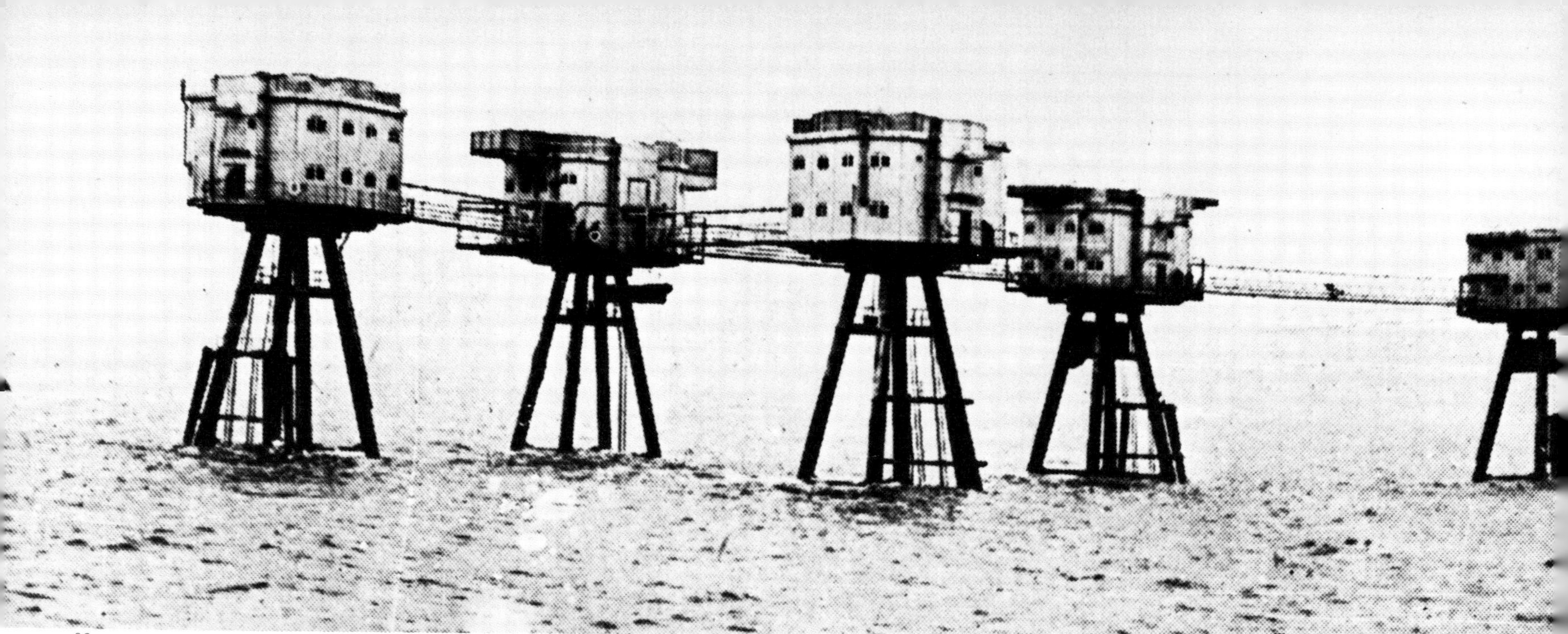

22

23

24

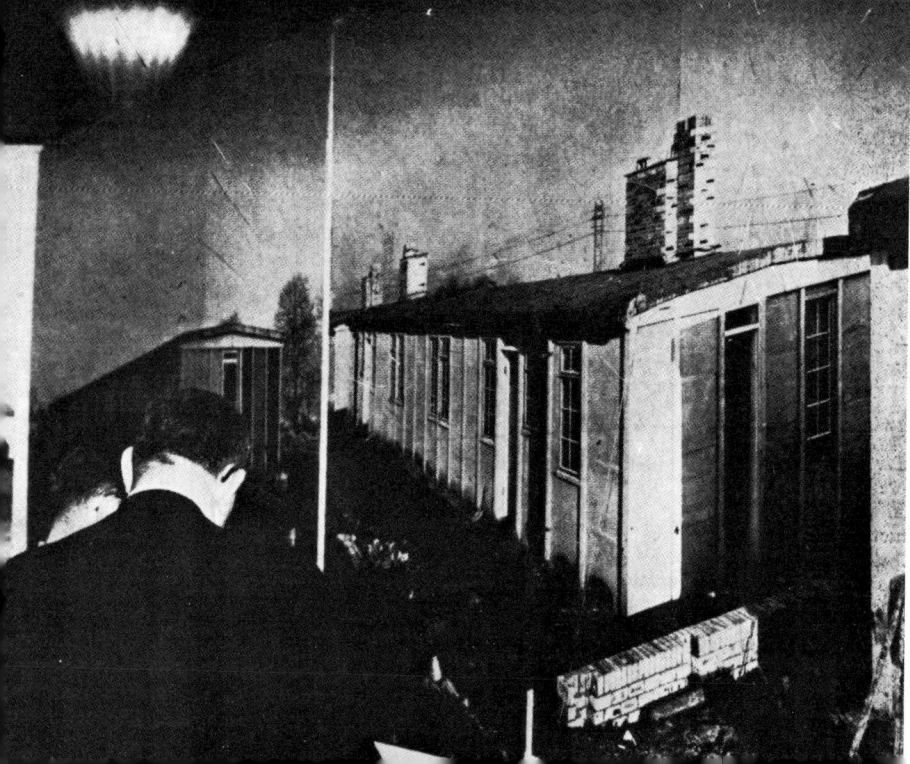

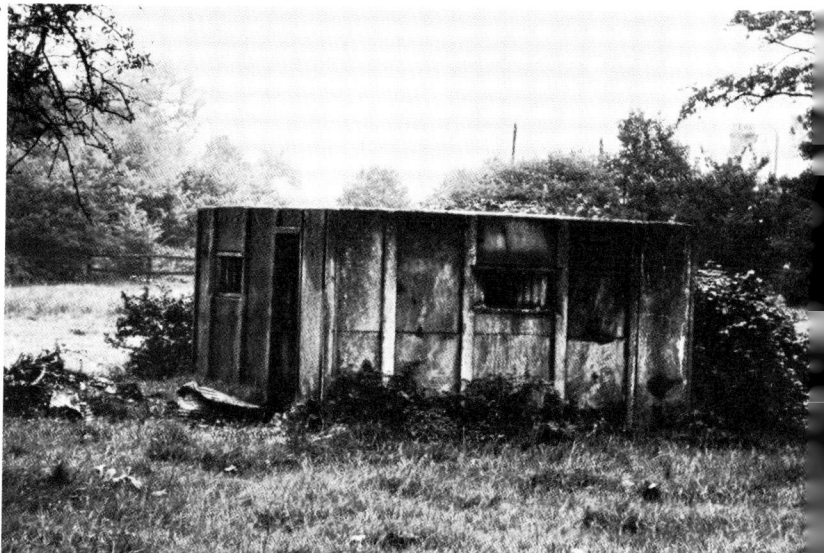

22. The Army Sea Forts

23. The BCF hut on exhibition in London 1942.

24. Prefabricated concrete pill box constructed during 1939–45 about 20 miles SW of London.

busier to make such a dramatic change in his attitude to concrete between 1939 and 1948 are uncertain. Admittedly he had started to tend towards natural materials before 1939, and his obsession with the 'honesty' of engineering cannot be discounted as an influence, but it is an interesting coincidence that this changeover period coincided with the construction of the Atlantic Wall along the French coast—defences in which the textural possibilities of rough boarded concrete were so clearly indicated in its bunkers.

The question of military influence on style is full of such unknowns. Only in a small number of cases can a positive relationship be established where an architect has consciously extracted military imagery. One example is the Red Sands Fort in the Thames Estuary, which inspired the Archigram group to devise 'The Walking City'. Another is the work of the French architects Paul Virilio and Claud Parent who, together with Michael Carrade and Morice Lipsi, have published a series of manifestos and projects as the 'Groupe Architecture Principe'. Their review No. 7 was devoted entirely to bunker archaeology. Entitled 'Architecture Cryptique' (20), their collection of sculptural photographs from the Atlantic Wall can perhaps be defined as a source of stylistic inspiration for their own work. Their Parish centre of Sainte Bernadette at Nevers, although described in the magazine 'Architectural Design' as 'cosy late-Corbusian' (21), can be related more positively with German coastal bunker design of the 40s.

But such links are exceptional and in trying to explain stylistic similarities between the two types of construction the conclusion must be drawn that such similarities are based primarily on the technical and material factors which they have in common. The stylistic comparison cannot be treated in any other way—as C. F. Voysey wrote in the Architects' Journal in 1935—when disputing an argument that modern architecture stemmed from *his* work:

"Steel construction and reinforced concrete are the real culprits responsible for the ultra modern architecture of today." (22).

The subject of the technical influence of military construction on civil architecture therefore assumes more importance than superficial stylistic influences, a fact which brings us back to the earlier classification of evolutionary and innovative. While German construction was primarily evolutionary it had limited influence on civil architecture, but Britain's tendency towards the innovative had very definite effects on her civil architecture.

At a recent conference at the RIBA on 'Innovation in the Construction Industry' the Conference Chairman, Professor D. A. Turin, made the following point in his foreword:

"Innovation is more than change—it demands a conscious attempt to move forward, to progress. It is imperative that innovation is not stultified either by ignorance of the market, or a credibility gap between the researcher, manufacturer, designer or user; possibly most important of all, that innovation is not held back by a social environment in which the products of innovation are restrained by economic, financial or psychological reasons." (23).

In civil construction, there always has been this gap created by the system between designer, manufacturer, researcher and client. There has been little attempt at market research in the building industry, nor is the social environment of peacetime conducive to innovation. Military construction and war, conversely, provide excellent conditions for innovation. Boundaries between research, design, use and manufacture are broken down. The matter of speed and of urgency provide an excellent climate for innovation and more important—its implementation. During the two wars ideas which had been rejected or not exploited in peacetime were developed within specific military programmes as part of the overall war effort. Examples such as the jet and atom bomb are well known but within the building industry, industrialization, prefabrication and site management are similar examples.

During the 1914–18 war the wide use made of reinforced concrete was a major feature; of lesser immediate post-war influence were the early British attempts at prefabrication such as the Nissen hut. Reinforced concrete had been used before 1914, but the war programme did help establish a wider use after 1918, particularly in Britain where such construction had met with much resistance within the designing professions. Now, not only had several contractors become experienced in its use, but a steel shortage after 1918 enforced the continued use of concrete as a building material until supply returned to normal (24). In at least this way the war influenced the civil architecture of 1918–39. The effect of the architecture of the Second World War on British civil construction was much more noticeable however. Marian Bowley, Professor of Political Economy at London University, writes:

"The gentle progress of innovation in building structures of the inter-

27

29

Arcon site at Great Yarmouth: some 39 000 Arcon units were supplied in the Temporary Prefabricated Housing programme and the Arcon became, together with the Aluminium (54 000) and Uni-Seco (29 000), one of the three main systems.

AIROH prefabricated aluminium houses in production at the Weston-Super-Mare works of the Bristol Aeroplane Company in 1946: the factory had originally been designed for the production of Beaufighter aircraft.

28. Units being despatched from the factory: four units were connected on site to form a single house. Units were fully finished in the factory and assembly on prepared foundations took only 40 man hours.

30. Recent photograph of an AIROH house and plan. The heating and plumbing unit was of a modified Portal type: a. site joints b. bedroom c. bath d. kitchen e. living room.

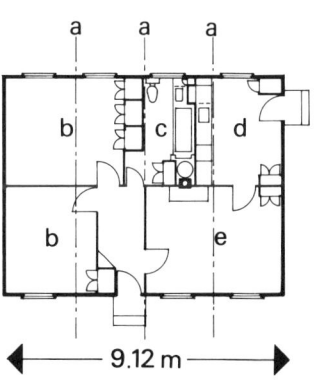

30 ← 9.12 m →

31

32

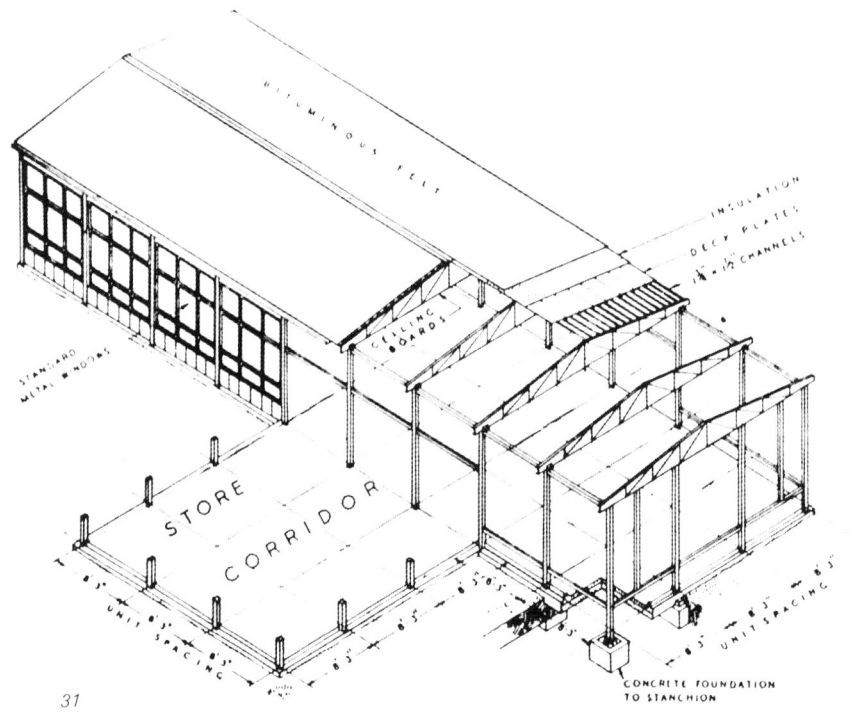

31. *Ministry of Works: Standardized School Construction 1944.*

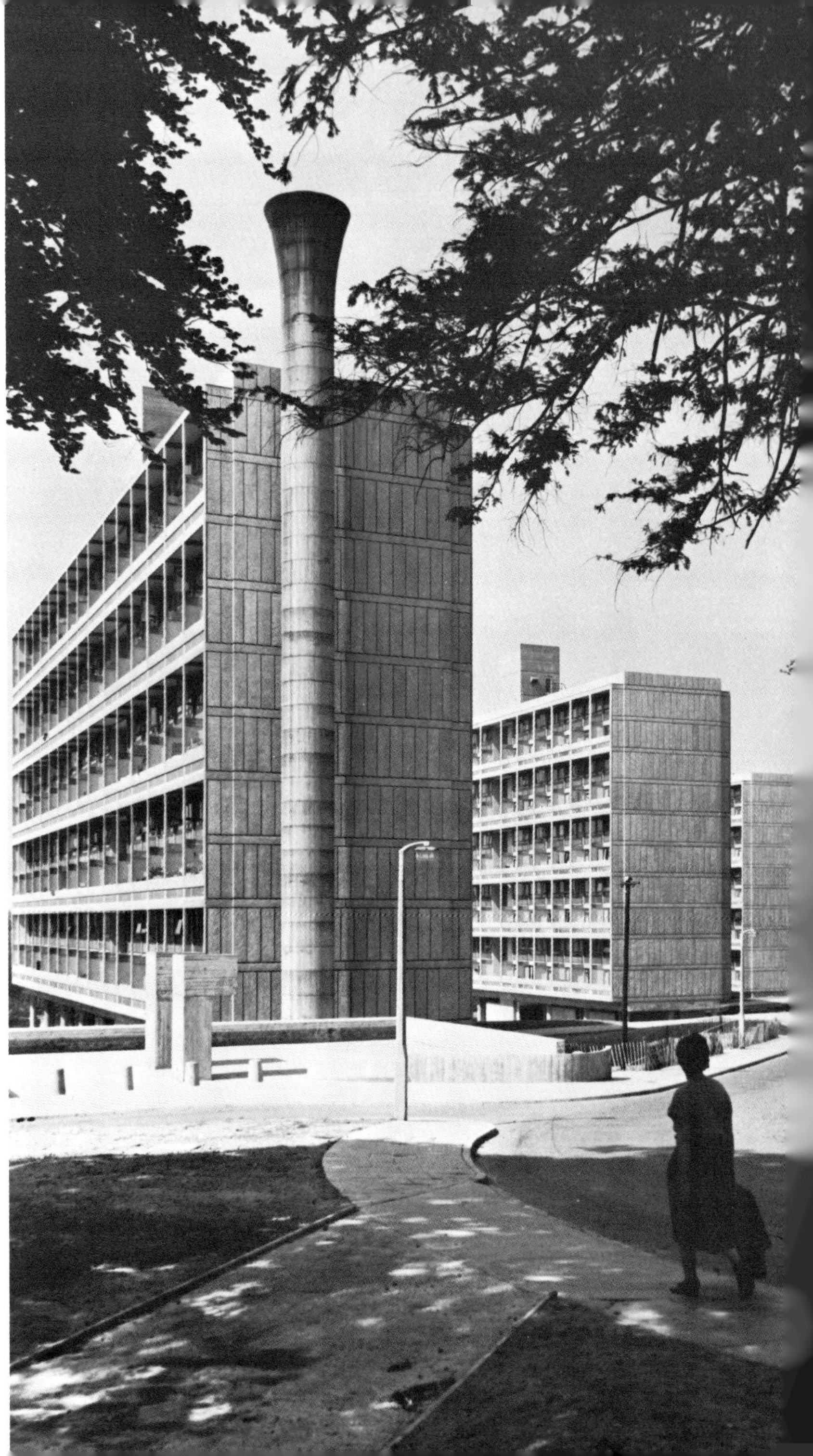

32. *British concrete prefabrication 1959: the LCC's Roehampton Estate constructed with precast exposed-aggregate facing slabs.*

war years was rudely interrupted by the Second World War. . . . Some building was still required, but the great multi-storey blocks, leisurely erected and of immense solidity, characteristic of the twenties and and thirties were no longer needed. Factories, factory extensions, temporary hostels and hosts of temporary buildings for both civil and military use (but ultimately part of the war effort) became of extreme urgency. The development of aerial warfare ushering in the air-age required not only aerodrome buildings but also some buildings that would withstand bombing. Strange new constructions such as the Mulberry Harbours came into existence. At the same time the resources most customarily used for such classes of work, unlimited steel, timber and labour were now of extreme scarcity. . . . Up to a point this had happened before, during the First World War, but the Second World War provided the opportunity for the exploitation of the new techniques and knowledge that had been used so hesitantly, or not at all, in the interwar period. . . . Instead of objections or indifference automatically obstructing innovations and enterprise in design and construction, there was at last a tendency to welcome them." (25).

The Sea Forts, Waller's concrete shell structures and the ingenuity of the hutting programme became as typical of British war construction as heavy concrete was in Germany. The point that they exhibited a national characteristic may be debatable, but certainly the technical effects of this construction programme on post-war British architecture were significant.

The most obvious example of this was in the housing shortage which resulted from the Blitz and the V-weapon attacks. As Marian Bowley points out, the opportunities to experiment with new construction methods have on several occasions been provided by national or local disasters. Bowley compares the destruction in Europe at this time with the Great Fire of London in 1666 which brought timber construction to an end in London (26).

With this urgent need for housing, industries and schools and the same restrictions which had dictated the hutting programme it was not surprising that the Government intended to use the same methods to solve the problem:

"I hope we may make up to half a million of these (industrial housing units)", Churchill told the Nation. "The whole business is to be treated as a military evolution . . ." (27).

Similarly the then Minister of Health stated that 'Housing should be tackled as one would tackle a military operation' (28). Named after the Minister of Works the Steel Portal House was designed to use the assembly lines which had previously been used for war materials. But when the house was unveiled at the Tate Gallery in mid-1944, it got a critical reception from architectural journalists. Nine months later the programme was cancelled as the continuing war made steel critical once more and, of the 500 000 Portal houses which were to be manufactured, little more than 150 000 were produced (29). During 1945–47 similar attempts were made in the Temporary Prefabricated Housing programme. Anthony Jackson writes:

"It had been hoped that prefabrication would solve the housing shortage. Aneurin Bevan admitted after a few months as Minister of Health that he had been eagerly looking for a system of mass production that would turn out houses in the same way as cars and aeroplanes, but so far it had eluded him . . . in 1947, over one hundred alternative prefabricated systems had been approved for development or use. The results were disappointing. With an unconvinced private industry reflecting public preferences, only 15 per cent of the houses built used industrialized methods." (30).

This belief in prefabrication was quite obviously derived from the success of wartime construction in such projects as the Sea Forts, Mulberry Harbour and the hutting programme, at least in technique. Even when the failure to solve the housing shortage quickly in this way was realised, the belief in prefabrication continued due to this success. Particularly in the field of pre-cast concrete panel construction the war marked an important change in British architecture. Prefabrication became a major feature of post-war Britain: at times excellent, at others deplorable—in both environmental and structural terms (31).

In addition to the direct innovation or developments brought about in the field of military construction, such as prefabrication, there were also the more indirect and numerous innovations of military technology in general: new materials, techniques and hardware. But while all of these greatly enriched civil architecture technically, two other factors should be considered: firstly that war-time conditions prompted the implementation of new ideas rather than their invention, secondly the huge energy potential which was absorbed by the wars.

In purely economic terms the First World War alone was calculated by

the economist E. L. Bogart to have cost £82 900 000 000 including the cost of munition, necessary machines, property losses and production losses but excluding the later billions paid in interest payment on loans, pensions and care for veterans. According to the Editor of Scholastic in 1934:

"It would have been enough to furnish: (I) every family in England, France, Belgium, Germany, Russia, the United States, Canada and Australia with a £625 house on a £125 one-acre lot with £250 worth of furniture, and (II) A £1 250 000 library for every community of 200 000 population in each of the countries. (III) A £2 500 000 university for each of these communities, and (IV) A fund that at five per cent interest would yield enough to pay indefinitely £250 a year to 125 000 teachers and 125 000 nurses and (V) Still leave enough money to purchase every piece of property and all the wealth in Belgium and France at a fair market price." (32).

History has unfortunately shown that this energy would only have reached civil architecture in a much diluted form if war had been avoided. Buckminster Fuller in fact points out that it is 'incomprehensible to society and economists' that in 1945, 'despite our having expended hundreds of billions of dollars, our economy was far richer than before the war'. Fuller terms this wealth as the 'techno-scientific literacy' or 'generalized toolbase' which developed inadvertently from the war programme (33).

Fuller has been one of the few post-war writers to identify architectural progress with military expenditure. He claims that fall out from the space programme will be more fundamental than that from previous 'weaponry':

"... now that scientific warfare has gone into space, men who handle the warfaring apparatus in space find no air to breathe and no water or food waiting to drink or eat. For the first time in history, it has been necessary for science to upgrade environmental and metabolic regeneration conditions of man and to package them for economic delivery by rockets ..." (34).

He sees the environmental 'black box' as the result: freeing earthbound man from existing limitations of enclosure and climate.

Since 1945 the strong relationship between military technology and civil architecture has certainly continued. In America the system of funding research established in World War II, the Research and Development Contract paid for by the Central Government, has dictated the type of research undertaken at universities and other institutions. Subsequently 80 per cent of academic science in 1970 relied on military and space budgets (35). In Europe the situation 25 years after the war, is far from this extreme but the conclusions made by Fuller in his recent book 'Utopia or Oblivion' are still applicable. Fuller claims that 'in order to be able to take care of 100 per cent of humanity in the shortest possible time, we are going to have to stop getting our technical advantage gains only as a secondhand event' (36). Theo Crosby made a more complete summary in his review of the book:

"He (Fuller) points out that we have reached our present relatively comfortable situation almost entirely through vast expenditure on defence, on weaponry, and that we have received all our domestic technology at second hand from this expenditure. How much could we have if we had made the investment directly in 'Livingry', through a comprehensive design science? Politicians and economists are ignorant and afraid of technology. Money can thus always be found for weapons. It is somehow never available for the application of scientific resources to housing or recreation, to the environment, for experimental cities rather than rocket installations we do not intend to use ..." (37).

33. Caquot Kite Balloon on the Western Front 1914–18.

REFERENCE NOTES

The reference notes for each chapter are self-contained: any book cited in a chapter is given full bibliographical reference upon first mention in that chapter.

CHAPTER 1: 1914 FORTRESS ZONE

1. BRUNNER, Major Moritz Ritter von. 'Permanent Fortification for the Imperial Military Training Establishments and for the Instruction of Officers of all Arms of the Austro-Hungarian Army'. Translated for the General Staff. HMSO. London 1910 (7th edition), para. 79.
2. *ibid.* para. 39.
3. *ibid.* para. 41.
4. CLARKE, Major G. Sydenham CMG. Royal Engineers 'Fortification: Its past achievements, recent developments and future progress', pub. Murray. London 1890. p. 88.
5. Brunner, para. 76.
6. Clarke, pp. 90–91.
7. HORNE, Alistair. 'The Price of Glory: Verdun 1916', pub. Macmillan. London 1962. p. 5.
8. MORIN, Lieut. Col. 'The Utility of Permanent Fortifications'. Journal of the Royal Artillery Vol. LVII 1930–31. pp. 491–505. (Translated from the Revue Militaire Français, May 1930, by Brig. Gen. W. Evans.)
9. EDMONDS, Brig. Gen. Sir James E. 'Minor French Fortresses and Barrier Forts in August–September 1914'. Royal Engineers Journal, Dec. 1939. p. 520.
10. RÉBOLD, Colonel J. 'La Guerre de Forteresse 1914–18', pub. Payot. Paris 1936. fig. 1.
11. Morin.
12. Rébold, pp. 19–20.
13. Morin.
14. BENOIT, Général. 'La Fortification Permanente pendant la Guerre'. Berger-Levrault Nancy-Paris-Strasbourg 1922, pp. 16–17.
15. SYDENHAM, Colonel, Lord of Combe. 'The Belgian Defences in 1914'. Royal Engineers Journal. January 1921. p. 58.
16. SCHRYVER, Colonel adjoint d'Etat-Major A. de. 'La Bataille de Liège (Août 1914)'. Imprimerie H. Vaillant-Carmanne. Liège 1922. p. 8. Sydenham 'The Belgian . . .'. p. 59.
17. Rébold, pp. 18–19.
18. Benoit, p. 18 (translation by the authors).
19. Sydenham, 'The Belgian . . .', p. 57.
20. *ibid.* (translation by the authors).
21. *ibid.*
22. Benoit, pp. 18–19.
23. Rébold, pp. 18–19.
24. ESSEN, Leon van der. 'The Invasion and the War in Belgium. From Liège to the Yser', pub. T. Fisher Unwin. London 1917. pp. 235–36.
25. Rébold, p. 19 (translation by the authors).
26. Morin.
27. Following the defeat of 1870 defensive thought had been predominant in France, the fortress zone was the result, but gradually her military leaders developed the opinion that the French defeat had been due to a lack of offensive spirit—'furia francese'. The result was Plan XVII. When war was declared, four out of the five French armies were to rush forward in a headlong offensive in the direction of Alsace-Lorraine.
28. FALLS, Cyril. 'The First World War', pub. Longmans. London 1960. p. 22.
29. DUFFY, Christopher. 'The Liège Forts' published in 'History of the First World War'. Purnell 1970. Vol. 1, No. 5. p. 131.
30. GOODSPEED, D. J. 'Ludendorff', pub. Rupert Hart-Davis. London 1966. p. 33.
31. HEYE, Oberst a. D. 'Panzer im Festungsbau'. Wehrtechnische Monatshefte. March 1939. p. 114.
32. Sydenham, 'The Belgian . . .', p. 59.
33. Goodspeed, p. 36.
34. Duffy.
35. 250 kg to the cubic metre as opposed to French 400 kg to the cubic metre (Benoit, p. 12).
36. Benoit, p. 12.
37. HAUTECLER, Georges. 'La Rapport du Général Leman sur la Défense de Liège en Août 1914'. Palais des Academies. Brussels 1960.
38. Heye and Duffy.
39. Heye, p. 115. Heye claims only four of the forts were under 420 mm attack. He claims that the 210 mm howitzers were sufficient on their own to capture the Belgian forts, breaking up the weak concrete structure and in this way making the forts' artillery unworkable.
40. 'The Times', October 6th, 1914, p. 7.

41. Duffy.
42. Heye, p. 114.
43. Hautecler, p. 37 (translation by the authors).
44. Duffy.
45. 'The Times', October 1st, 1914, p. 7.
46. Essen, p. 239.
47. LIDDELL HART. 'History of the World War 1914–18', pub. Chivers. Bath 1968. p. 290 (originally pub. as 'The Real War', Faber & Faber Ltd. London 1930).
48. Morin.
49. Horne, p. 36.
50. *ibid.* pp. 47–48.
51. PETAIN, Marshal H. P. 'Verdun', pub. Elkin Mathews & Marrot. London 1930. pp. 242–43.
52. MACKSEY, Kenneth. 'Fort Douaumont' published in 'History of the First World War'. Purnell 1970. Vol. 3, No. 14. p. 1266.
53. Macksey.
54. Benoit, pp. 11–12.
55. Horne, pp. 107–8.
56. Morin.
57. Horne, p. 327.
58. Pétain, pp. 250–51.
59. Clarke, p. 105.

CHAPTER 2: DEFENCE IN DEPTH

1. ESSAME, Major-General H. 'The New Warfare', published in 'History of the First World War'. Purnell 1970. Vol. 3, p. 1181.
2. KEEGAN, John. 'Trench Warfare', published in 'History of the First World War'. Purnell 1970. Vol. 2, pp. 576–87.
3. WAR OFFICE, General Staff. 'Field Service Regulations, Part I: Operations'. 1909. (Reprinted 1914.) Quote from Chapter VII: 'The Battle', Section 105: 'The General Conduct of the Attack', Sub-section 6, p. 142.
4. PARSONS, Colonel William Barclay, 11th US Engineers. 'The American Engineers in France'. New York 1920. p. 318.
5. Outposts and listening posts were built in 'No Man's Land', forming an intelligence support, which could give reports after dark. The main obstacles in front of the Front trench, the main defensive position, were formed by barbed wire.
6. Concrete gun positions did not appear until after 1916.
7. BRITISH INTELLIGENCE. 'German Methods of Trench Warfare'. 1916.
8. In the vicintiy of the Somme during 1915 the second position comprised only a number of detached posts. Similarly, south of the La Bassée canal there was only a single line of entrenchments until March 1915.
9. ROYAL MONMOUTHSHIRE R.E. MILITIA. Regimental History (unpublished).
10. CLARK, Alan. 'The Donkeys'. London 1961. p. 106.
11. PRUSSIAN WAR MINISTRY. 'Manual of Position Warfare for All Arms: Part I: The Construction of Field Positions (Stellungsbau). Issued Berlin 1916. (Translation of a captured German Document Ia/20934.) p. 6, para. 11.
12. *ibid.*
13. McKEE, Alexander. 'Vimy Ridge'. London 1966. p. 114. McKee is quoting Captain H. U. S. Nisbet, 5th Imperial Division.
14. SIXT VON ARMIN. 'The Construction of Defensive Positions'. 4th Army HQ. July 1917. p. 6. (Translation of a captured German Document Ia/38494.)
15. LIDDELL HART. 'History of the World War 1914–18'. Chivers Ltd. Bath 1968. p. 331. (First published as 'The Real War'. Faber & Faber 1930).
16. Prussian War Ministry 'Manual of . . .'
   "It is a ruling principle that the ground to be held must be fortified in such a way that an obstinate defence by sectors is obtained, and to such a depth, that the loss of, or withdrawal from parts of the positions does not endanger it as a whole. . . . A strongly constructed first position with plenty of depth. . . . Behind the first position at least one rearward position should be prepared . . . at such a distance from the first position that a simultaneous artillery attack on both is not possible. . . . The bulk of the garrison (including machine guns) must be accommodated in the rearward lines, in the ground between the lines, in the communication trenches, and in the country behind the first position. . . . Similarly, the strength of construction employed in the rearward lines, and particularly of the emplacements, dug-outs, etc., must be appreciably greater than in the front line."
17. In December 1916 a manual called 'Die Führung der Abwehrschlacht' (Conduct of the Defensive Battle) was published, putting forward more complete principles of 'in depth' based on the experience of the Somme, to be followed in January 1971 by 'Instructions on the counter-attack in depth' and in August 1917 by 'Allgemeines über Stellenbau' giving the constructional details of 'in depth'.
18. GENERAL STAFF (British Intelligence). 'Diagram showing the Organization and Defences of a Divisional Sector'. Ia/32971 German Document captured April 1917 and published by British GHQ May 1st, 1917.
19. FALLS, Cyril. 'The First World War'. Longmans. London 1960. p. 174.
20. VON LOSSBERG, Fritz. 'Meine Tätigkeit im Weltkrieg, 1914–18'. Berlin 1939. p. 283.
21. SIXTH ARMY. German 'Supplementary Instructions as to the Construction of Defences, Sept. 27, 1916'. Published as a Translation of a captured

German Document by British General Staff (Intelligence). GHQ. May 5th, 1917. SS 577-la/32517.
22. The full name was 'Mannschafts-Eisenbeton-Unterstände'.
23. Sixth Army 'Supplementary Instructions . . .'.
24. Information based on:
    (i) THURLOW, Colonel E. G. L. 'The Pill-Boxes of Flanders', pub. Nicholson & Watson 1933.
    (ii) Letter from Dr. A. Caenepeel, author of 'Ieper en de Frontstreek 1914–18'.
25. Prussian War Ministry 'Manual of . . .', para. 36.
26. Figures correlated from:
    (i) Prussian War Ministry 'Manual of . . .';
    (ii) SEETZELBERG, Friedrich. '1914–18 Der Stellungskrieg'. Berlin 1926;
    (iii) WILSON, B. T. 'Study of German Defences near Lille', *circa* 1918.
27. Prussian War Ministry 'Manual of . . .', para. 24.
28. Wilson (Survey drawings of reinforced steeple and chimney).
29. Prussian War Ministry 'Manual of . . .', para. 110.
30. Based on:
    (i) Wilson (Survey of Fin de la Guerre Brigade HQ);
    (ii) Seetzelberg (plate 8);
    (iii) Caenepeel letter to authors . . .
31. Prussian War Ministry 'Manual of . . .', para. 145, and: 6th BAVARIAN DIV. 'Order of the 6th Bavarian Division regarding Machine Guns'. 3.9.16.
32. 'Notes on the Construction of Positions on the Ypres Battle Front for the Coming Winter'. 5.9.17. (Captured German Document la/40342.)
33. Based on:
    (i) Prussian War Ministry 'Manual of . . .' (para. 73, fig. 32, para. 68);
    (ii) Wilson (Survey of standard trench shelter for two machine gun crews at Pont-a-Vendin, Aubers Ridge);
    (iii) Seetzelberg, p. 165.
34. Wilson, 21A.
35. Prussian War Ministry 'Manual of . . .', para. 14.
36. *ibid.* paras. 17, 79, 80.
37. Thurlow. *See also:* 'Six Examples of the German use of Concrete in Farm Buildings', a series of photographs taken by 21 Squadron RFC June–Sept. 1917 in the Julien-Frezenberg sector.
38. Thurlow and McKee.
39. Sixt von Armin. 'The Consruction of . . .'.
40. *ibid.*
41. LUDENDORFF. 'The Defence of the Outpost Zone'. Issued by German GHQ 6.7.18. (Translation of Captured German Document la/53917.)
42. LUDENDORFF. 'Experiences of the Recent Fighting'. Issued by German GHQ 22.7.18. (Translation of a Captured German Document.)
43. BARNETT, Correlli. 'The Swordbearers—Studies in Supreme Command in the First World War'. Eyre & Spottiswoode. London 1963. pp. 346–53.

CHAPTER 3: WIDER SCOPE OF DEFENCE

1. The resistance of the Verdun forts was later to be the main impetus behind the building of the Maginot Line but the forts, such as Douaumont, were all built pre-1914 (*see* Chapter 1).
2. CLARK Alan. 'The Donkeys'. Hutchinson. London 1961. p.42. Clark is quoting from 'GHQ' by Brig.-Gen. Charteris—pub. Cassell 1931.
3. 'The Times History of the War'. London 1916. Vol. IX, p. 483.
4. BRITISH INTELLIGENCE. 'German Methods of Trench Warfare'. 1915–16.
5. LIDDELL HART. 'Europe in Arms'. Faber & Faber Ltd. London 1937. p. 273.
6. *ibid.* p. 274. *See also:* OGORKIEWICZ, R. M. 'Design and Development of Fighting Vehicles'. MacDonald. London 1968. pp. 26–27.
7. PITT, Barrie. '1918—The Last Act'. Cassell. London 1962. p. 8.
8. BARNETT, Correlli. 'The Swordbearers—Studies in Supreme Command in the First World War'. Eyre & Spottiswoode. London 1963. p. 249.
9. A Company would move into a position, start constructing trenches based on their own ideas, disregarding what was already there. Before they had finished the company would move on and the process was repeated. As no division trusted the defensive power of the neighbouring company, most of the initial effort would be expended in the construction of switch lines along company borders.
10. PRIDHAM, Colonel G. R. 'Field Defences'. Royal Engineers Journal. March 1938. p. 19.
11. The Ypres Salient was held by the British throughout the war and the defences there were of a more complete nature than those on other parts of the front.
12. Information based on:
    (i) THURLOW, Colonel E. G. L. 'The Pill-Boxes of Flanders', pub. Nicholson & Watson 1933.
    (ii) Letter from Dr. Caenepeel;
    (iii) ENGINEER IN CHIEF, GHQ France. 'File on Recent Fighting, Field Fortification Notes and Field Fortification Plates'.
13. ROYAL MONMOUTHSHIRE R.E. MILITIA: Regimental History (unpublished).
14. GENERAL STAFF. 'Notes on Trench Warfare for Infantry Officers'. Dec. 1916. (Revision of notes of March 1916.) Figs. 49 and 50.
15. Engineer in Chief, GHQ France. 'File on Recent . . .'.
16. WAR OFFICE. 'Manual of Field Works (All Arms)'. 1921 (Provisional).

HMSO. London. Plate 22.

17. Engineer in Chief, GHQ France. 'File on Recent . . . .
18. Pridham, p. 19.
19. See reference notes for Chapter 2.
20. Reading between the lines, the criticism levelled against the British home defences, the GHQ line (Chapter 8), it seems the British were still uncertain of the difference between 'in depth' and dispersion during World War II. In France Pétain was opposed by his contemporaries when he wanted to introduce 'in depth' in 1918 and also later when he wanted to build the Maginot Line on these principles (Chapter 5).
21. CHURCHILL, Winston S. 'The World Crisis, 1911–14'. Thornton Butterworth. London 1923. pp. 154–55.
22. *ibid.* p. 391.
23. HANKEY, Lord. 'The Supreme Command, 1914–18'. Vol. I. George Allen & Unwin. London 1961. p. 69.
24. *ibid.* p. 215.
25. CRUTTWELL, C. R. M. F. 'A History of the Great War, 1914–18'. Oxford 1934. p. 68.
26. 'Field Defences: Thames and Medway Garrison'. Vol. I (Chatham land front) and Vol. II (Sheppey section). Both volumes of unmounted photographs—date uncertain. Copies held at the Imperial War Museum, London.
27. MAURICE JONES, Colonel K. W. 'The History of Coastal Artillery in the British Army'. Royal Artillery Institution. London 1959. pp. 180 and 218.
28. SYDENHAM CLARKE, Major G. Royal Engineers, 'Fortification: Its past achievements, recent development and future progress'. John Murray. London 1890. pp. 78 and 199–204.
29. As Germany had started the defence of her coastline at the very end of the 19th century, all her coastal fortifications were on this principle until the building of the Atlantic Wall. In Britain, however, the project proposed by the Royal Commission along the south coast had been started in 1860, and at first thus incoporated the principles of casemating. Similarly the theories of protection against fire through heavy constructions rather than through lessening the size of both emplacements and quarters led to the building of the massive Verne Citadel at Portland and Garrison point at Sheerness. *See:* BUCHANAN, Colonel A. G. B. 'Coast Defences 75 Years Ago'. Royal Engineers Journal. Dec. 1938.
30. LIDDELL HART. 'History of the World War 1914–18', this edition by Chivers. Bath 1968. p. 457. (Originally published as 'The Real War' by Faber & Faber 1930).
31. MURRELL, H. F. Architect. 'German War Construction: Submarine Shelters and Zeppelin Sheds'. pub. in 'Journal of the Royal Institute of British Architects'. Oct. 23rd, 1920. pp. 485–90.
32. RAWLINSON, A. 'The Defence of London 1915–18'. pub. 1923. pp. 258–59.

CHAPTER 4: SUPPORT STRUCTURES

1. LIDDELL HART. 'Europe in Arms'. Faber & Faber. London 1937. p. 21.
2. LIDDELL HART. 'The War in Outline, 1914–18'. Faber & Faber. London 1936. p. 74.
3. LIDDELL HART. 'History of the World War, 1914–18'. pp. 55, 62 and 385.
4. The variety of building types increased as the war developed, including specialized structures such as 'delousing plants'. A bizarre list of building types is descirbed in 'Historical Report of the Chief Engineer: Allied Expeditionary Forces 1917–19'. pub. Washington 1919.
5. INSTITUTION OF ROYAL ENGINEERS. 'Work of the Royal Engineers in the European War, 1914–1919: Work under the Director of Works (France)'. Chatham 1924. p. 1.
6. *ibid.* p. 6.
7. *ibid.* p. 10.
8. 'Report of the (RIBA) Council for the official year 1916–17'. The RIBA Journal April 1917.
9. Institution of Royal Engineers 'Work of . . .', p. 64
10. *ibid.* p. 53.
11. *ibid.* p. 172.
12. Vice-versa if weather-boarding was used.
13. Institution of Royal Engineers. 'Work of . . .', pp. 172–73.
14. 'The Nissen Hut on the Western Front'. The Architects' and Builders' Journal. Feb. 14th, 1917. p. 91.
15. *ibid.* p. 92.
16. Institution of Royal Engineers. 'Work of . . .', pp. 173–74.
17. BRITISH INTELLIGENCE. 'German Methods of Trench Warfare' (probably written in Winter of 1915–16).
18. NEUMANN, G. P. 'The German Airforce in the Great War'. Hodder & Stoughton. London 1921. This edition Chivers, London 1969. p. 32.
19. *ibid.* p. 20.
20. HAENIG, A. 'Luftschiffhallenbau'. Rostock 1910. pp. 7, 8, 9 and 90.
21. Neumann, p. 8.
22. MURRELL, H. F. 'German War Construction, Submarine Shelters and Zeppelin Sheds'. RIBA Journal. Oct. 23rd, 1920. p. 488.
23. *ibid.* p. 490.
24. Neumann, p. 32.
25. JOHNSON, Air Vice-Marshal J. E. 'The Story of Air Fighting'. Chatto and Windus. London 1964. pp. 68–69 and 80–81.

26. MASON, John. 'A Report on Structural Engineering in Germany'. The Structural Engineer. London. June 1946. pp. 331–32.
27. 'These War Buildings were Significant'. Engineering News-Record (NY). Oct. 19th, 1944. p. 111.

CHAPTER 5: MAGINOT LINE

1. ROWE, Vivian. 'The Great Wall of France'. Putnam. London 1959. p. 48.
2. BARNETT, Correlli. 'The Swordbearers'. Eyre and Spottiswoode. London 1963. pp. 254–55.
3. BARRIE and ROCKLIFF. 'The ancient art of warfare'. Crescent Press. p. 448.
4. EASTWOOD, James. 'The Maginot and Siegfried Lines'. Pallas Publishing Co. Ltd. London 1939. pp. 10–11.
5. LIDDELL HART, Captain B. H. 'A History of the World War 1914–18', (Originally 'The Real War'. Faber and Faber 1930.) Ref. from former p. 290.
6. EIS, Egon. 'The Forts of Folly'. Oswald Wolff Ltd. London 1959. p. 226.
7. The Rhine defences were started as part of the Maginot Line defences but, as the Rhine itself provided a natural barrier, only minor shelters and casemates were considered necessary. The construction of defences along the Belgian border, started when the Nazis renewed fortification fever, never really got off the ground. The Government felt that it would endanger Franco-Belgian relationships and the military staff always maintained that the French could only defend themselves in the north by taking up positions in Belgium. This strategy would probably have had some success had not the hinge of this planned forward move into Belgium, the Ardennes, been considered to be impregnable. Any suggestions that it was not were crushed by Pétain when, in 1934, he called the Ardennes Forests 'inpenetrable': 'the enemy will not deploy there'. The Alpine position, however, did meet with military success. When the armistice was signed in 1940 the Italians had hardly moved from their first position. The 'Maritimes Fortified Sector', as it was called, was strengthened under the guidance of General Degoutte in 1936. Concrete bunkers, ranging from large casemates to simple pill-boxes were built in order to take the fullest advantage of the existing natural obstacles of the Alps.
8. Eastwood, p. 23. *See also:* BELPERRON, Pierre. 'Maginot of the line'. Translated by H. J. Stenning. Williams and Norgate. London 1940. p. 104.
9. Fortification construction in the Low Countries—Belgium and Holland—had not started till late in the 1930s as a policy of absolute neutrality had been insisted upon by the governments. The Belgian Fort Eben Emael, because of its late date of construction, was therefore often considered to be the ultimate in modern fortress design. Both lines of Belgian defence, the Albert Canal and the KW line, ran between Antwerp, Liège and Namur. Fort Eben Emael thus formed a cornerstone: the first serious point of resistance to the German invasion. Ample warning of attack was expected as the Germans would have to cross the Dutch frontier before reaching the fort. Yet, due to the lack of cover given to the fort by neighbouring artillery, the Germans were able to overpower this strongpoint by one ingenious operation carried out by a glider detachment. Eighty men descended on it by this means and bottled up the 1200 strong garrison while the German advanced ground troops came up in relative safety to capture the fort.
10. *See* Chapter 6, pp. 118–119; and Chapter 6, reference note 17.
11. Eastwood, p. 39.
12. Rowe, p. 73.
13. *ibid.* p. 72.
14. Rowe.
15. Eastwood, p. 13.
16. LIDDELL HART, Captain B. H. 'Europe in Arms'. Faber & Faber. London 1937. p. 47.
17. Eastwood, p. 26.
18. *ibid.* p. 16.
19. Eis, p. 224.
20. BRYANT, Arthur. 'The Turn of the Tide'. Collins. London 1957. p. 72.
21. LIDDELL HART, Captain B. H. 'Strategy'. Faber & Faber. London 1954. p. 232.

CHAPTER 6: THE AUTOBAHN AND WEST WALL

1. As late as 1939 Todt was invited to give a lecture on motorway transport to the Civil Engineers' German Circle. At that time the first English flyover was still only at the design stage.
2. MASON, John. 'A Report on Structural Engineering in Germany', published in The Structural Engineer, The Journal of the Institute of Structural Engineers. London. June 1946, pp. 300–301.
3. WEBER, J. B. 'Aspects of National Socialist Architecture'. The Architectural Association Quarterly. 1969. April, Vol. I, No. 2. pp. 46–55.
4. These bridges were ofted designed by fairly well known architects of the time. In 'Neue Deutsche Baukunst' (SPEER, Albert. Berlin 1941), the official appreciation of Nazi architecture, there were illustrations of several bridges by such architects as Fritz Tamms, Hans Freese, Wilhelm Haerter and Paul Bonatz.
5. Among the Allies the West Wall became known as the Siegfried Line. This should not be confused with the Siegfried Line of the First World War and for this reason is referred to only as the West Wall in the text.
6. EIS, Egon. 'The Forts of Folly', pub. Oswald Wolff Ltd. London 1959. p. 228.
7. SHIRER, William. 'The Rise and Fall of the Third Reich', first published 1959, this edition BCA. 1970. p. 318.

8. MACDONALD, Charles B. 'US Army in World War II: The European Theatre of Operations: The Siegfried Line Campaign', published US Dept. of Army Office of the Chief of Military History. 1963. p. 31.
9. PÖCHLINGER, Josef. 'Das Buch vom West Wall'. Otto Elsner. Verlagsgesellschaft. Berlin 1940. p. 65.
10. LIDDELL HART. 'Strategy'. Frederick A. Praeger (NY 1954), originally published Faber and Faber (London 1954). p. 230.
11. O'NEILL, Robert J. 'The German Army and the Nazi Party'. Cassell and Co. Ltd. London 1966. p. 124.
12. This was the name given by American Intelligence in 'German Doctrine of the Stabilized Front'.
13. Rows of eight dragon's teeth were also used but these were only two consecutive rows of four and followed the same design.
14. US MILITARY INTELLIGENCE DIVISION. 'The German Doctrine of the Stablized Front'. US War Dept. 1943. p. 24.
15. *See* Chapter 12.
16. *ibid.* p. 68.
17. The Czech fortifications were started in 1933 when Hitler became German Chancellor. France, in attempting to form a ring of allies around Germany, greatly assisted in the construction by making her experience from the construction of the Maginot Line available and by providing French engineers and technicians to assist in the design. The detailed result was therefore a replica of the Maginot Line but it did not try to make Czechoslovakia impregnable: instead it was constructed so as to force a German attack to follow an eastern course along the length of the country where the topography would slow down such an attack until French assistance could be brought in. But the theory was of little use for Hitler had chosen to use political methods, not the army, what he called 'intellectual weapons'. When it came to a political showdown France and Britain gave in to German pressure. The very areas where the massive defences had been built were given to Germany with their fortifications intact. Only after the Munich agreement did Gamelin point out to Daladier and the Foreign Minister Bonnet that the area given away without Prague's consent contained nearly all the Czechoslovakian defences. The subsequent unopposed invasion of the rest of the country was the inevitable conclusion.
18. KÜHNE, Rudolf Theodor. 'Der West Wall', published J. F. Lehmans. Verlag. (München 1939) gives figures for the West Wall. On p. 22 there is the following table which gives a detail growth pattern for the number of men working under Todt's direction on the West Wall in 1938:

July 20	35 000 men	Aug. 31	170 000 men	Sept. 21	241 000 men
July 27	45 000 men	Sept. 7	191 000 men	Sept. 28	278 000 men
Aug. 3	77 000 men	Sept. 14	213 000 men	Oct. 6	342 000 men
Aug. 10	93 000 men				
Aug. 17	121 000 men				
Aug. 24	145 000 men				

19. As part of these statistics were issued by the Nazi government they are probably not conservative.
20. MIRS. 'Handbook of the Organization Todt' (OT). London 1944. p. 5.
21. Though several officers were arrested, the jealousy between Army and SS as well as Hitler's unfavourable view of officers who did not support the Nazi cause might have increased the number beyond those actually involved.
22. Kühne, p. 39.
23. EASTWOOD, James. 'Walls of Death'. Pallas. London 1939. p. 53.
24. WILMOT, Chester. 'The Struggle for Europe'. Collins. London 1952.
25. MacDonald, p. 35.

CHAPTER 7: FORTRESS BRITAIN

1. MOOREHEAD, Alan. 'Churchill and his World'. Thames and Hudson. London 1960. p. 79.
2. COLLIER, Basil. 'History of the Second World War: The Defence of the United Kingdom'. HMSO. London 1957.
3. FRANKLAND, Noble. 'The Bombing Offensive against Germany, Outline and Perspective'. Faber and Faber. London 1965.
4. During 1914–18 the total number of civilian deaths caused by air raids was 1413. Of the total of 103 raids, 51 were by airships, and the rest by planes. The most successful single raid was on June 26th, 1917, when 14 twin-engined Gotha aeroplanes bombed London causing 162 deaths. The fact that these raids caused considerable alarm and panic is quite remarkable when one compares them with the huge losses being suffered on the Western Front at this time where, for example, British losses over August and September 1917 alone are estimated at almost 240 000 (FALLS, Cyril. 'The First World War'. London 1960. p. 285).
5. *See* LIDDELL HART. 'Europe in Arms'. Faber and Faber Ltd. London 1937, for discussion on the subject of 'Would another war end civilisation?'.
6. Collier, p. 13.
7. *ibid.* p. 16.
8. *ibid.* p. 33.
9. *ibid.* p. 43.
10. *ibid.* p. 119.
11. MAURICE-JONES, Colonel K. W. 'The History of Coastal Artillery in the British Army'. Royal Artillery Institution. London 1959. pp. 142, 180, 218, 228.
12. Maurice-Jones (p. 228) gives the following break-down of the construction

of 'emergency batteries':

During June	1940, 1st instalment,	47 batteries finished
June	1940, 2nd instalment,	24 batteries finished
July	1940, 3rd instalment,	8 batteries finished
August	1940, 4th instalment,	28 batteries finished
September	1940, 5th instalment,	5 batteries finished
November	1940, 6th instalment,	15 batteries finished
January	1941, 7th instalment,	26 batteries finished

13. Maurice-Jones, p. 223.
14. *ibid.* p. 223.
15. *ibid.* p. 234.
16. Letter to the authors from T. Kent.
17. The standardized pill-box designs used can be directly compared with those built in Britain during 1914–18. It is probable, however, that they also owed some of their design features to the pill-boxes built by the British Expeditionary Force, BEF, along the Belgian border during 1939–40. It had been suggested that the BEF should build defences something like the Maginot line along their section of the border, but, with little time available, they instead built a more flexible defence incorporating some 400 pill-boxes. The pill-boxes consisted of five standard designs which, using re-usable steel shuttering were to be 'mass produced'. *See* PAKENHAM-WALSH, Major General R. P. 'The Pill-Box Row: the Defensive System of the BEF in France 1939–40'. Royal Engineers Journal, Vol. 74. 1960. p. 496.
18. VIVIAN, Brigadier J. A. 'Coast Artillery'. The Royal Artillery Commemoration Book 1939–45'. The Royal Artillery Institution.
19. Although on the Atlantic Wall, in comparison, Major General J. L. Moulton points out (in a letter to the authors) that on D-Day the houses which had been converted into strongpoints at St. Aubin and elsewhere did, in his opinion, 'more damage to the assault than the more imposing gun emplacements at Merville and elsewhere'. The camouflaged strongpoints were relatively untouched by the preliminary bombing.
20. *See also:* PAWLE, Gerald. 'The Secret War 1939–45', published Harrap. London 1965, who describes some of the incredible inventions which took place in Britain at this time.
21. Collier, p. 129.
22. IRONSIDE, Major General Sir Edmund. 'Role of the Fortress in Modern War'. Journal of the Royal Artillery. 1929. (A lecture delivered at the RA Institution February 15th, 1927).
23. Collier, p. 130.
24. *See* the captured German documents which are referred to in Chapters 2 and 3.
25. Collier, p. 142.
26. BRYANT, Arthur. 'Turn of the Tide'. (A study based on the diaries and autobiographical notes of Field Marshal the Viscount Alanbrooke). Collins. London 1957.
27. *See* WARLIMONT, General Walter. 'Inside Hitler's Headquarters'. Weidenfeld and Nicolson. London 1964. p. 109, where Warlimont refers to 'Das Unternehmen Seelöwe' by K. KLEE, pp. 39–43.
28. POSFORD, John Albert. 'The Construction of Britain's Sea Forts'. The Civil Engineer in War. Vol. 3.
29. The Mystery forts, also known as 'Nab towers', would have each cost over five times as much as a Maunsell tower and taken four times as long to build. The Mystery fort design called for a cellular concrete base and steel superstructure.
30. Moorehead, p. 92.
31. CHURCHILL, Winston S. 'The Second World War'. Vol. II. 'Their finest Hour', published Cassell. London 1949. p. 216. *See also* Chapter 10, pp. 201–3.
32. In 1921 the Institute of Civil Engineers published a detailed paper on the wartime development of concrete in ship building by Prof. Thomas Bertrand Abell, 'Reinforced Concrete for Ship Construction', Paper No. 4351, 'Minutes of Proceedings of the Institution of Civil Engineers with other selected papers'. London 1921.
33. Posford, p. 132.
34. Civil Engineer in War. Vol. 3. p. 166: discussion on the foregoing papers.
35. DALZIEL, Lance. 'Secrets of the Sea Forts'. Coastal Artillery Journal. 1945. Vol. LXXXVIII, No. 2. p. 33.
36. *See* Chapter 13, pp. 279–84.
37. As the units were interdependent, only five had mooring facilities and the fuel was stored in tanks suspended from two of the units.
38. The technical information on the sea forts is based primarily on Posford.
39. Posford, pp. 160–61.

CHAPTER 8: ATLANTIC WALL

1. LEBRAM, Hans Heinrich. 'Kritische Analyse der Artillerie des Atlantikwalles'. Marine Rundschau, Zeitschrift für Seewesen, 1955. No. 2. p. 29.
2. WARLIMONT, Walter. 'Inside Hitler's Headquarters'. Weidenfeld & Nicolson. London 1969. p. 192.
3. Warlimont, p. 198.
4. MEAD, Hillary P. 'The Martello Towers of England'. Mariners Mirror. Vol. 34. London 1948.
5. MAURICE-JONES, Colonel K. W. 'The History of Coastal Artillery in the British Army'. Royal Artillery Institution. London 1959. p. 107.
6. *See* Chapter 12, for Dorsch's interest in underground construction.

7. SPEIDEL, Lieutenant-General Hans. 'We defended Normandy'. Herbert Jenkins. London 1951. p. 51.
8. MAJDALANY. 'Fall of Fortress Europe'. Hodder and Stoughton. London 1968. p. 35. The extent of German coastal fortifications during 1939–45 is shown by a series of maps in VON HARNIER, Wilhelm. 'Artillierie im Küstenkampf' (Munich 1970), which shows coastal defences extending from the northern tip of Norway along the coast of Europe, including the Baltic, along the coast of the Mediterranean into the Black Sea. Only part of this can be called the Atlantic Wall proper.
9. Majdalany, pp. 30–31. The original publication from which Majdalany quotes is Walter Hubatsch's 'Hitlers Weisungen für die Kriegsführung 1939–45'.
10. Majdalany, p. 39.
11. SPEER, Albert. 'Inside the Third Reich'. Weidenfeld and Nicolson. London 1970. p. 352.
12. In Autumn 1942, the use of such a tower was suggested by the naval Commander in Chief of Fortifications, France, at OKM in Berlin. The experiments were started in Gennevilliers outside Paris on a ball bearing seven metres diameter from the German naval ship 'Provence', using two metres thickness of concrete. By the time it was ready for testing, not only the Navy but also Rommel was interested. Time had run out, however, it was April 10th, 1944, initial Allied bombing had already started. For further information on rotating artillery emplacements see STJERNFELD, Bertil. 'Alarm i Atlantervallen'. Hörsta Förlag AB. Stockholm 1953. p. 126.
13. *See also.* LEPOTIER, A. 'L'Atlantikwall de Brest' in 'Revue Maritime'. No. 105. 1955.
14. Speer, p. 352.
15. Majdalany, p. 31.
16. Speer, p. 352.
17. *See* Speer's 'Inside the Third Reich' for a complete description of Hitler's architectural aspirations.
18. Information on the OT is based primarily on: MIRS. 'Handbook of the Organisation Todt'. London 1945. *See* p. 25 in particular.
19. YOUNG, Desmond. 'Rommel'. Collins. London 1950. p. 192.
20. Majdalany, p. 40.
21. Warlimont, p. 400.
22. Young, p. 191.
23. This was in fact proved to be the case following D-Day.
24. In the navy 93 out of 132 coastal guns had been placed in casemates in the Pas de Calais area, and 27 out of 47 in the Normandy naval command. The practice of providing only half of the coastal guns with overhead protection was not due to lack of time, but experience from the Mediterranean, where loss of two-thirds of the field of fire had been strongly criticized.
25. Stjernfeld, p. 103.
26. Lebram, p. 37.
27. *ibid.*
28. US ARMY, Office of the Chief Engineer, Headquarters, European Theatre of Operations. 'Report on German Concrete Fortifications'. September/October 1944.
29. Quoted from: EIS, Egon. 'The Forts of Folly'. Oswald Wolff Ltd. London 1959. p. 233. This description by Shulman was, in the light of information available today, an extremely accurate one. Diagrams of Harnier's 'Artillerie im Küstenkampf', when linked into one map of Europe, bear an uncanny resemblance to 'knots in a piece of string'.
30. Young, p. 204.
31. Speidel, p. 71.

CHAPTER 9: THE ARMED CAMP

1. BARNETT, Correlli. 'Britain and her Army, 1509–1970'. Allen Lane. The Penguin Press. London 1970. p. 423.
2. The C'tesiphon hut will be described in detail later, but *see also:* BILLIG, Kurt. 'Concrete Shell Roofs with Flexible Moulds'. Paper No. 5506. ICE Journal 1945–46. January. pp. 228–31.
3. FITZMAURICE, R. 'Wartime Building'. RIBA Journal. June 1940. p. 187.
4. AALTO, Alvar. RIBA Journal. March 17th, 1941. p. 78.
5. Compare the building of the West Wall (Chapter 6) and the GHQ line (Chapter 7).
6. 'Transportable Timber Huts'. The Architects' Journal. April 29th, 1943. pp. 286–87.
7. SINGER, Major C. M. 'Notes on Alternative Materials and Methods of Construction for War Hutting'. Royal Engineers Journal. June 1940. p. 180. (Based on an investigation carried out by Directorate of Fortifications and Works during 1939–40.)
8. *See*, for example, the design patented by the architect George Coles—which was approved by the War Office as an alternative design to their standard Army hut. Illustrated in: 'Precast Unit Construction for Hutting'. The Builder. January 17th, 1941. pp. 90–91.
9. 'Hutments in Concrete, A design eliminating steel and shuttering'. The Builder. February 16th, 1940. pp. 221–22.
10. 'Concrete Shutter Plates'. Concrete and Constructional Engineering. March 1940. p. 128.
11. 'Rapid Construction of Hut Camps: an all concrete system'. The Builder. December 27th, 1940. pp. 625–27.

12. 'Rapid Construction of Hut Camps', p. 625.
13. 'War-Time Buildings. The "Stancon" System'. Designed by Stanley H. Hamp, FRIBA. Consulting Engineers: L. G. Mouchel & Partners. The Architect & Building News. 3.5.40. p. 92.
14. *ibid.*
15. BOWLEY, Marian. 'The British Building Industry'. Cambridge University Press. 1966. p. 60.
16. 'The BCF Hut'. The Architects' Journal. April 9th, 1942. pp. 257–58. *See also:* 'BCF Concrete Hostels'. Building. April 1942. p. 88.
17. 'Prefabricated Units'. RIBA Journal. June 1942. p. 129.
18. 'The MOWP Standard Hut'. RIBA Journal. September 1942. p. 194.
19. 'Concrete Hutments'. The Architect & Building News. 22.12.39. pp. 267–68.
20. Bowley, p. 90.
21. Bowley, pp. 88–89.
22. Singer, pp. 183–84.
23. 'War-Time Buildings. The Tarran System'. The Architect & Building News. 10.5.40. p. 120.
24. Tarran Industries of Hull were also involved in the construction of defences on the North East coast.
25. 'Tarran Improved Construction' Building. April 1942. pp. 86–87. *See also:* developments shown in 'A Wartime Hospital at Hull'. The Builder. March 6th, 1942. p. 218.
26. Billig.
27. DAWBARN, Graham R. 'Blisters'. RIBA Journal. April 1941. pp. 108–110.
28. DAWSON, Donald S. 'A Study of the Development of Airport Architecture'. Thesis for final 1948. RIBA Library. London. p. 11.
29. Dawbarn.
30. *ibid.* p. 110.
31. 'A New Constructional Method'. The Architects' Journal. June 12th, 1941. pp. 387–90.
32. 'New Building Techniques: Development During War Time'. The Architect & Building News. June 27th, 1941. p. 184.
33. 'Hy-rib' construction was later used as permanent shuttering for the building of the Mulberry Harbours.
34. 'Cleansing Station'. Architectural Design & Construction. February 1943. pp. 40–41.
35. Billig.
36. ELLIS, Major L. F. 'History of the Second World War: Victory in the West'. Vol. I. HMSO. London 1962. p. 29.
37. THOMPSON, R. W. 'D-Day: Spearhead of Invasion'. MacDonald. London 1968. p. 6.
38. 'Camps'. The Architects,' Journal. November 19th, 1942. p. 328.
39. BOWMAN, Waldo G. 'Military Huts and Structures at American Installations in Britain'. Engineering News Record (NY). October 21st, 1943. p. 98.
40. BUTLER, R. Cotterell. 'War Time Building Practice. 18—Current Technical Considerations: Substitutes for timber and metal for non-structural building components'. The Builder. May 24th, 1940. pp. 611–14.
41. *See:* 'These War Buildings were Significant'. Engineering News Record. October 19th, 1944. pp. 109–18. And: DIETZ, Albert G. H. 'War Time Innovations in Timber Design'. Engineering News Record. October 18th, 1945. pp. 120–3.
42. 'Aircraft Plant has 150 feet Timber Trusses'. Engineering News Record. October 21st, 1943. pp. 114–19.
43. The USA had a tradition and expertise in prefabricated timber hutting which stretched back into the 19th century. See Chapter 13.
44. 'The Harvard Field Hospital'. RIBA Journal. August 1941. p. 172.
45. Bowman.
46. JACKSON, Anthony. 'The Politics of Architecture'. The Architectural Press. London 1970. p. 164.
47. Bowley, p. 200.
48. Ellis, p. 29.

CHAPTER 10: BIRTH OF MULBERRY HARBOUR AND FALL OF ATLANTIC WALL

1. WILMOT, Chester. 'The Struggle for Europe'. Collins. London 1952. p. 171.
2. MAJDALANY, Fred. 'Fall of Fortress Europe'. Hodder and Stoughton. London 1968. pp. 65–76.
3. MORGAN, Sir Fredrick. 'Overture to Overlord'. Hodder and Stoughton London 1950. p. 262.
4. *ibid.* p. 261.
5. CHURCHILL, Winston S. 'The Second World War, Vol. II: Their Finest Hour'. Cassell. London 1949. p. 216. (*See also:* Chapter 7 for development of sea forts.)
6. CHURCHILL, Winston S. 'The Second World War, Vol. I: Closing the Ring'. Cassell. London 1948. p. 66 and p. facing 78.
7. EISENHOWER, Dwight D. 'Crusade in Europe'. William Heinemann. London 1948. p. 258.
8. PAWLE, Gerald. 'The Secret War'. Harrap. London 1956. p. 254.
9. *ibid.* pp. 256–57.
10. *See:* LOCHNER, R., FABER, O., PENNEY, W. G. 'The Bombardon Floating Breakwater'. pub. in The Civil Engineer in War: a symposium of papers on War-Time Engineering Problems. London 1948. Vol. 2. pp. 256–91.

11. Pawle, p. 268. Once the decision on a specific site had been taken in 1943, the design responsibility had been given to the Directorate of Transportation at the War Office headed by D. J. McMuller who again allocated the job to the Director of Ports and Inland Waterways, Brigadier Bruce White. Each different part of the harbour was the responsibility of large numbers of designers, including several civilian engineers and contractors. To pick out any in particular is difficult, except perhaps in the case of Colonel A. S. Rolfe, who seems to have had much of the responsibility for concrete design both for the Phoenix units and the floating roadways, and Captain W. J. Hodge who was in charge of the War Office drawing office.
12. WOOD, C. R. J. 'Phoenix', pub. in The Civil Engineer in War: a symposium of papers on War-Time Engineering Problems. London 1948. Vol. 2. pp. 336–69.
13. *See:* PAVRY, R. 'Mulberry Pierheads', pub. in The Civil Engineer in War: a symposium of papers on War-Time Engineering Problems. London 1948. pp. 369–85, quoted pp. 369–70.
14. *See:*
    (i) BECKET, A. H. 'Some Aspects of the Design of Flexible Bridging, including "Whale" Floating Roadways', (pub. as above). pp. 385–401.
    (ii) WOOD, C. R. J. 'Reinforced-Concrete Pier Pontoons and Intermediate Pierhead Pontoons', (pub. as above). pp. 401–26.
15. Pawle, p. 199 and p. 212.
16. STJERNFELD, Bertil. 'Alarm i Atlantervallen'. Hörsta Förlag AB. Stockholm 1953.
17. Wilmot, p. 269.
18. The special armour used on D-Day is illustrated in: THOMPSON, R. W. 'D-Day: Spearhead of Invasion'. MacDonald. London 1968. pp. 36–40. The main types of special armour were as follows:
    (i) Churchill Mk. III 'Bobbin' which was designed to lay a canvas mat over soft sand for the special tanks which were to attack the defences;
    (ii) Churchill Mk. VIII 'Crocodile' which was a flamethrower tank with a trailer carrying 400 gallons of fuel;
    (iii) Churchill Mk. III AVRE (SBG) which carried either brushwood or the Small Box Girder bridge to neutralize anti-tank walls and ditches;
    (iv) Churchill Mk. III AVRE (Armoured Vehicle Royal Engineers) designed to knock out concrete emplacements at short range with heavy mortar bombs ('flying dustbin');
    (v) Sherman M4—A4 DD (Duplex Drive), which was amphibious and kept afloat by a collapsible canvas screen, was intended to be the first weapon to reach the shore;
    (vi) Sherman M4—A4 'Crab' Flail was intended to clear a path through minefields on the beach.
19. Wilmot, p. 17.
20. The layout of the Bombardons was changed at the last minute, however. Instead of a double line as successfully tested, it was decided only to use a single line and anchorage proved deeper than initially calculated.
21. Pawle, p. 285. *See also:* JELLET, J. H. 'The Lay-out, Assembly, and Behaviour of the Breakwaters at Arromanches Harbour (Mulberry 'B')', pub. in The Civil Engineer in War: a symposium of papers on War-Time Engineering Problems. London 1948. Vol. 2. pp. 291–313.
22. Wilmot, p. 387.
23. SPEER, Albert. 'Inside the Third Reich'. Weidenfeld and Nicolson. London 1970. p. 352–53.

CHAPTER 11: THE SHELTER CONTROVERSY

1. O'BRIEN, Terence. 'History of the Second World War: Civil Defence'. HMSO. London 1955. p. 19.
2. Experiments with gas, which could be carried out with reasonable secrecy were made fairly extensively in Britain. By 1932 provision had been made for the protection of individuals and for the inclusion of anti-gas features in the new Underground stations. The superiority of British gas protection, the personal gas mask, was one of the reasons why gas was not used against civilians in 1939–45. Germany had no comparable protection and only her large bunkers had gass filters. Hitler consequently feared any possibility of reprisal.
3. LE CORBUSIER. 'The Radiant City'. Faber and Faber. London 1967. pp. 60–61. Originally published 'La Ville Radieuse'. Paris 1933.
4. O'Brien, p. 148.
5. The real value of the dispersal policy lay in the evacuation of people from densely populated or otherwise vulnerable areas, not dispersal within these areas. In 1939 over 3.5 M people were evacuated from such areas but the population gradually drifted back. Plans were made for similar evacuation of the Government. *See:* Calder 'The People's War, Britain 1939–45', pp. 3–37.
6. Much of the design by Dr. David Anderson.
7. CLERKE, R. W. G. 'Constructional Work on Air Raid Shelters and Other Protective Works.. Journal of ICE. April 1939. pp. 573–84.
8. Standard trench design and specification eligible for grant issued to Local Authorities November 1938. Revision issued January 9th, 1939, as original specification and design did not agree. Even then trenches were of poor quality, and the prefabricated concrete lining proved vulnerable.
9. The results of the investigation were published in:
    (a) TECTON architects. 'Planned ARP'. Architectural Press. London 1939.
    (b) ARUP, Ove. 'Design, Cost, Construction and relative safety of Trench, Surface, Bomb-proof and other air-raid shelters'. Concrete Publications. London. March 1939.
10. LEE, Donovan H. 'Design and Construction of Air Raid Shelters'. Concrete Publications Ltd. London 1940. p. 6.

11. Arup, p. 1.
12. *ibid.*, p. 20.
13. ARUP, Ove. 'London's Shelter Problem'. London. October 1940. p. 2.
14. No accurate tests had yet taken place, but the committee defined bomb-proof as minimum 18–24 metres underground or 900 mm of concrete with an air expansion chamber below followed by 750 mm inch concrete roof, preferably with steel linings.
15. CALDER, Angus. 'The People's War: Britain 1939–45'. London 1969. pp. 338–39.
16. O'Brien, p. 192.
17. During 1917 the public had sheltered in police stations, public buildings and the London Tube stations.
18. The raids on Britain ultimately called for much greater thickness than those laid out in 5A. Based on a bomb size of 500-lb these figures might have been quite relevant during the Blitz of 1940–41 when only a few bombs were over this size (most being of the 50-kg and 250-kg type) but in 1942 heavier bombs were more frequently used, and in the 'Little Blitz' of 1944 bombs of up to 2500-kg were being dropped.
19. For discussion on figures *see*: Donovan H. Lee, p. 14 and ANDERSON, David. 'The Design of Bomb-proof Shelters'. Journal of ICE. October 1939. pp. 242–44.
20. Strength of steel framed buildings later confirmed by Government Research and Experiment Dept. for use as public shelters.
21. Home Office and other ARP recommendations for industrial and commercial use made obligatory in Civil Defence Bill of 1939 (July).
22. O'Brien, p. 282.
23. ARP Co-ordinating Committee's memorandum to Sir John Anderson quoted from 'The Builder'. December 22nd, 1939. London. p. 852.
24. *ibid.*
25. Information from Arup's 'London's Shelter Problem'. It is worth noting that the compartmental shelters at Finsbury reacted very favourably during the Blitz. Apart from the fact that they offered the facilities which were soon going to be necessary in other shelters, the effect of bomb damage was as had been calculated by the designers. A bomb penetrating one of the compartments of a 426-person shelter caused 12 casualties in that compartment, but none of the occupants of the other compartments were affected.
26. Anderson shelters were put on sale to the public in October 1939, with prices ranging from £6-14-0 to £10-18-0. Only one thousand were sold.
27. Calder, pp. 112–13.
28. O'Brien, p. 507.
29. PAWLEY, Martin. 'The Other Shelter Problem'. Architectural Design. September 1970. London.
30. Approximately 184 000 Anderson shelters were manufactured during the last two months of 1940 and the first three months of 1941. In March 1941 production stopped due to steel shortage. Extra curved sheets distributed to extend smallest shelter to a size long enough to sleep in.
31. Calder, p. 199.
32. Two designs had originally been produced in 1940, one of which had a curved top and proved difficult to produce. The Morrison shelter gave less protection than the Anderson shelter. It gave no lateral protection and was intended for ground floor use. It could carry the debris from two floors. Selling price £7-12-6.
33. Designed for larger families by Research and Experiments Dept. in April 1941 and produced in the autumn, it had received 10 000 orders by March 1942.
34. For further details *see*: JOPLING, G. S. 'Mass Production of Reinforced Concrete Shelters constructed by Direct Labour, using a Mobile Type of Shuttering'. The Journal of the Institute of Municipal and County Engineers. June 23rd, 1942. p. 400.
35. The Builder, November 24th, 1939. p. 736.
36. Arup 'London's Shelter . . .', p. 18.
37. O'Brien, p. 519.
38. Calder, p. 167.
39. Calder, p. 224.
40. SEGAL, Walter. 'Family Shelters'. Building. January 1941. pp. 12–16.
41. Calder, pp. 414–15.
42. Arup 'London's Shelter . . .', p. 15.

CHAPTER 12: GERMAN BOMB-PROOF MANIA

1. SPEER, Albert. 'Inside the Third Reich', Weidenfeld and Nicolson. London 1970. p. 473.
2. US STRATEGIC BOMBING SURVEY. 'Public Air Raid Shelters in Germany'. 1945. p. 1. The figures are based on interrogation of Prof. F. Kristen, who was in charge of laboratories for development of German air raid protection.
3. CALDER, Angus. 'The People's War: Britain 1939–45', originally published by Jonathan Cape. London 1969. This edition by Literary Guild. p. 229.
4. US Strategic Bombing Survey, 'Public . . .', pp. 1–14.
5. *ibid.* pp. 2, 6 and 12.
6. British calculations of 'bomb-proof' were more pessimistic.
7. US Strategic Bombing Survey, 'Public . . .', p. 5.
8. The pamphlets were based mainly on research carried out in the ARP laboratories under Prof. Kristen.

9. The general method of foundation was a type of free-floating reinforced concrete slab, usually about the same thickness as the walls except for a deeper perimeter section. This type of foundation 'floated' and occupants experienced a rocking motion as the bunker settled back into equilibrium after near or direct hits. On some occasions piles had to be used in order to support the foundation slab but in any case a bomb-proof or 'bomb-actuating apron' was specified, extending out from the bunker's external walls for a distance of up to 10 metres. This apron was comparable to the foundations used for the naval casemates of the Atlantic Wall.

10. Bunker design was co-ordinated by specifications and orders as early as July 1941.

11. GLOVER, Charles. 'Civil Defence', London 1938. p. 149.

12. US Strategic Bombing Survey, 'Public . . .', pp. 6 and 14.

13. LANE, Barbara Miller. 'Architecture and Politics in German 1918–45', pub. Harvard University Press: Cambridge, Massachusetts 1968. See photograph of Weather Broadcasting Station, p. 199.

14. Note: In the early 30's the National Socialists found the issue of the 'new style' excellent campaign material and their eventual rise to power in 1933 committed them to a positive architectural policy. The Nazis attached great ideological importance to architecture and this aspect was given extensive use as propaganda material. Two publications in this context were: 'DEUTSCHLAND BAUT. DIE BAUTEN UND BAUVORHABEN DER PARTEI UND DES REICHES, DER ARBEITSFRONT, DER HITLER-JUGEND, DER LUFTWAFFE, DES HERRES UND DER MARINE'. Stuttgart 1939. And: 'NEUE DEUTSCHE BAUKUNST'. Albert Speer. Berlin 1941. Both of these show divergent sytlistic preferences rather than a single Nazi 'style', Hitler's own stylistic preferences being mainly limited to the 'monumental' buildings directly commissioned by him and in which he indulged his own artistic ambitions. The best explanation of the variations in Nazi architecture is in Lane's 'Architecture and Politics in Germany 1918–45'. Miss Lane writes (pp. 185–86) that the:

". . . diversity in Nazi architecture reflected the widely differing views of the party's leaders, who, after Goebbels had prevented the establishment of centralized mechanisms of stylistic control in 1933, assumed initiative, individually, in deciding questions of architectural style. Despite Hitler's many pronouncements on the subject, Feder, Schirach, Ley, Goering, and other officials who became the regime's principal architectural patrons never agreed on a consistent theory of what Nazi architecture should be. Some of these men favoured the modernized neoclassicism of Speer's Party Congress buildings, conceived as a new gathering place for the masses of the party faithful; others, the neo-Romanesque style of the castle-like Ordensburgen, quasi-military schools for a 'heroic' party leadership; still others, the rustic appearance of government housing projects, intended to symbolize the repatriation of urban workers to the soil. A few commissioned buildings as radically modern as any constructed in the twenties; these were advertised as evidence of the revolutionary and 'modern' character of the Nazi regime."

15. US Strategic Bombing Survey, 'Public . . .', pp. 6–7. The translated letter ordering this new type of design is in Exhibit L of the survey.

16. MIRS. 'Handbook of the Organization Todt (OT)'. London 1945. para. 32.

17. Information on the Berlin towers from: Royal Engineers Journal, Vol. LXIII. March–December 1949. p. 257. The largest of the Berlin towers—the Tiergartentower—was a six-storey building of monolithic construction without expansion joints to weaken it and was carried on a 'dish-shaped' reinforced concrete raft coated with a thin layer of bitumen.

18. The heavy monumental style which Hitler planned for Berlin is best illustrated by the triumphal arch Hitler designed himself. See chapter 13 illustration 15 and Speer, pp. 118–19.

19. Lane, p. 186.

20. 'L'Architecture d'Aujourd'hui'. August–September 1970. No. 151. pp. 47–50. Article: 'FLAKTÜRME' (Tours de DCA, Vienne).

21. IRVING, David. 'The Destruction of Dresden'. William Kimber. London 1963. p. 32. Also Calder, pp. 229–30.

22. Irving, 'The Destruction of . . .', p. 32 and: FRANKLAND, Noble. 'The Bombing Offensive against Germany, Outline and Perspective'. Faber and Faber. London 1965. pp. 59–61.

23. US Strategic Bombing Survey, 'Public . . .', p. 2.

24. Irving, 'The Destruction of . . .', p. 45.

25. The Dresden raid, the precautions taken in the city, and the results are described fully in Irving, 'The Destruction of . . .'.

26. HEZLET, Vice Admiral Sir Arthur. 'The Submarine and Sea Power', pub. Peter Davies. London 1967. p. 107.

27. See Chapter 3 and: MURRELL, H. F., Architect. 'German War Construction: Submarine Shelters and Zeppelin Sheds', pub. in 'Journal of the Royal Institute of British Architects'. October 23rd, 1920. pp. 485–90.

28. Figures from: WHITEHOUSE, A. 'Subs and Submariners'. 1963.

29. US STRATEGIC BOMBING SURVEY. 'Submarine Pens—Deutsche Werft, Hamburg, Germany'.

30. MASON, John. 'Report on Structural Engineering in Germany', published in 'The Structural Engineer—Journal of the Institute of Structural Engineers'. June 1946. pp. 297–334. An English translation of Freyssinet's paper 'A Revolution in the Technique of the Utilization of Concrete' was published in the 'Structural Engineer', May 1936, and about this time Freyssinet introduced his system into Germany where there was much wider interest in and use of the system. Wayss and Freytag were appointed licensees of the system in Germany by Freyssinet, although other firms were very active in the field and there was considerable competition.

31. Hezlet, p. 167.
32. LEPOTIER, A. 'L'Atlantikwall de Brest', pub. in 'Revue Maritime'. No. 105. 1955. p. 61.
33. *ibid.*
34. Information on the Brest pen based on:
    (i) US STRATEGIC BOMBING SURVEY. 'Brest Submarine Pens'.
    (ii) COMBINED INTELLIGENCE OBJECTIVES SUB-COMMITTEE. 'German Submarine Pens in France'. G-2 Division, SHAEF (Rear) APO 413. Report by Lieut. G. R. Wernisch, June 1945.
    (iii) Personal visit by the authors Spring 1970 (pens still used by French navy).
35. CIOS, 'German Submarine Pens . . .'.
36. *ibid.*
37. Hezlet, p. 187.
38. US STRATEGIC BOMBING SURVEY. 'Submarine Assembly Plant at Farge, Germany'.
39. As an example, a Central Control Station built at Deptford in 1939, following this report (Hailey Report), had a two metre reinforced concrete shell and was designed to resist 500-lb bombs. Illustrated in 'The Builder'. December 15th, 1939. p. 823.
40. PAWLEY, Martin. 'The Other Shelter Problem'. Architectural Design. September 1970. p. 465.
41. Information from Calder, p. 161. A small underground factory built for a London firm asked to move out of London is described in detail by its architect in: LOBB, H. Vicars. 'The Construction of an Underground Factory', in the Journal of the Royal Institute of British Architects. November 1945. pp. 3–7.
42. The main British effort in industrial protection was put into making existing factories safer and the provision of shelter for factory workers. New factory construction was designed to reduce the effects of high explosive and incendiaries rather than to eliminate them altogether. A number of 'Wartime Building Bulletins' were issued by the Research and Experiments Department of the Ministry of Home Security on this subject. An example is Bulletin c.12 which is described in: BAKER, Prof. J. F. 'Single Storey Wartime Factory Design'. Journal of ICE. March 1941. pp. 73–79. 'Shelter factories, with splinter-proof protection, baffle walls and 'safety-valve' construction (blow-out walls) were constructed or adapted allowing workers to remain at their workplace as long as possible. *See:* BUTLER, R. Cotterell. 'War-Time Building Practice' The Builder Sept. 27th, 1940. pp. 304–306.
43. Speer, p. 336.
44. Dorsch was one of Todt's key men during the building of the West Wall (letter from General Warlimont to the authors). He was appointed at OTZ by Speer in 1942 and had probably been aiming at further independence for a long time.
45. Speer, p. 336.
46. *ibid.* p. 553.
47. MIRS. OT Handbook. para. 21. Speer kept Dorsch in check by appointing him as his representative in the OT under the title Generalbevollmächtigter Bau (Plenipotentiary General for Construction).
48. COMBINED INTELLIGENCE OBJECTIVE SUB-COMMITTEE. 'Underground Factories in Central Germany', G-2 Division, SHAEF (REAR) APO 413. CIOS Team 13. 1945. p. 7.
49. COMBINED INTELLIGENCE OBJECTIVES SUB-COMMITTEE. 'Messerschmitt Bombproof Assembly Plant, Landsberg'. G-2 Division, SHAEF (REAR) APO 413. Reported by Lieut. R. E. Erikson June 1945.
50. IRVING, David. 'The Mare's Nest', pub. William Kimber. London 1964. p. 142.
51. CIOS, 'Underground Factories . . .', p. 12.
52. CIOS, 'Underground Factories . . .'.
53. Irving, 'The Mare's . . .', p. 238.
54. *ibid.* p. 143.
55. Dimensions based on CIOS 'Underground Factories . . .'.
56. Irving, 'The Mare's . . .', p. 28.
57. *ibid.* p. 142.
58. *ibid.* p. 247.
59. *ibid.* p. 304.
60. *ibid.* pp. 213–14.
61. SHIRER, William. 'The Rise and Fall of the Third Reich'. 1959. This edition by Book Club Associates 1970, pp. 1105 6. Foreword to MIRS, 'OT Handbook' of 5/1945 makes the same point. 'To-day OT is indispensable in any protracted resistance the Nazis may intend to offer. Their experience in making the most of terrain in the building of field fortifications, in the building of underground tunnels, depots of all kinds, hide-outs, shelters, in fact, of regular subterranean living and operating quarters of vast proportions is unique'.
62. Irving, 'The Mare's . . .', pp. 301–5 and CIOS, 'Underground Factories in . . .'. The latter gives diagram of complete 'last stand' complex and detail of 'Kuckuck I'.

CHAPTER 13: CONTEXT

1. In the Pacific theatre during the inter-war period the enormous concrete fort known as Fort Drum was constructed in Manila Bay by the Americans. With thicknesses of between 5 and 9 metres this artificial island was literally a huge concrete battleship in appearance. Large tunnelled underground installations such as those on Corregidor in Malinta Bay, also in the

Philippines, were similarly constructed by the Americans. This work was continued during the war by the Japanese and included large underground installations such as those at Guam, dug-outs, and numerous field defences. For further information on building activity in the Pacific theatre *see:* DOD, Karl C. 'The Corps of Engineers : The War Against Japan'. US Army in World War II, The Technical Services.

2. In America, unlike Britain, timber was comparitively plentiful and her military hutting programme reflected this availability. The US War Production Board asked the Institute of Paper Chemistry to develop an inexpensive, portable paper house that could be mass produced' and prompted similar developments in that field. *See:* 'Paper House Review'. Architectural Design. October 1970. p. 499. Early in the war programme, while steel was not critical, several huge blimp hangars were constructed for the Navy and Buckminster Fuller's steel housing unit, the 'Dymaxion Deployment Unit', was produced in limited numbers for the use of American and Russian mechanics assembling US fighters in the Persian Gulf. *See:* McHALE, John. 'R. Buckminster Fuller'. Prentice-Hall International Inc. And : MARKS, Robert W. 'The Dymaxion World of Buckminster Fuller'. Reinhold. Some energy was directed into concrete prefabrication, as in Britain, but the major advances were in the field of reinforced concrete barrel arches. For a detailed description of the innovative nature of US war construction at this time, and some suggestion of its influence, *see:*
   (i) DIETZ, Albert G. H. 'Wartime Innovations in Timber Design'. Engineering News Record. October 18th, 1945. pp. 120–23.
   (ii) 'Aircraft Plant has 150 feet Timber Trusses'. Engineering News Record. October 21st, 1943. pp. 114–19.
   (iii) 'These War Buildings were Significant'. Engineering News Record. October 19th, 1944. pp. 109–18.
   (iv) MICHAELS, Leonard. 'Contemporary Structure in Architecture'. Reinhold. New York. 1950.

3. For the full story of Zeppelin's struggle to convince the armed forces that his airship was of military significance *see:* ROBINSON, Douglas H. 'The Zeppelin in Combat', pub. G. T. Foulis. London 1962.

4. LIDDELL HART. 'Europe in Arms', pub. Faber and Faber. London 1937. pp. 273–74.

5. 'Cardington Killingry'. Architectural Design. April 1970. p. 173.

6. SCHWARZ, Walter. 'Israelis confident Suez missiles could be by-passed'. The Guardian. November 5th, 1970. p. 2.

7. *ibid.*

8. EIS, Egon. 'Forts of Folly', pub. Oswald Wolff. London 1959. p. 246.

9. FOLEY, Charles. 'Carnage by People Sniffers', article in The Observer. June 20th, 1971.

10. *ibid.*

11. SPEER, Albert. 'Inside the Third Reich', pub. Weidenfeld and Nicolson. London 1970. p. 159.

12. BLAKE, Peter. 'German Architecture and American'. Architectural Forum. New York, August 1957. p. 135.

13. SMITH, G. E. Kidder. 'The New Architecture of Europe', pub. Pelican 1961. p. 113.

14. SANTIAGO, Michel. 'Bellicose Prophecies'. The Architectural Review. July 1963. pp. 59–61.

15. 'Culture Bunkers'. Architectural Design. July 1967.

16. CHALK, Warren. 'South Bank Arts Centre, London'. Architectural Design. March 1967. p. 122.

17. The 380 mm concrete was for purposes of sound insulation :
   (i) author's conversation with Norman Engleback of the GLC Architects Department, Group Leader for the project, Spring 1969.
   (ii) BANHAM, Reyner. 'The Architecture of the Well-tempered Environment'. The Architectural Press. London 1969. p. 257.

18. Chalk, 'South Bank . . . .

19. BANHAM, Reyner. 'The New Brutalism'. The Architectural Press. London 1966. p. 16.

20. 'Bunker Archaeology'. Architectural Design. May 1967. Photographs form 'Architecture Cryptique' are used in the Atlantic Wall chapter : Illustrations 38 and 39.

21. 'Architecture Principle'. Architectural Design. August 1966. p. 374.

22. 'C. F. A. Voysey'. Architectural Design. January 1971. p. 7.

23. This conference, sponsored by Imperial Chemical Industries, was held at the RIBA on November 11th and 12th, 1970.

24. BOWLEY, Marian. 'The British Building Industry : Four Studies in Response and Resistance to Change'. Cambridge University Press 1966. p. 397.

25. *ibid.* p. 83.

26. *ibid.* p. 11.

27. JACKSON, Anthony. 'The Politics of Architecture'. The Architectural Press. London 1970. p. 164.

28. PAWLEY, Martin. 'The Other Shelter Problem'. Architectural Design. September 1970. p. 465.

29. Jackson, p. 164.

30. *ibid.* p. 169.

31. After the war confidence grew in precast panel construction despite the failure of the 'prefab' programme. By the mid-fifties the public sector was constructing ten storey blocks using precast panels. The growing use of this form of construction had a severe set back with the Ronan point collapse—but ironically the potential weakness of precast panel construction had

already been forecast in the military sector five years earlier. In 1963 the Officers Mess at Aldershot, designed on the G.80 system, had collapsed in the course of construction.

32. SNYDER. Louise L. 'The War, A concise History, 1939–45', pub. Robert Hale. London 1962. p. 11.
33. FULLER, R. Buckminster. 'Utopia or Oblivion'. Allen Lane. The Penguin Press. London 1970. p. 169 and pp. 273–74.
34. *ibid*. pp. 404–405.
35. CROOME, Angela. 'America cuts back on Science'. Daily Telegraph. August 15th, 1970. p. 8.
36. Fuller, p. 291.
37. CROSBY, Theo. Review of 'Utopia or Oblivion', in The Sunday Times 1970.

CREDITS

The authors wish to thank the individuals and organizations who have allowed reproduction of illustrations in their copyright. Photographs and drawings not listed are by the authors:

CHAPTER 1
3, 4, 5, 6, 7, 9 from 'Permanent Fortification for the Imperial Military Training Establishments and for the Instruction of Officers of all Arms of the Austro-Hungarian Army' by Major Moritz Ritter von Brunner, pub. HMSO London 1910; 8, 10 from 'Fortification: Its past achievements, recent developments and future progress' by Major G. S. Clarke, pub. John Murray, London 1890; 13 Musée royal de l'Armée et d'Histoire militaire, Bruxelles, cliché Ed. R. Laffont, Paris; 15 Bibliothek für Zeitgeschichte, Stuttgart; 16 from 'La Fortification Permanente pendent la Guerre' by Général Benoit, pub. Berger-Levrault, Nancy-Paris-Strasbourg 1922; 18 from 'History of the First World War' Vol. 3 No. 14, pub. Purnell, BPC, London 1969; 19 Bayer Hauptstaatsarchiv Abt IV Kriegsarchiv, München; 20 BBC Publications, London.

CHAPTER 2
1 from 'Bayerische Pioniere im Weltkrieg' by Karl Lehman, München 1918; 2 Bapty, London; 4 from 'The American Engineers in France' by Col. W. B. Parsons, New York 1920; 5 Süddeutscher Verlag, München; 6, 11, 21 Imperial War Museum, London; 8, 10, 17, 18, 19 Ordnance Survey, Southampton; 9 by permission of Institution of Royal Engineers, UK; 13, 14, 16 from '1914–18 Der Stellungskrieg' by F. Seetzelberg, Mittler & Sohn, Berlin 1926.

CHAPTER 3
1, 4 Imperial War Museum, London; 2, 5, 8 by permission of Institution of Royal Engineers, UK; 3, 7 by permission of the Controller of Her Britannic Majesty's Stationery Office; 6 the Royal British Legion; 10 November Books, London; 11 from Country Life 1961, photograph by Andrew Saunders; 12, 13, 14, 15, 16 from 'Thames and Medway Garrison'; 17, 18 RIBA Journal, London.

CHAPTER 4
1 from 'Bayerische Pioniere im Weltkrieg' by Karl Lehman, München 1918; 2, 3, 4, 5, 6, 7, 8, 9, 12 by permission of Institution of Royal Engineers, UK; 11 Architectural Press; 14 from 'Luftschiffhallenbau' by A. Haenig, pub. C. J. E. Volckmann Nachfolger, Rostock 1910; 13, 15, 16 Douglas H. Robinson MD, New Jersey USA; 18, 19 from 'Von Richthofen and the Flying Circus' by Nowarra and Brown, pub. Harleyford, UK 1964; 20, 21 by kind permission of the Council of the Institution of Structural Engineers, London; 22 Engineering News-Record, New York.

CHAPTER 5
1 photograph by Roger Cowles; 2 from Daily Express-Photonews, London 1936; 4, 5, 7, 8, 10, 11, 13, 14, 16, 24 photographs by Terence Fowler; 9, 20, 22, 25 from 'Männergegen Stein und Stahl' by Joachim Backhausen, pub. Schützen-Verlag Berlin; 15 LEA, The Illustrated London News and Sketch Ltd.; 17 Franklin Watts Ltd., London; 18 Imperial War Museum, London; 19, 21 Vivian Rowe; 23 Ministère d'État chargé de la Défense Nationale, Paris.

CHAPTER 6
1 from 'Neue Deutsche Baukunst' by Albert Speer, pub. Volk und Reich Verlag, Berlin 1941; 3, 4, 5, 9, 10, 11, 14 Department of Defense, USA; 6, 7, 12, 13 by permission of Institution of Royal Engineers, UK; 15, 16, 17 J. F. Lehmanns Verlag, München; 18 Imperial War Museum, London.

CHAPTER 7
1, 2, 7, 8, 9, 10; 11, 12, 17, 18, 20, 23 Imperial War Museum, London; 6 photograph by Terence Fowler; 15, 16, 25 from 'The Civil Engineer in War: a symposium of papers on War-Time Engineering Problems' Vol. 3, pub. London 1948, paper by J. A. Posford on 'The Construction of Britain's Sea Forts', courtesy the Council of the Institution of Civil Engineers; 21, 26 Keystone Press Agency, London; 28 photograph by 'Skyfotos Ltd.' of Ashford Airport, Hythe, Kent, UK.

CHAPTER 8
1 Keystone Press Agency, London; 2, 3, 4, 7, 8 by permission of Institution of Royal Engineers, UK; 5 Staatsbibliothek Preussischer Kulturbesitz Bildarchiv, Berlin; 13 Société Jersiaise, Jersey; 16 based on material from Priaulx Library, Guernsey; 17 Carel Toms, Guernsey; 29 photograph by Robert Pickering; 35 from 'Alarm i Atlantervallen' by Bertil Stjernfeld, pub. Hörsta Förlag, Sweden 1953; 36 Imperial War Museum, London; 38, 39 from 'Architecture Cryptique' by Paul Virilio.

CHAPTER 9
1, 8, 22, 25 Engineering News-Record, New York; 2, 14, 15, 16 Architectural Press; 3, 12 Building, London; 4, 5, 10, 11, 17 The Architect, London; 9, 21 RIBA Journal, London; 19, 20 Architectural Design, London; 23 Barchild Construction Ltd., UK.

CHAPTER 10
2, 4, 5, 6, 7, 8, 9, 10, 11, 12, 13, 14 from 'The Civil Engineer in War: a symposium of papers on War-Time Engineering Problems' Vol. 3, pub. London 1948 (details of authors and papers given in full in the reference notes to chapter 10), courtesy

the Council of the Institution of Civil Engineers; 15, 17 Imperial War Museum, London.

CHAPTER 11
1, 14, 21 from 'Civil Defence' by Charles Glover, pub. Chapman & Hall 1938; 2, 13 Architectural Press; 5, 6, 7 Cement and Concrete Association, London; 9, 12 Fox Photos., London; 10, 20 Building, London; 15 Heal and Son Ltd., London; 16, 17 from 'Mass Production of Reinforced Concrete Shelters constructed by Direct Labour, using a Mobile Type of Shuttering' by G. S. Jopling, pub. in Vol. LVXIII No. 13, June 23rd 1942, of The Journal of Municipal and County Engineers, these illustrations are reproduced by permission of the Council of the Institution of Municipal Engineers, London; 19 Imperial War Museum, London.

CHAPTER 12
1 by permission of Institution of Royal Engineers, UK; 2, 3, 19, 20, 24, 25, 26, 28, 32 Department of Defense, USA; 6 J. F. Lehmanns Verlag, München; 7 Imperial War Museum, London; 8 photograph by Cecil Newman, Belfast; 10, 11, 12 L'Architecture d'Aujourd'hui, Boulogne; 21, 22, 23 Tore Brantenberg/ Paal Didrik Holm, Norway; 29 Daisy Hill Real Estates Ltd., Jersey; 30, 31 from 'The Mare's Nest' by David Irving, published William Kimber, London 1964.

CHAPTER 13
1 photograph by John Donat, London; 2, 8, 9 R. Buckminster Fuller, Illinois, USA; 3, 10 Intercontinental Literary Agency, London–New York; 4 photograph by US Navy; 5 Engineering News-Record, New York; 6, 7 photograph by Vasari, Rome; 11, 12 National Aeronautics and Space Administration, USA; 13 US Government; 14 from 'Inside the Third Reich' by Albert Speer, pub. The Macmillan Company, New York; 15 The Library of Congress, Washington; 17 LEA, The Illustrated London News and Sketch Ltd.; 20 photograph by Brecht-Einzig, London; 21 Claude Parent photograph by Gilles Ehrmann, Paris; 22 from 'The Civil Engineer in War: a symposium of papers on War-Time Engineering Problems' Vol. 3, pub. London 1948, courtesy the Council of the Institution of Civil Engineers; 23 Architectural Press; 25 Andrew Reid/Martin Pawley, London; 26 The Aluminium Federation, UK; 27, 28 RIBA Journal, London; 31 by permission of the Department of the Environment, London; 32 GLC Photo Unit, London; 33 Imperial War Museum, London.

INDEX

Figures in italic refer to illustrations

A

acoustic mirrors 129
air bases: Allied, *84*; German airship, *80*; German, mobile, of J.G.2, *84*
AIROH prefabricated aluminium houses 283
air raid shelters, *see* shelters
air-space theory 59, *60*
Aluminium housing system 283
American camps in Britain 195–9
Anderson, Sir John 217, 225, 229
Anderson shelter 215, 217, *218*, 225, 227, 229, 237
anti-tank: defences on British coast, *126*; wall, *212*
Antwerp fortress 13, *14*, *22*, 23, 27
Arcon housing system *282*
Armstrong Hut 73, 77
army forts 143, *144*, *145*, *146*; construction of, 147–8
artillery block 90, *274*; in Czech fortifications, *118*
Arup, Ove 219, 221, 225, 233, 235, 237
Asbestos Arch hut 197
Atlantic Wall 149–78; map of, *158*, *210*
Autobahn: bridges, *108*, 111; building of, 111
avant-poste *98*, 99
Aylwin Hut 73, *75*

B

Bar-Lev line 269
'barrier' fort 13–15, 17
batteries: British anti-aircraft, *128*; coastal, in Britain, *66*, *67*, 131; 'Emergency batteries', 131; long-range, 131
battle headquarters: German, 51
BCF hut *186*, 189, 197, *280*
beach defences *210*, 211
Beetles bridge pontoon *200*, 207, *208*
Blister hangars *190*, 191, 199
blockships *202*, 211

blockwork *176*, *248*
Bofors tower: of army fort *144*, *145*
Bombardons *202*, *204*, 205, *207*, 211
bomb deflecting shelter *56*, 241, *242*
Bonatz, Paul 108
breakwaters: bubble, 205; pneumatic, *see* Lilo
Brialmont, General Henri 17, 21, 23, 25, 26, 27
Bruges pens 69, *70*
Brunner, Major Moritz Ritter von 15; his training manual, *14*, 15, *16*
bunkers: ant-hill, 241; bomb-deflecting, 241, *242*; camouflaged, 241, *243*; effect of fire storms on, 249; German air raid, *240*, 241, *242*, 243
burster: principle, 59; slab, *248*, 251, *253*

C

camouflage: Allied fieldworks built within existing buildings, 59, *60*; German fieldworks built within existing buildings, *48*, 49, 59; German bunkers, 241, 243; Germany in World War II, 134; pillboxes in Britain, *132–3*, 134
Caquot Kite Balloon *286*
casemates: armoured, *18*; Atlantic Wall, 167, *170*, 171, *172*, *174*, 175, *177*, *212*, *277*; beach-flanking, *172*; British coastal artillery, *128*, 131; coupled, 99; de Bourges, *30*; in Britain, *66*, 69; Maginot Line, *96*, *97*, *98*, *99*, *100*, *274*; of Honfleur Naval Battery, *166*; of Lindemann Battery, *150*, *154–5*; of Longues Naval Battery, *164*, *165*; West Wall, 119
cellular shelter *224*, 225
Channel Islands: under Nazi control, 157, 158, *159*, *160*, *161*, *260*, 279
Churchill, Winston 63, 127, 137, 201, 203, 285
coastal defence: British, 63, *64*, 65, *128*, 129, *130*, 131, *136*; faulty siting of, in Britain, 134; German, *64*
command post *47*, *168*
compartmental shelter *224*, 227

Committee for Scientific Survey of Air Defence 129
concrete: air raid shelters, 219, *222*, 223, *224*, 225, 227, 228, 229, 233; Allied use of, 59; army forts, 147; Atlantic Wall casemates, 167, 213, *277*; bomb deflecting shelter, *56*; bomb-proof assembly plants, *256*, 257; early use of in Belgian fortresses, 21, 23, 25; early use of in French fortresses, 29, 33; flak towers, 244; German bunkers, *240*, 241, *243*; German dressing station, *42*; hangars, *271*; hutting, 181, 183, *185*, *186*, 187, 189, *192*, 195, 197; in West Wall, 121; machine-gun emplacement, *56*; Maginot Line casemates, *96*, *97*, *98*; 'Moir' pill-box, *61*; naval forts, 143; post-war replacement of, 275; post-war use of, *278*, 279, *284*, 285; quick-setting, 131; reinforced, early use of in German fieldworks, *45*, *46*, *48*, 49; reinforced, importance of on Western Front, 51; replacement of masonry with, 13, 15; submarine pens, 69, 253, *254*, *255*; trenches, dugouts and emplacements, *34*, *41*, *44*, 45; vessels, 137
'concretitis' 103
control tower *176*
corrugated iron in huts 77, 78, 81
C'tesiphon hut 183, 197, *198*
cupolas *28*; artillery, *20*, *30*, *101*; fixed, *101*, 117; hydraulic vanishing, 21; machine-gun, *20*, *30*, *101*; mobile armoured, 17, *19*; observation, *30*; of iron, 15, 21; of steel, 15, 21; retracting, 17, *19*, 29, *101*; rotating, *18*, *19*, 23
Czech fortifications *118*, 119

D

'Dachs IV' underground factory *256*, 259
deep shelters: *see* shelters; deep shelter mentality, 223, 233, 239
detached fort *16*, 17, 21, 23

305

DEW (Distant Early Warning) 269, 275
'Dora I' submarine pen 251, 252
Dorsch, Xaver 257, 261
Douaumont, Fort 29, 30, 31, 32, 93
dragon's teeth 112, 117, 122
dugouts 54; German, 39, 40, 43, 53; German concrete, 34, 41, 43, 49
Dymaxion Deployment Units 270

E

early warning station 268
'Electronic Battlefield' 275
'elephant iron' 59
emplacements: Allied, 56, 67; French, 69; German, 45, 46, 49, 51; open artillery, 170; range-finder, 162–3, 165, 166; Tobruk, 168, 169, 170, 172, 173, 212
'en barbette' artillery emplacement 69, 131, 153

F

Falkenhayn 27, 29, 35, 37
Farge Assembly Plant 254
fire control tower 161, 279
fire storms 249
flak towers 243, 244, 245, 246, 247, 248, 255, 275; Harburg, 238, 247; Tiergarten, 245
Forest Hut 77
fort, early circular sea 66; see also 'Sea Forts' project
Fuller, Buckminster 269, 271, 272, 287
Fylingdales early warning station 268, 275

G

geodesic domes 269, 272
GHQ Line 135, 136
'girdle' fort 13, 14, 17, 22, 23
Glover precast concrete 228
Gooseberry harbours 203, 211
Grosser Kurfürst Battery 151, 156

H

Hailey Committee 223; Hailey Report 255
hangars: Blister, 190; concrete, 271; demountable, 85, 86; dirigible, 80, 83, 85, 191; dirigible, circular, 80, 83; dirigible,
double revolving, 80; dirigible, revolving, 83; dirigible, swimming, 83; Hünnebeck, 86; lightweight transportable, 86; mobile, 84, 85; steel, to house blimps, 270
'Hindenburg Line' 35, 45
Hitler, Adolf 119, 153, 167, 239, 257, 259, 261, 274; builds West Wall 113
Honfleur Naval Battery 166
Hughes-Hallett, Commodore John 201
huts, prefabricated 74, 75, 76, 77, 78, 79, 81, 185, 186, 187, 189, 190, 191, 193, 196
hutting: British, 78, 79, 81, 181, 184; circular concrete, 192, 195; concrete, 181, 183, 185, 186, 187, 189, 195, 197; German, 72; steel, 181, 183, 191, 195, 197; timber, 182, 183, 191, 195, 196, 197
'Hy-rib' construction 195, 207

I

Iris huts 181, 197
Ironside, General 127, 134–5

J

Joffre, Marshal 27, 35, 93

K

Kammler, Hans 259, 261, 265
Kennedy Space Centre 273

L

Landsberg plant 256, 257
Le Corbusier 215, 217, 279, 281
Leman, General Gérard 25, 27
Liddell Portable Hut 77
Liège fortress 13, 22, 23, 25, 26, 27, 28, 35
Lignocrete 191
Lilo 204, 205; Hard, see Bombardons
Lindemann Battery 150, 151, 152, 154–5, 178
Lochner, Lieutenant-Commander 205
Longues Naval Battery 164–5, 209
Loos, Adolf 279
Lossberg, Colonel Fritz von 35, 43, 45, 53
Ludendorff, Major-General Erich von 23–5, 35, 37, 43, 53

M

Maginot, André 91
Maginot Line 90–107, casemates, 96, 97, 98; evolution of its design, 93–5; map of, 94; popular conception of, 92
Maunsell, G. A. 137, 143
'Mebu' fieldworks 45
Meuse fortifications 21–3
Mirus Battery 157, 161
'Moir' pill-box 59, 61
Moltke 23, 27, 37
Mopin system 185, 189
Morrison shelter 215, 228, 229, 237
Moselstellung 13, 21, 23
Mougin, M., 15, 17, 20
MOWP hut 188, 189, 197
Mulberry Harbour: Mulberry A, 201, 203, 204, 211; Mulberry B, 200, 201, 202, 203, 211; start of project, 201, 203
Mutzig fortress 21
'Mystery Forts' 137

N

Namur fortress 13, 22, 23, 25, 26, 27
naval forts 138–42, 143; structure of, 138–9, 143
Nervi, Pier Luigi 271
Nevers, parish centre of Sainte Bernadette 278, 281
Nissen Hut 73, 75, 77, 81, 183, 191, 197, 217; Bow, 77, 78, 81; Hospital, 77, 79, 81, 196

O

observation posts 47, 60, 153, 157; naval, 162–3, 165, 166, 176, 275; on Channel Islands, 157, 159, 160, 276
ouvrages 99, 100, 102, 103, 105, 106; conditions in, 103, 104; coupled, 99; electrical supply, 105

P

Padmos hut 194, 195
parabolic arch 191
Patrick Portable Building 193
Pétain, Marshal 29, 33, 63, 93, 215

Phoenix units *202*, 203, 205, *206*, 207, 211
pierheads, *see* Spud pierhead
pill-boxes *52*; construction of, in Britain, 131;
    in Britain, 65, *68*, *130*, *132–3*, *136*, *280*;
    'Moir', *see* 'Moir' pill-box; type machine-gun,
    50, 51
Plycrete system *185*, 189
pontoon, bridge *see* Beetle; steel, *208*
Portal House 285
prefabrication: in British hutting programme,
    59, 73, *74*, 75, *76*, *77*, *78*, *79*, 81, *185*, *186*,
    187, 189, *190*, 181, *196*; in construction of
    British army and naval forts, 147; post-war
    concrete flats, *284*; shelter construction, 177;
    Temporary Prefabricated Housing programme,
    282, 285

Q

Queen Elizabeth Hall *278*, 279

R

radar: stations *268*, 269, *272*, 275; towers,
    245, *246*, 247
Radiant City 215, 217
radomes 269, *272*, 275
Railway Battery 151, *156*
ramps to sunk shelters 219
redoubt, earthwork, *68*
'ring' fortress, *see* 'girdle' fortress
Rivières, General Séré de 13, 17, 22, 23
roadblocks, in Britain 135, 137
Roehampton Estate *284*
Rommel, Field-Marshal 175, 177–8; 'Rommel
    Bar', 151, 177
Romney hut *180*, 181, 197
Rundstedt, Field-Marshal von 171; his second
    line of defence in Pas de Calais area, *174*, 175
'Ryes' O.P. Plates 59, *62*

S

Schlieffen, General Graf von 13, 22, 23;
    Schlieffen Plan, 23, 35

Schumann, Lieutenant-Colonel 15, 19
'Sea Forts' project *136*, 137–48, 269, *280*
searchlight platform *170*
'secret weapon': factories, 259, 261, *264*, 265;
    launching of, *262*, 263, 265
Seeckt, General von 111
Sereth Line 17, 19
shell-bursting slabs 59
shelters: bed, *228*; brick and concrete, *222*;
    circular, 219, *221*; deep shelter controversy,
    219, 223, 225, 235; designs, *220–1*, *224*,
    228, *236*, 241; experimentation, 223; German
    policy, 239; lack of amenities in, 229, 233;
    mobile shuttering construction, *230*; personnel,
    168, 177; sanitation in, 235; semi-detached
    family, *234*; steel, *216*, *217*, *and see* Anderson
    shelter *and* Morrison shelter; surface, *226*,
    227; trench, 227, 229
shuttering: concrete blockwork, 178; corrugated
    steel, *60*; timber, 178
Siegfried Stellung, *see* 'Hindenburg Line'
sleeping accommodation, *74*, *75*, *76*, *77*, 81, *270*
Speer, Albert 151, 171, 173, 213, 239, 243,
    255, 257, 259, 261, 274, 275
Spit Bank Fort *66*
Spud pierhead *202*, *206*, 207, 211
Stancon system *185*, 187, 189
Steel-Bartholomew Committee 127, 129
steel, used in hutting 181, 183, 191, 195, 197,
    270
submarine pens 55, 69, *70*, 249, 253, 255;
    Bordeaux, *251*, 253; Brest, *248*, *250*, 253,
    255, *274*; Trondheim, *251*, *252*, 253
subterranean fort 15, *20*, *30*
Swiss Roll *202*, 207

T

Tarran hut *190*, 191
Tarrant Light Portable Sleeping Hut *74*, 77
Texas Towers 269, *272*
timber, used in hutting *182*, 183, 191, 195, *196*,
    197

Todt, Fritz 109, 111, 113, 119, 121, 151, 153,
    157, 173
Todt Battery 151, *156*, 178
trenches: Allied, on Western Front, *36*, 37, 39,
    57, *58*; concrete, *44*, 45; 'Front', 37, 39;
    German, on Western Front, *36*, 37, *38*, 39,
    43, *44*; in Britain, 65; 'Reserve', 37, 39;
    'Support', 37, 39; 'Wege', 37

U

Underground, used as public shelter 223, 229,
    232, 233, 235
underground: factories 255, *256*, *257*, *258*,
    259, 261, *264*, 265; hospital, *260*
Uni-Seco Housing System 283

V

V-weapons, *see* 'secret weapons'
Vauban fortress *12*
Vaux, Fort *30*
Verdun defences 13, 17, 22, 27, 29, *30*, *31*,
    33, 35, 36, 93, 95

W

Wall Shelters 233
Waller, Major J. H. de W. 191, 192, 193, 195,
    197
Watten bunker 261
Weblee Hut 77; erection of, *76*
Wehrmachtsgefolge 151, 153
Weingut II *256*, 257
Western Front: map of fortresses in 1914, *22*
West Wall: broken by U.S. Army, *122*, 123;
    compared with Maginot Line, 115; defences,
    114, *116*, 117, 119, *122*; map of, *110*;
    propaganda pictures, *120*
Whale bridges *202*, 207

Z

Zeppelinfeld *274*
Zeppelin sheds, *see* hangars, dirigible

ABOUT THIS BOOK

During the first half of this century astronomical sums of money were spent on war and the preparation for war, including military construction. Yet the architecture of war has been largely ignored by historians even though it easily matches in importance the resources dedicated to it. As this deeply researched and marvelously entertaining book shows, military architecture in its various manifestations both reflected and influenced the course of warfare to a surprising degree. Thus, in World War I fixed fortifications were first discredited and then, with growing sophistication in weaponry and military theory, rehabilitated to the extent that they dominated defense budgeting between the wars. The notion of the impregnable fortress, a twentieth-century myth which has its origins in the Battle of Verdun, found its ultimate expression in the Maginot Line—a masterpiece of misapplied ingenuity. Nor did it lose its appeal for the Germans, who proceeded, with their own Atlantic Wall, to fall into the illusion that they themselves had dispelled in 1940.

Apart from orthodox fortifications, World War II gave rise to a host of novel and experimental structures, which are described—concrete bunkers for civilians, underground arms factories, temporary harbors, flak towers, and deep pens for submarines. All of them demonstrate the essential quality that differentiates military from civilian architecture: adaptability to rapidly changing events and new technologies. Keith Mallory and Arvid Ottar, in this original work, afford wonderful insight into the startling and amusing contradiction between the applications of sophisticated technology and weaponry and the misapplications of discredited military strategies, the consequences of Hitler's failed architectural ambitions, the politics of defense budgeting, and the over-all impact of military architecture helping to shape civil architecture itself.

Keith Mallory studied architecture at Northern Polytechnic, London, and at the University of Bath. The recipient of an Architects' Journal Award, Mr. Mallory has lectured on military architecture at the Institute of Contemporary Arts, London, and is working in a private firm in England.

Born in Oslo, Arvid Ottar studied architecture at the University of Bath. One of the winners of design awards for two architectural complexes in Norway, Mr. Ottar is with the Oslo firm of Eliassen and Lambertz-Nilssen, where he is primarily concerned with the social implications of architecture.